Springer Proceedings in Physics 36

Springer Proceedings in Physics

Managing Editor: H. K. V. Lotsch

Volumes 1 – 29 are listed on the back inside cover

Nonlinear Optics of Organics and Semiconductors

Proceedings of the International Symposium
Tokyo, Japan, July 25-26, 1988

Editor: T. Kobayashi

With 236 Figures

Springer-Verlag Berlin Heidelberg GmbH

Professor Takayoshi Kobayashi
Department of Physics, University of Tokyo, 7-3-1 Hongo, Bunkyo-ku, Tokyo 113, Japan

ISBN 978-3-642-93428-5 ISBN 978-3-642-93426-1 (eBook)
DOI 10.1007/978-3-642-93426-1

Originally published by Springer-Verlag Berlin Heidelberg New York in 1989
Softcover reprint of the hardcover 1st edition 1989

2154/3150-543210 – Printed on acid-free paper

Preface

Just after the International Quantum Electronics Conference, which took place in Tokyo, July 18–21, 1988, a symposium on "Nonlinear Optics of Organics and Semiconductors" was held on the Hongo campus of the University of Tokyo on July 25 and 26, 1988.

Recently, nonlinear optical materials have started to attract the attention of many scientists, not only in the field of quantum electronics but also in organic chemistry, and solid-state physics, especially semiconductor physics. Therefore this area of investigation is really interdisciplinary and necessitates understanding from different viewpoints for continuous growth of the research activity in the field.

This book contains many papers written by scientists active in the field of nonlinear optics of organic materials and semiconductors. The first two parts of the book discuss mainly the basic physics of nonlinear optics of semiconductors, including bulk and quantum well structure semiconductors. The papers in Parts III–VIII concern mainly organic materials. They include the calculation of molecular polarizability, synthesis, the measurement of optical nonlinearity and characterization. The final part is on potential device applications.

The symposium was organized in collaboration with Prof. Shinsuke Umegaki of Tokyo Technical Institute, and Dr. Hachiro Nakanishi of the Research Institute of Polymers and Textiles. Preparations for the conference were made with the help of Dr. Masayuki Yoshizawa, Miss Nanae Indou, Mrs. Yoko Sasaki and graduate students (Mrs. T. Hattori, M. Terauchi, K. Ichimura, A. Terasaki, K. Misawa, and M. Taiji and Miss K. Minoshima) and Mr. K. Ueki, Dr. U. Stamm and Mr. X. Chen. Without their help, the symposium could not have been held. The symposium was sponsored by the Kobunshi Sozai Center. I would like to gratefully acknowledge this organization, especially Mrs. Kenichi Ohno and Yuzuru Oide, for providing us with the opportunity to hold the symposium.

Tokyo, *T. Kobayashi*
November 1988

Contents

Part VI Molecular Crystals – Single Crystals

Part I

Fundamental Physics of Semiconductors

Phase Conjugation Enhanced by Weak Localization of Exciton Polaritons *

E. Hanamura

Department of Applied Physics, University of Tokyo,
7-3-1 Hongo, Bunkyo-ku, Tokyo 113, Japan

Generation of a phase-conjugated light wave is shown to be enhanced by weak localization of polaritons as a precursor to Anderson localization. The phase-conjugated wave which is time-reversal propagation of the probe light is shown to be enhanced by constructive interference between elastic multiple-scatterings of the polariton at impurities and its time-reversed processes. We list several characteristics of generation of phase-conjugated waves due to the weak localization of polaritons. Observations of these will be used to check the present theory.

1. Introduction

Phase conjugation is defined as the process in which the phase of an output wave is the complex conjugate of the phase of an input wave. If the phase-conjugated output propagates in the backward direction with respect to the corresponding input wave, then it can be used to correct aberration due to phase distortion experienced by the input wave [1]. This is one example of degenerate four-wave mixing and can be understood from the following physical picture. Under three input light waves with a common frequency ω_0, *i.e.*, forward ($\mathbf{K}_0$) and backward ($-\mathbf{K}_0$) pump waves and probe wave ($\mathbf{K}_i$), the signal wave ($\mathbf{K}_s$) is possible in three directions $\mathbf{K}_s = \mathbf{K}_i \pm 2\mathbf{K}_0$ and $\mathbf{K}_s = -\mathbf{K}_i$. However, as $|\mathbf{K}_s|$ is not in general equal to $\omega_0\sqrt{\epsilon_\infty}/c$, the generation in the directions $\mathbf{K}_i \pm 2\mathbf{K}_0$ may not be phase-matched but that in $\mathbf{K}_s = -\mathbf{K}_i$ is always phase-matched. Here ϵ_∞ is an optical dielectric constant and c is light velocity in vacuum. This is the phase-conjugated wave and time-reversal propagation of the probe wave $\mathbf{K}_i$. This is also characteristic of an open system to which the energy is supplied by the forward and backward pump waves. Here arises a question about the microscopic understanding of generation of the phase-conjugated wave.

---

* A paper "Weak Localization of Exciton Polaritons and Generation of Phase-Conjugated Light Wave" with same contents was submitted to Solid State Communications.

Springer Proceedings in Physics, Vol. 36
Nonlinear Optics of Organics and Semiconductors
Editor: T. Kobayashi

We will point out in this paper that the weak localization of exciton-polaritons plays a key role in enhancing the phase-conjugated wave for the cases of using exciton-polaritons as the nonlinear optical medium. We will describe microscopic mechanism for formation of the population grating from one of the pump waves and the probe wave and reflection of the other pump wave by the grating in terms of multiple elastic scatterings of polaritons at impurities. This will be shown to be related to elementary excitations which are precursors of the Anderson localization and are also characteristic of the time-reversal invariance of the material system. These elementary excitations in metals are called cooperons and diffusons [2,3].

2. Hamiltonian

We consider a crystal much larger than the relevant optical wavelength. The exciton-photon coupling is much larger than other interactions in many cases, to which we will confine ourselves in the present work. Therefore first the Hamiltonian is rewritten in terms of the hybridized modes of polaritons [4]. The exciton is scattered elastically by impurity potential $\Sigma_i V(\mathbf{r}-\mathbf{r}_i)$, and we keep only the elastic scattering on the polariton dispersion in the lower branch. Therefore we start from the Hamiltonian

$$H=\sum_{\mathbf{k}}\omega(\mathbf{k})c_{\mathbf{k}}^{\dagger}c_{\mathbf{k}}+\sum_{\mathbf{kq},i}V_0(\mathbf{q})\exp(i\mathbf{q}\cdot\mathbf{r}_i)c_{\mathbf{k+q}}^{\dagger}c_{\mathbf{k}}\ , \tag{1}$$

where $V_0(\mathbf{q})$ is described by the Fourier transform of $V(\mathbf{r}-\mathbf{r}_i)$ and exciton-to-polariton transformation matrix elements. Here and hereafter we use a unit $\hbar=1$. In the following, we have to consider the effect of external radiation fields on the material system. The ground state of a material system $|0\rangle$ is a vacuum of any kinds of polaritons. The radiation field was considered to be divided into those in two spaces inside and outside the crystal. The former was taken into account already in forming the polaritons. The external radiation fields $\mathbf{E}_{ext}$ excite the polaritons at the crystal surface. This process is described by the interaction Hamiltonian H':

$$H'=\sum_{\alpha,j,\mathbf{k}}(c_{\mathbf{k}\alpha}+c_{\mathbf{k}\alpha}^{\dagger})\mu_{\alpha}\cdot\mathbf{E}_j\ , \tag{2}$$

where μ_α is the polarization vector of the polariton (α) and is evaluated in terms of the exciton dipole moment and the exciton-to-polariton transformation matrix elements. External field $\mathbf{E}_j$ with an angular frequency ω_j can excite the polariton with the same frequency $\omega(\mathbf{k})=\omega_j$ with the wave-number vector $\mathbf{k}$ which is determined by the incident angle and the polariton dispersion. The wave-number vector $\mathbf{k}$ and μ_α can be also determined classically in terms of refractive index and the incident frequency. The signal polaritons are produced through nonlinear processes as shown in the following. These polaritons can be taken outside the crystal as signal lights. This process is described also by Eq.(2) and can be also classically determined in terms of the refractive index and the angle of hitting the surface from inside the crystal. Therefore we will pursue the generation of phase-conjugated polariton waves through the excitations of the time-invariant collective modes in impure crystals.

3. Phase Conjugation

The amplitude of the phase-conjugated wave is derived in two steps. First, we derive third-order perturbations of the density matrix in the polariton-external field interaction H' and obtain the expression for an amplitude of signal polariton through four-wave mixing. Second, in terms of this expression, we take into account effects of elastic multiple-scatterings of polaritons at impurities. Then we will be able to give microscopic expressions for formation of population gratings due to two input polaritons and reflection of the third input polariton by the gratings. The amplitude of the signal polariton is expressed as [5,6]

$$\langle c_{\mathbf{K}_s}(\omega_s)\rangle \equiv \int_{-\infty}^{\infty} dt\ e^{i\omega_s t}\langle\langle c_{\mathbf{K}_s}|\rho(t)\rangle\rangle = \int_{-\infty}^{\infty} dt\ e^{i\omega_s t}\langle \mathrm{Tr} c_{\mathbf{K}_s}\rho(t)\rangle\ ,$$

where $\rho(t)$ is a density matrix of the total system $H_T = H + H'$, and the double brackets mean taking the ensemble average over the impurity distribution as well as the quantum mechanical average. The density matrix $\rho(t) = \exp(-iH_T t)\rho_0\exp(iH_T t)$ is expanded to the third order of H', where ρ_0 denotes the crystal ground state $|0\rangle\langle 0|$. Then we have a familiar expression [7,8]

$$\langle c_{\mathbf{K}_s}(\omega_s)\rangle = (-i)^3\int_{-\infty}^{\infty} dt\ e^{i\omega_s t}\int_{-\infty}^{t} dt_3\int_{-\infty}^{t_3} dt_2\int_{-\infty}^{t_2} dt_1$$

$$\times\ \langle\langle c_{\mathbf{K}_s}[H'(t_3),[H'(t_2),[H'(t_1),\rho_0]]]\rangle\rangle\ . \tag{3}$$

Here we apply three input optical fields as the external field $\mathbf{E}_{ext}$: (1) forward pump field $(\omega_0, \mathbf{K}_0)$, (2) backward pump field $(\omega_0, -\mathbf{K}_0)$, and (3) probe field $(\omega_i, \mathbf{K}_i)$. For the case of the degenerate four-wave mixing, we put $\omega_i = \omega_0$. Here and hereafter we consider a case under nearly resonant pumping of the lowest exciton so that a rotating wave approximation is well justified. Three commutators in Eq.(3) bring about eight terms but only four kinds of diagrams drawn in Fig.1 contribute to the four-wave mixing under a rotat-

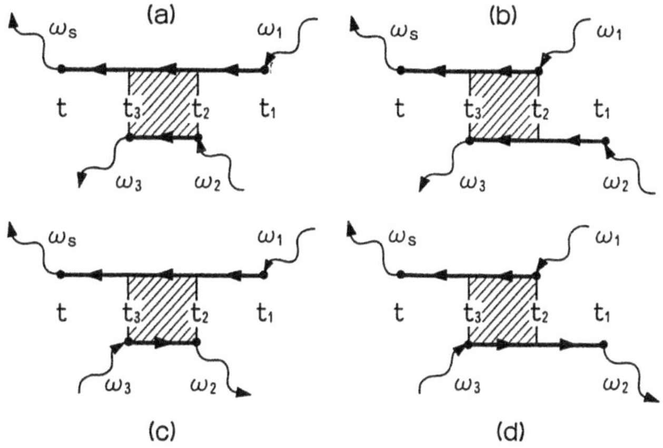

Fig.1. Feynman diagrams describing four kinds of contributions to four-wave mixing under a rotating wave approximation. The leftward and rightward arrows denote, respectively, the corresponding propagations of states in the density matrix. Time development in the diagram is leftward. ω_1, ω_2 and ω_3 represent three input fields and ω_s the signal field at time t.

ing wave approximation. In the diagrams in Fig.1, $(\omega_1, \mathbf{K}_1)$, $(\omega_2, \mathbf{K}_2)$ and $(\omega_3, \mathbf{K}_3)$ denote polaritons generated by input fields at times t_1 (or t_2), t_2 (or t_1) and t_3, respectively. There are six ways of combining these three input fields per each diagram in Fig.1 so that we have a total of 24 contributions as shown in Table 1. Then we get three kinds of signal polaritons ($\omega_s = \omega_i$, $\mathbf{K}_s = \mathbf{K}_i \pm 2\mathbf{K}_0$) and ($\omega_s = 2\omega_0 - \omega_i$, $\mathbf{K}_s = -\mathbf{K}_i$). The last wave corresponds to generation of a polariton wave phase-conjugate to the probe field $(\omega_i, \mathbf{K}_i)$ for the degenerate case $\omega_0 = \omega_i = \omega_s$. As shown in Table 1, the phase-conjugated wave is generated in cases of I-c and -d, *i.e.*, diagrams (c) and (d) of Fig.1 with $\omega_1 = (\omega_0, \mp\mathbf{K}_0)$, $\omega_2 = (\omega_i, \mathbf{K}_i)$ and $\omega_3 = (\omega_0, \pm\mathbf{K}_0)$, and in cases of III-a and -b, *i.e.*, diagrams (a) and (b) of Fig.1 with $\omega_1 = (\omega_0, \mp\mathbf{K}_0)$, $\omega_2 = (\omega_0, \pm\mathbf{K}_0)$ and $\omega_3 = (\omega_i, \mathbf{K}_i)$. We will show evaluation of $\langle c_{\mathbf{K}_s}(\omega_s)\rangle$ for the case I-c as an example:

$$\langle\langle c_{\mathbf{K}_s}(t) H'(\mathbf{K}_0, t_3)\rho_0 H'(\mathbf{K}_i, t_2) H'(-\mathbf{K}_0, t_1)\rangle\rangle = A\langle G(t, t_3, t_2, t_1)\rangle \ , \quad (4)$$

where

$$A = \langle 0|c_{\mathbf{K}_s}|\omega_s \mathbf{K}_s\rangle\langle\omega_0 \mathbf{K}_0|c^\dagger_{\mathbf{K}_0}\mu\cdot\mathbf{E}(\mathbf{K}_0)|0\rangle$$
$$\times\ \langle 0|c_{\mathbf{K}_i}\mu\cdot\mathbf{E}(\mathbf{K}_i)^*|\omega_i\mathbf{K}_i\rangle\langle\omega_0 - \mathbf{K}_0|c^\dagger_{-\mathbf{K}_0}\mu\cdot\mathbf{E}(-\mathbf{K}_0)|0\rangle \ , \quad (4a)$$

and

Table 1. The 24 cases contributing to four-wave mixing of ω_1, ω_2 and ω_3 input waves. The cases a, b, c and d correspond, respectively, to the processes of Feynman diagrams of Figs.1a, 1b, 1c and 1d, with described combinations of forward and backward pump waves and probe wave. ω_s and $\mathbf{K}_s$ denote, respectively, the signal angular frequency and signal wave-number vector. $\mathbf{E}_s$ describes the polarization dependence of the signal wave on the polarizations of the three input waves.

Case		ω_1	ω_2	ω_3	ω_s	$E_s \sim$
I	a, b	$\omega_0 \mp K_0$	$\omega_l K_l$	$\omega_0 \pm K_0$	$\omega_s=\omega_l$, $K_s=K_l \mp 2K_0$	$(E_l \cdot E_\mp) E_\pm$
	c, d				$\omega_s=2\omega_0-\omega_l$, $K_s=-K_l$	
II	a, b	$\omega_l K_l$	$\omega_0 \mp K_0$	$\omega_0 \pm K_0$	$\omega_s=\omega_l$, $K_s=K_l \mp 2K_0$	$(E_l \cdot E_\mp) E_\pm$
	c, d				$\omega_s=\omega_l$, $K_s=K_l \pm 2K_0$	
III	a, b	$\omega_0 \mp K_0$	$\omega_0 \pm K_0$	$\omega_l K_l$	$\omega_s=2\omega_0-\omega_l$, $K_s=-K_l$	$(E_\mp \cdot E_\pm) E_l$
	c, d				$\omega_s=\omega_l$, $K_s=K_l \mp 2K_0$	

$$\begin{aligned}\langle G(t,t_3,t_2,t_1)\rangle \doteq\ & (-i)^3\, \mathrm{e}^{-i(2\omega_0-\omega_i)t}\mathrm{e}^{i\omega_s(t-t_3)} \\ & \times \langle\langle \omega_s \mathbf{K}_s|\mathrm{e}^{-iH(t-t_3)}|\omega_s' \mathbf{K}_s'\rangle_+\rangle \mathrm{e}^{i(\omega_0-\omega_i)(t_3-t_2)} \\ & \times \langle\langle \omega_s' \mathbf{K}_s'|\mathrm{e}^{-iH(t_3-t_2)}|\omega_0' \mathbf{K}_0'\rangle_+ \langle \omega_i \mathbf{K}_i|\mathrm{e}^{iH(t_3-t_2)}|\omega_0 - \mathbf{K}_0\rangle_-\rangle \\ & \times \mathrm{e}^{i\omega_0(t_2-t_1)}\langle\langle \omega_0' \mathbf{K}_0'|\mathrm{e}^{-iH(t_2-t_1)}|\omega_0 \mathbf{K}_0\rangle_+\rangle\ . \end{aligned} \tag{4b}$$

Here the time interval (t,t_1) is divided into three parts (t,t_3), (t_3,t_2) and (t_2,t_1) as shown in Fig.1. Then an ensemble average of G over impurity distribution for the time interval (t,t_1) is approximated by a product of three ensemble averages for (t,t_3), (t_3,t_2) and (t_2,t_1). Then a contribution of the case I-c in Table 1 to $\langle c_{\mathbf{K}_s}(\omega_s)\rangle$ is expressed in terms of three propagators as

$$\begin{aligned}\langle c_{\mathbf{K}_s}(\omega_s)\rangle_c =\ & A\int_{-\infty}^{\infty} \mathrm{d}t\mathrm{e}^{i(\omega_s-2\omega_0+\omega_i)t}\int_0^{\infty} G_s(\tau_3)\mathrm{d}\tau_3 \\ & \times \int_0^{\infty} \Xi(\tau_2)\mathrm{d}\tau_2 \int_0^{\infty} G_f(\tau_1)\mathrm{d}\tau_1\ . \end{aligned} \tag{5}$$

Here

$$G_s(\tau_3) = -ie^{i\omega_s\tau_3}\langle\langle\omega_s\mathbf{K}_s|e^{-iH\tau_3}|\omega_s'\mathbf{K}_s'\rangle_+\rangle \ , \tag{6a}$$

$$\Xi(\tau_2) = -ie^{i(\omega_0-\omega_i)\tau_2}\langle\langle\omega_s'\mathbf{K}_s'|e^{-iH\tau_2}|\omega_0\mathbf{K}_0\rangle_+ \times \langle\omega_i\mathbf{K}_i|e^{iH\tau_2}|\omega_0-\mathbf{K}_0\rangle_-\rangle \ , \tag{6b}$$

$$G_f(\tau_1) = -ie^{i\omega_0\tau_1}\langle\langle\omega_0'\mathbf{K}_0'|e^{-iH\tau_1}|\omega_0\mathbf{K}_0\rangle_+\rangle \ , \tag{6c}$$

where we have used the abbreviation $|\omega_s'\mathbf{K}_s'\rangle\langle\omega_s'\mathbf{K}_s'| = \Sigma_{\omega_s'\mathbf{K}_s'}|\omega_s'\mathbf{K}_s'\rangle\langle\omega_s'\mathbf{K}_s'|$, which is equal to one. Propagators $G(t)$ describe propagation of polaritons $(\omega_0, \mathbf{K}_0)$ and $(\omega_s, \mathbf{K}_s)$ in a crystal with elastic scatterers. It is vertex part Ξ that represents formation of population gratings of polaritons as well as generation of the conjugated wave $(\omega_s, \mathbf{K}_s)$ due to reflection of one of the pump fields by the population grating. Suffixes + and − of the inner brackets in Eqs.(6) mean the leftward and rightward propagations of the states in the density matrix as also realized in Fig.1c.

Contribution of the diagram in Fig.1d is obtained by replacing the last propagator $G_f(\tau_1)$ in Eq.(5) by $G_b(\tau) = -ie^{-i\omega_0\tau}\langle\langle\omega_0' - \mathbf{K}_0'|e^{-iH\tau}|\omega_0-\mathbf{K}_0\rangle_-\rangle$. Contributions to $\langle\langle c_{\mathbf{K}_s}(\omega_s)\rangle\rangle$ from I-c and -d in Table 1 with $\mathbf{K}_0$ and $-\mathbf{K}_0$ exchanged are obtained from the above results by exchanging $\mathbf{K}_0$ and $-\mathbf{K}_0$. These give the same results besides the factor A given by Eq.(4a) as long as $\omega(-\mathbf{K}_0) = \omega(\mathbf{K}_0)$.

4. Multiple Elastic Scattering of Polaritons

Effects of elastic scatterings of polaritons by impurity centers are evaluated. For the diagram of Fig.1c with a combination I-c in Table 1, the forward pump polariton $(\omega_0, \mathbf{K}_0)$ propagates for the initial interval (t_2, t_1) while the signal polariton does in the last interval (t, t_3). For the intermediate interval (t_3, t_2), polaritons $(\omega_0, \mathbf{K}_0)$ and $(\omega_i, \mathbf{K}_i)$ propagating, respectively, in the leftward and rightward directions in the density matrix form the population grating with the difference wave-number vector $\mathbf{K}_0 - \mathbf{K}_i$ and oscillating in time with the difference frequency $\omega_0 - \omega_i$. Then the third polariton $(\omega_0, -\mathbf{K}_0)$ is reflected by this grating and the phase-conjugated wave $(\omega_s = 2\omega_0 - \omega_i, \mathbf{K}_s = -\mathbf{K}_i)$ is generated as a signal polariton. Therefore the vertex part plays the key role in generation of the phase-conjugated wave. Integrals in times in Eq.(5) are performed using perturbational expansions of the propagators in the polariton-scatterer interaction V_0 of Eq.(1).

Here two points should be noted. First the polariton amplitude decays with a decay rate γ_0 through surface transmission and/or inelastic scatterings, *e.g.*, by phonons. This effect is taken into account by replacing e^{-iHt} and e^{iHt} in Eq.(6b) by $e^{-i(H-i\gamma_0)t}$ and $e^{i(H+i\gamma_0)t}$, respectively, and $\omega(\mathbf{K}_0)$ and $\omega(\mathbf{K}_i)$ by $\omega(\mathbf{K}_0)+i\gamma_0$ and $\omega(\mathbf{K}_i)-i\gamma_0$, respectively. This comes from the fact that our system of polaritons is a quasi-stationary state far from the thermal equilibrium and a polariton once scattered inelastically cannot contribute again to generation of the phase-conjugated wave. This is in contrast to the Anderson localization in metals in thermal equilibrium. Second, we take an ensemble average over a uniform spacial distribution of scattering centers, so that the ensemble average of products of V matrix elements is divided into products of ensemble averages of a pair of V's. Here we note that $\langle V_{\mathbf{qq}'} V_{\mathbf{kk}'} \rangle \doteq n_i |V(0)|^2 \delta_{\mathbf{q}-\mathbf{q}',\mathbf{k}'-\mathbf{k}}$ with the impurity concentration n_i. First, the ensemble averages of scatterings within the leftward propagation of a polariton (on the upper lines in all the diagrams of Figs.1) and within the rightward one (on the lower lines of the diagrams of Figs.1c and 1d) in the density matrix are described, respectively, by the retarded and advanced Green's functions (Fig.2a):

$$G^R(\mathbf{q}) = \frac{1}{\omega(\mathbf{K}_0) - \omega(\mathbf{q}) + i(\gamma_0 + \gamma)} ,$$

$$G^A(\mathbf{q}') = \frac{1}{\omega(\mathbf{K}_i) - \omega(\mathbf{q}') - i(\gamma_0 + \gamma)} .$$

Diagrams of Figs.2b and 2c are shown to have singularly dominating contributions to the vertex part Ξ. Note that the self-energy described by Fig.2a is taken into account in each propagator in the diagrams of Figs.2b and 2c. Then the vertex part $\Xi(\omega_0 - \omega_i)$ is written in terms of contributions from the ladder diagram Γ_d (Fig.2b) and the maximally crossed diagram Γ_c (Fig.2c) as

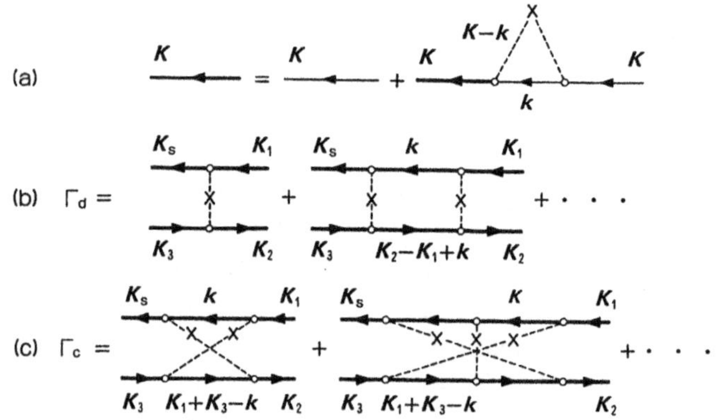

Fig.2. (a) Diagram describing the lowest-order contribution to the self-energy of polariton $\mathbf{K}$. The crosses and dotted lines mean scattering centers and elastic scattering interactions of the polariton with an impurity. (b) Ladder-like scatterings of the leftward and rightward polaritons. (c) Maximally crossed scattering of polaritons at impurity centers.

$$\Xi(\omega_0 - \omega_i) = \frac{-i}{2\gamma_0} \frac{1}{[\omega(\mathbf{K}_0) - \omega(\mathbf{K}_s) + i(\gamma_0 + \gamma)]} \times (\Gamma_d + \Gamma_c) \frac{1}{\omega(\mathbf{K}_i) - \omega(-\mathbf{K}_0') - i(\gamma_0 + \gamma)} .$$

The ladder-like contribution of Fig.2b, which is characteristic of a diffusive motion of the polariton, is summed up as $\Gamma_d = U_0/(1 - U_0 X)$ where $U_0 \equiv n_i |V(0)|^2$ and

$$X = \sum_{\mathbf{q}} G^R(\mathbf{q}) G^A(\mathbf{K}_i - \mathbf{K}_0 + \mathbf{q}) \doteq \frac{\pi N(\omega_0)}{\gamma_0 + \gamma} \left[1 + \frac{i(\omega_0 - \omega_i)}{2(\gamma_0 + \gamma)} - \frac{|\mathbf{K}_i - \mathbf{K}_0|^2 v_g^2}{12(\gamma_0 + \gamma)^2} \right] .$$

Here summation over $\mathbf{q}$ was replaced by an integral in energy with a state density $N(\omega_0)$ and we introduced the group velocity of the polariton at the energy $\omega_{\mathbf{q}} = \omega_0$ as $\mathbf{v}_g = \nabla_{\mathbf{q}} \omega_{\mathbf{q}}|_{\omega_{\mathbf{q}} = \omega_0}$. $v_g \equiv |\mathbf{v}_g|$. We used the fact that $2(\gamma_0 + \gamma) \gg |\omega_0 - \omega_i|$ and $|\mathbf{K}_i - \mathbf{K}_0| v_g$. Then we have

$$\Gamma_d(\mathbf{K}_i - \mathbf{K}_0) \doteq \frac{2(\gamma_0 + \gamma) U_0}{2\gamma_0 - i(\omega_0 - \omega_i) + D|\mathbf{K}_i - \mathbf{K}_0|^2} .$$

Here we used the fact $\gamma \gg \gamma_0$ and a diffusion constant $D \equiv v_g^2/6(\gamma_0 + \gamma)$. Next we evaluate a contribution Γ_c from the series of maximally crossed diagrams of Fig.2c. This is done very similarly to the case of Γ_d and is expressed as $\Gamma_c = U_0^2 Y/(1 - U_0 Y)$, where $Y = \sum_{\mathbf{q}} G^R(\mathbf{q}) G^A(\mathbf{K}_0 - \mathbf{K}_0' - \mathbf{q})$. Then we have

$$\Gamma_c(\mathbf{K}_o - \mathbf{K}_0') \doteq \frac{2(\gamma_0 + \gamma) U_0}{2\gamma_0 - i(\omega_0 - \omega_0') + D|\mathbf{K}_0 - \mathbf{K}_0'|^2} .$$

Here we used the fact $U_0 Y \doteq 1$ as $2(\gamma_0 + \gamma) \gg |\omega_0 - \omega_0'|$ and $D|\mathbf{K}_0 - \mathbf{K}_0'|^2$ and denoted the backward pump polariton as $(\omega_0', -\mathbf{K}_0')$. As a result we can conclude that the strongest singularity comes from the maximally crossed diagram for small $2\gamma_0$ under the usual case ($\omega_0 = \omega_0'$, $\mathbf{K}_0' = \mathbf{K}_0$) while the ladder contribution is much smaller, as the value of $D(\mathbf{K}_i - \mathbf{K}_0)^2$ is usually much larger than $2\gamma_0$. The other contribution to the generation of the phase-conjugated wave comes from the diagrams of Figs.1a and b with combinations of input waves IIIa and IIIb in Table 1. These have the same form as those of Ic and Id but a sum of Γ_d and Γ_c is replaced by U_0, because contributions of ladder-type diagrams (Fig.3b) and maximally crossed diagrams (Fig.3c) vanish. Therefore these give the conventional expression for generation of a

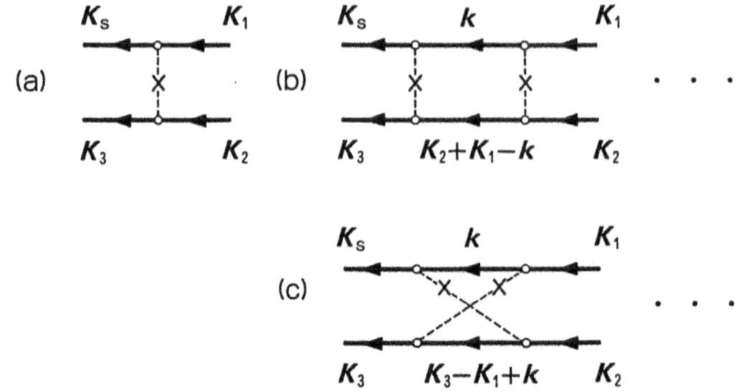

Fig.3. (a) The lowest-order scattering mechanism of two polaritons which are leftward propagating in the density matrix. (b) Ladder-like scatterings of these two polaritons. (c) Maximally crossed scatterings of these two polaritons by impurities.

phase-conjugated wave and are smaller by a factor $(\gamma_0 + \gamma)/\gamma_0$ than those of Ic and Id in Table 1. Therefore we may conclude that the diagrams of Figs.1c and 1d with the combinations of input polariton Ic and Id in Table 1 have the dominant contribution from the maximally crossed diagrams given by Fig.2c to generation of the phase-conjugated wave:

$$\langle c_{\mathbf{K}_s}(\omega_s)\rangle \doteq 2\pi A\delta(\omega_s - 2\omega_0 + \omega_i)\frac{1}{\gamma_0(\gamma_0+\gamma)^3} \times \frac{2U_0}{2\gamma_0 - i(\omega_0 - \omega_0') + D(\mathbf{K}_0 - \mathbf{K}_0')^2} \,. \tag{7}$$

5. Discussion and Conclusion

Effects of the weak localization were clearly demonstrated in the d.c. resistance of thin disordered metallic films and MOS inversion layers as a function of temperature [9-11]. These linear responses could pick up the singularity at $\omega = 0$ and $\mathbf{q} = 0$ of the time-reversal collective modes "cooperons"[2]. The linear optical response, however, is insensitive to these collective modes even though the exciton-polaritons suffer from weak localization. On the other hand, the weak localization of polaritons has been shown in this communication to play a key role in nonlinear optical responses, especially in generation of the phase-conjugated wave. This is because we can control the difference of two input frequencies $\omega_0 - \omega_0'$ and a combination of two wave vectors near the singular point of the collective excitations, *i.e.*, $\omega_0 - \omega_0' = 0$ and $\mathbf{q} = \mathbf{K}_0 - \mathbf{K}_0' = 0$.

The generation of the phase-conjugated wave, which is time-reversal propagation of the probe light, has been shown here to originate from the singularity due to coherent superposition of the forward- and backward-propagating waves of polaritons under elastic multiple scatterings at impurities, reflecting the time-reversal symmetry of the system. We will list some characteristics of the phase-conjugated wave generated by the present mechanism. The phase-conjugation has already been observed near the exciton-frequency region of CuCl [12] and the polaritons of this material have been studied very well [13] so that we will study especially the case of CuCl.

(1) The first characteristic of the present theory is the singular dependence of $\langle c_{\mathbf{K}_s}(\omega_s)\rangle$ through Γ_c on $\omega_0 - \omega_0'$ and $\mathbf{K}_0 - \mathbf{K}_0'$. For degenerate four-wave mixing $\omega_0 = \omega_i = \omega_s$, the intensity $|\langle c_{\mathbf{K}_s}(\omega_s)\rangle|^2$ of the phase-conjugated wave is very sensitive to the angle $(\pi - \theta)$ between the two pump fields $\mathbf{K}_0$ and $-\mathbf{K}_0'$. As long as $\theta > 2.2 \times 10^{-2}$ radian, the signal intensity is proportional to $(1 - \cos\theta)^{-2} \doteq 4\theta^{-4}$.

(2) On the other hand, when we use nondegenerate four-wave mixing, we will be able to determine the dispersion of the collective mode from the dependence $|\langle c_{\mathbf{K}_s}(\omega_s)\rangle|^2 \propto [(\omega_0 - \omega_0')^2 + \{2\gamma_0 + D(\mathbf{K}_0 - \mathbf{K}_0')^2\}^2]^{-1}$.

(3) The diffusion constant $D \equiv v_g^2/6(\gamma_0 + \gamma)$ can be varied over several orders of magnitude by changing the input frequency ω_0 as the group velocity v_g depends sensitively on the input frequency over the polariton dispersion [14]. This brings about the incident frequency dependence of the phase-conjugated wave intensity even for the degenerate case with fixed θ.

(4) The cut-off frequency $2\gamma_0$ in Eq.(7) is sensitive to the lattice temperature $T > 40$ K [15].

(5) The signal depends also on the polarizations of the three input fields as shown in Table 1. This comes from the expression A in Eq.(4a). These dependences were already observed by MIZUTANI and NAGASAWA [12]. As the vertex part Γ_c shows, the formation of polariton gratings and reflection of one pump wave into the conjugated wave are enhanced in proportion to the square of impurity concentration n_i through $2(\gamma_0 + \gamma)U_0$. However, the intensity of the phase-conjugated wave depends inversely on the impurity concentration n_i for the case of $\gamma \gg \gamma_0$ as follows:

$$\langle c_{\mathbf{K}_s}(\omega_s)\rangle \sim n_i^{-2} \text{ for } D(\mathbf{K}_0 - \mathbf{K}_0')^2 \ll 2\gamma_0 \,, \tag{8a}$$

$$\sim n_i^{-1} \text{ for } D(\mathbf{K}_0 - \mathbf{K}_0')^2 \gg 2\gamma_0 \,. \tag{8b}$$

This is because the propagation of input as well as signal polaritons is incoherently scattered by impurities although the grating formation as well as the reflection due to the grating are enhanced by the impurity scatterings through Γ_c. Note that we cannot extrapolate the results of Eqs.(8) to the limit of $n_i \rightarrow$ zero as these are restricted to the case $\gamma \equiv \pi N(\omega_0) n_i |V(0)|^2 >$ γ_0. Of course $\langle c_{\mathbf{K}_s}(\omega_s)\rangle$ vanishes in the limit of $n_i|V(0)|^2 \equiv U_0 \rightarrow 0$ as Eq.(7 shows.

The present theory has clarified microscopic mechanisms of the phase-conjugation processes, in which the weak localization of polaritons plays important roles. This will also be able to clarify a mystery of polariton dynamics observed by KUWATA and NAGASAWA [16]. They observed th generation of backward propagating polaritons under pumping of the forwar wave, and the strong coupling among the forward and backward polaritons. This will also be understood by the weak localization of polaritons presente in this communication.

Acknowledgement

The author thanks Dr. N. Nagaosa and N. Taniguchi for fruitful discussions. This work was partially supported by Grant-in-Aid No. 63604516 for Scientific Research on Priority Area, New Functionality Materials -- Design, Preparation and Control -- by the Ministry of Education, Science and Culture of Japan.

References

1 For a general review of optical phase-conjugation see A. Yariv, IEEE J. Quantum Electronics **QE-14**, 650 (1978); C. Giuliano, Physics Today April, 27 (1981); and Y.R. Shen, *The Principles of Nonlinear Optics*, (John Wiley & Sons, 1984), Chap.14.

2 H. Fukuyama, J. Phys. Soc. Jpn. **50**, 3407 (1981); Prog. Theor. Phys. Suppl. **84**, 47 (1985).

3 For a general review of Anderson localization see J.G. Bergmann, Phys. Rev. **107**, 1 (1984); P.A. Lee and T.V. Ramakrishnan, Rev. Mod. Phys. **57**, 287 (1985); *Anderson Localization*, ed. by Y. Nagaoka, Prog. Theor. Phys. Suppl. **84**, (1985); and *Anderson Localization*, ed. by Y. Nagaoka and H. Fukuyama (Springer-Verlag, 1982).

4 J.J. Hopfield, Phys. Rev. **112**, 1555 (1958).

5 S. Mukamel and E. Hanamura, Phys. Rev. **A33**, 1099 (1986).

6 E. Hanamura, in *Advances in Multi-photon Processes and Spectroscopy* vol.3 ed. by S.H. Lin (World Scientific, 1987) p.1.

7 E. Hanamura, Phys. Rev. **B37**, 1273 (1988).

8 E. Hanamura, Phys. Rev. **B38**, 1228 (1988).

9 G.J. Dolan and D.D. Osheroff, Phys. Rev. Lett. **43**, 721 (1979).

10 S. Kobayashi, F. Komori, Y. Ootuka, and W. Sasaki, J. Phys. Soc. Jpn. **49**, 1635 (1980).
11 L. Van den dries, C. Van Haesendonck, Y. Bruynseraede, and G. Deutscher, Phys. Rev. Lett. **46**, 565 (1981).
12 G. Mizutani and N. Nagasawa, J. Phys. Soc. Jpn. **52**, 2251 (1983).
13 M. Ueta, H. Kanzaki, K. Kobayashi, Y. Toyozawa, and E. Hanamura, *Excitonic Processes in Solids*, (Springer-Verlag, 1986), Chaps. 2 and 3.
14 Y. Masumoto, Y. Unuma, Y. Tanaka and S. Shionoya, J. Phys. Soc. Jpn. **47**, 1844 (1979).
15 T. Itoh, T. Katohno, T. Kirihara, and M. Ueta, J. Phys. Soc. Jpn. **53**, 854 (1984).
16 M. Kuwata and N. Nagasawa, Solid State Commun. **45**, 937 (1983).

Dimensionality and Scaling in Nonlinear Optical Susceptibilities

C. Flytzanis

Laboratoire d'Optique Quantique, Ecole Polytechnique,
F-91128 Palaiseau, France

1. INTRODUCTION

New developments in nonlinear optics and implementation of nonlinear effects in devices are conditioned by a substantial improvement of several characteristics of the nonlinear optical materials in their transparency region, the one extending below the lowest electronic transitions and above the vibrational frequencies . The absorption losses and laser-induced irreversible modifications there can be safely avoided and the recovery time of the optical nonlinearities can be considered instantaneous for all practical purposes. The linear and nonlinear susceptibilities then are fashioned by the overall characteristics of the valence electron distribution /1/ and in particular by the effective size and form of the space available to each valence electron ; the quantity that matters is the *optical nonlinearity per valence electron.*

It is a striking fact for instance that the value of the third-order susceptibility $\chi^{(3)}$ in the transparency region of crystalline Ge, an optically isotropic material, is three orders of magnitude larger than that of the polydiacetylene crystal PTS when measured perpendicular to its chain axis but equal to that measured parallel to this axis /2,3/ and this for widely different valence electron densities (an order of magnitude larger in Ge than in PTS) ; at first sight such drastic differences are not expected when one compares their linear refractive indices. Also the value of the second-order susceptibility $\chi^{(2)}$ of crystalline GaAs, also an isotropic compound isoelectronic to Ge but noncentrosymmetric, is substantially larger than that of the noncentrosymmetric polydiacetylene recently synthetized /4/ and this irrespective of its crystal axes directions.

Such features for each compound separately can certainly be derived and calculated from detailed microscopic equations but with considerable computational effort ; such an approach is most often cumbersome and in the final analysis does not reveal the general trends and interrelations between the different macroscopic coefficients which facilitate comparison of different compounds. The above examples, however unrelated they might appear at first sight, on closer inspection actually hint at the working of some simple but powerful functional dependence of the susceptibilities on an effective parameter whose identity can be extracted by simple models that contain the gross features of the valence electron density, the ones that really persist after all summations and averages over the quantum states.

Below we show that the nonlinear susceptibilities, like the linear one $\chi^{(1)}$, can be cast in the form of scaling laws, essentially power laws of an effective parameter, with exponents that critically depend on the dimensionality of the material system for infinitely extended systems ; the effective parameter is amenable to a rigorous definition within a quantum mechanical description of the material system but also possesses a direct physical meaning, since it turns out to be a measure of the valence electron delocalization and can also be expressed in terms of measurable macroscopic quantities. The main approach was outlined in /5/, where also the main results were derived.

Springer Proceedings in Physics, Vol. 36
Nonlinear Optics of Organics and Semiconductors
Editor: T. Kobayashi

This approach does not rely on any detailed microscopic description in particular but on the topological character of the critical points (or van Hove singularities) in the joint density of states in extended periodic systems (crystals) , see for instance /6/. This approach was used by CARDONA and POLLACK /7/, van VECHTEN et al./8/, AGRAWAL and FLYTZANIS /9/, and AGRAWAL et al. /3/ to derive the functional dependence of $\chi^{(3)}$ on certain effective parameters related to these singularities. The results were subsequently generalized to all odd-order susceptibilities $\chi^{(2n+1)}$ using a deductive approach /5/. A key consequence of the critical point analysis is that the effective parameter, which as we anticipated measures the valence electron delocalization, can be simply expressed in terms of experimentally accessible quantities that are characteristic of the k-space (global quantities) rather than the r-space (local). The situation is strikingly different for even-order susceptibilities $\chi^{(2n)}$, as will be discussed below. We shall also summarize some results obtained for finite size systems, also called *zero-dimensionality* systems, that concern the relationship between the nonlinear polarizabilities and the size of the system /10,3/ ; here the approach is not as straightforward as for infinitely extended systems and the derived relationships can be strongly affected by pronounced boundary effects *(quantum confinement)* or local field redistribution *(dielectric confinement)* when the small systems are embedded in a different matrix as is quite often the case in practice /11-14/.

In a certain sense the first hint that the nonlinear susceptibilities can be expressed in terms of other macroscopic parameters was Miller's empirical rule, which in its initial form states /15/ that

$$\chi^{(2)}_{ijk} = \delta_{ijk}\chi^{(1)}_{ii}\,\chi^{(1)}_{jj}\,\chi^{(1)}_{kk} \tag{1.1}$$

and δ_{ijk} was conjectured to be approximately a constant by simple survey of the existant experimental values of $\chi^{(2)}$ for some twenty different crystals ; thus one could relate $\chi^{(2)}$ to $\chi^{(1)}$ where a large body of information exists. The first analysis of Miller's empirical rule using dimensionality arguments /16/ based on detailed quantum mechanical calculations revealed that quite generally δ is not a constant but is roughly proportional to the dipole moment of the electronic density distribution within a bond and can even reverse its sign for a givon axis convention. As we will see below the critical point analysis is not that relevant for even-order susceptibilities.

The laws we summarize below for infinitely extended systems are only valid in their transparency regions, the one extending below the lowest electronic transition and above the vibrational frequencies ; this ensures that the detailed energy spectrum or the excited-state wave functions and their dynamics will not be needed. We also assume that the energy separation between the highest occupied and lowest unoccupied electronic states is much higher than kT, so that no thermodynamic considerations will be needed. For finite size systems, also termed zero-dimensionality systems, like short polymer chains, semiconductor or metal microcrystallites, the situation in certain cases may be different as quite often there one may exploit resonances and the analysis must take into account their dynamics.

2. NONLINEAR OPTICAL SUSCEPTIBILITIES

For extended periodic systems the electron states are represented in terms of Bloch band states and the expressions of the optical susceptibilities $\chi^{(n)}$ using the Genkin-Mednis approach /17,1/, the most appropriate for our purpose, are

$$\chi^{(1)}_{xx}=\frac{4e^2}{hV}\int_{B.Z.}\Omega_{vc}S_{cv}dk \tag{2.1}$$

$$\chi^{(3)}_{xxxx}=\frac{8e^4}{h^3V}\int_{B.Z.}\left[\frac{1}{\omega_{cv}}\left(\frac{\partial S_{cv}}{\partial k}\right)\left(\frac{\partial S_{vc}}{\partial k}\right)-\Omega_{vc}S_{cv}S_{vc}S_{cv}\right]dk\ , \tag{2.2}$$

where $h\omega_{cv} = \varepsilon_c - \varepsilon_v$, Ω_{vc} is the transition dipole moment matrix element between the highest valence (v) and lowest conduction (c) bands, $S_{vc} = \Omega_{vc}/\omega_{vc}$, and ε_v and ε_c are the band energies for the v and c bands, respectively ; all quantities are functions of the wavevector k, which is let to vary over the entire Brillouin zone (B.Z.) and V is the normalization volume. In (2.1) and (2.2) we have introduced the two-band approximation, which is quite sufficient for our purpose. Furthermore, we have assumed that the system possesses inversion symmetry so that in particular $\chi^{(2)} \equiv 0$; for noncentrosymmetric systems, the expression of $\chi^{(3)}$ is slightly more lengthly /3/ and the second-order susceptibility is

$$\chi^{2}=\left(\frac{6e^3}{h^2V}\right)\frac{i}{2\pi}\int_{B.Z.}\left[S_{cv}S_{cv}(\Omega_{vv}-\Omega_{cc})-\frac{1}{2}\left(S_{vc}\frac{\partial S_{cv}}{\partial k}-\frac{\partial S_{vc}}{\partial k}S_{cv}\right)\right]dk \tag{2.3}$$

where Ω_{vv} and Ω_{cc} are intraband transition dipole-moment matrix elements. One may formally define also the polarizabilities $\alpha^{(n)}$ for such infinitely extended systems by the relation

$$\chi^{(n)}\equiv\frac{\alpha^{(n)}}{v} \tag{2.4}$$

where v is the repeat unit-cell volume ; for localized systems of extension much smaller than the optical wavelength the above expressions for $\alpha^{(n)}$ reduce to the well-known ones /1/.

As can be seen from (2.1) to (2.3) the susceptibilities are expressed as integrals over the Brillouin zone. The main contribution to these integrals is in general expected to come from some few nonoverlapping critical regions in the joint density of states, e.g. /6/ ; these are *points, lines* or *surfaces* depending on the spatial extension of the electronic density distribution and are defined by the condition

$$\nabla_k\,\omega_{cv}(k) = 0. \tag{2.5}$$

This is indeed the case in $\chi^{(1)}$ (see, for instance, /7/) and is also the case for odd-order susceptibilities $\chi^{(2n+1)}$. This directly establishes a relation between the optical susceptibilities $\chi^{(2n+1)}$ and the topology and dimensionality of the joint density of states. More importantly, it expresses $\chi^{(2n+1)}$ in terms of the values Ω_{vc} and ω_{cv} at these critical regions ; the latter can also be related to the characteristics of the electron distribution and calculated within a quantum mechanical description. This functional dependence is precisely the origin of the scaling laws that will be summarized in the following section for the odd-order susceptibilities ; it also reveals the crucial role played by the dimensionality of the electron distribution. For the even-order susceptibilities $\chi^{(2n)}$, as will become evident, the critical points do not play such a prominent role and scaling laws cannot be derived.

Before we move on to derive the scaling laws using the above expressions, we wish to address our attention to another important feature in the expressions of the nonlinear susceptibilities. As can be seen from (2.2), for instance, the magnitude and sign of the nonlinear optical sysceptibilities $\chi^{(n)}$ is determined by the competition of two terms, an *intraband* one, the first term in the integrand in (2.2), that arises from field mixing of Bloch band states within a band, and *an interband* one, the second term in the integrand in (2.2), that arises from mixing of states across the gap with wave vector conservation ; in contrast, the linear susceptibility $\chi^{(1)}$ involves only interband terms and this has crucial qualitative and quantitative implications.This competition can also be interpreted as displacements of opposite signs of the effective band gaps at the critical points when an electric field is applied. The interband term can be identified as the analogue of the Stark shift for atomic levels and consists in a repulsion between valence and conduction states at each k and in particular at the critical regions ; this leads to an increase of the effective energy gap there or a negative contribution to $\chi^{(3)}$ as can also be inferred from the negative sign in front of this term in the integrand in (2.2). The intraband term on the other hand, which is also responsible for the Franz-Keldysh effect, consists of a repulsion of the states within a single band, which results in a net repulsion of the critical points from all other states in the same band ; this leads to a decrease of the effective energy gap there or a positive contribution to $\chi^{(3)}$. It is quite evident that the intraband term will be dominant whenever the bands vary strongly with k (wide bands), which will be the case for very delocalized systems or strong overlap between wave functions of neighboring units in the periodic structure ; on the other hand, this contribution will be negligible for highly localized systems (very narrow bands) and the interband term then becomes dominant. Accordingly, the sign of $\chi^{(3)}$ gives a crucial indication of the degree of electron delocalization.

3. PERIODIC ONE-, TWO- AND THREE-DIMENSIONAL SYSTEMS

As previously stated, the essential contributions to the integrals in (2.1) to (2.3) come from some few nonoverlapping critical regions in the joint density of states ; these are the van Hove singularities where (and only there) an infinite joint density of states accumulates. Their features, in particular their dimensionality and relative positions, are shaped by the topology of the k-space.

For a one-dimensional semiconductor one can show on quite general topological grounds that this infinite accumulation of states takes place at a *point* located at the edge of the Brillouin zone, ka = π, where a is the unit cell length, where in addition the transition dipole moment $|\Omega_{cv}|$ attains its maximum value

$$|\Omega_{cv}(\pi)| \equiv L_d \equiv aN_d \qquad (3.1)$$

and the energy gap is smallest

$$h\omega_{cv}(\pi) \equiv E_o\,. \qquad (3.2)$$

The coincidence of these three features at a single point is an essential characteristic of the one-dimensional systems and is reminiscent of a metallic-type behavior : this point is the only one that is preserved from the metallic state when the Peierls transition (bond alternation) sets in and opens a gap to form the semiconducting state. One can show /5/ that as long as $N_d > 1$

$$\chi^{(2n-1)} \approx \chi_o^{(2n-1)}\left(\frac{E_F}{E_o}\right)^{4n-2} \quad (3.3)$$

where E_F is the Fermi energy and in particular

$$\chi^{(1)} \approx \chi_o^{(1)}\left(\frac{E_F}{E_o}\right)^{2} \quad (3.4)$$

$$\chi^{(3)} \approx \chi_o^{(3)}\left(\frac{E_F}{E_o}\right)^{6}. \quad (3.5)$$

which contains quantities directly accessible to measurement. This form in particular allows one to compare organic semiconductors, e.g.polydiacetylene, and inorganic ones, e.g.SbSI , and more importantly it directly relates $\chi^{(2n-1)}$ to the electron delocalization parameter defined by (3.2), which becomes

$$N_d \approx \frac{E_F}{E_o}. \quad (3.6)$$

Expression (3.4) was initially derived by PENN /18/ and expression (3.5) by CARDONA and POLLACK /7/ using critical point analysis, by SAUTERET et al. /2/ using a simple model of a one-dimensional semiconductor and by AGRAWAL et al. /3/ using the tight binding approximation : the latter approach allows a direct analytical treatment for a wide class of one-dimensional systems and indicates that the scaling law (3.5) for $\chi^{(3)}$ is intrinsically associated with the critical point E_o of the one-dimensional system. Similarly one expects that the scaling law (3.5) for higher-order susceptibilities is universally valid for one-dimensional systems.

The detailed analysis of AGRAWAL et al. /3/ also shows that the main contribution to $\chi^{(3)}$, and actually to all odd-order susceptibilities, comes from the intraband term in (2.2) as long as $N_d > 1$; this contribution is positive, in contrast to the interband one, which has a negative sign.

Before moving into the higher-dimensionality critical regions, we wish to say a few words about the second-order susceptibility $\chi^{(2)}$, given by (2.3). Here the interband and intraband terms can each have either sign, so that they may add or subtract. Careful analysis of (2.3) reveals that contributions to the integrand only come from regions where Ω_{cv} is complex /3/ ; it vanishes whenever Ω_{vc} becomes real or pure imaginary and this is precisely what happens at the edge of the Brillouin zone of a one-dimensional system namely at the critical point E_o. Thus, in contrast to $\chi^{(3)}$, $\chi^{(2)}$ does not take full advantage of the infinite densities of states there and the highly delocalized character of these states. In particular, with the critical point contribution being suppressed in $\chi^{(2)}$, integration over the whole B.Z. is now required to extract the behavior of $\chi^{(2)}$. This implies that $\chi^{(2)}$ is sensitive to local properties of the electron density. Indeed this quantity is sensitive to the charge asymmetry that is localized within a molecular entity (bond or molecule) and may interfere with the electron delocalization. Miller's empirical rule (1.1) hints that the two effects can be factorized, but more work is still needed to obtain useful scaling laws of universal validity /3/.

For a two-dimensional system the critical regions are of two kinds : a point, at energy E_o at the edge of the Brillouin zone, reminiscent of one-dimensional patterns in the electron density distribution, with its contribution to $\chi^{(2-1)}$ given by the E_o contribution (3.3) ; and a line, at energy E_1, intrinsic to the two-dimensional character of the system. In general, $E_1<E_o$, and furthermore the impact of E_o is now reduced as E_1 attracts a large density of states. Using plausible simplifications in the model density of states, one finds /7/ that the E_1 contributions to $\chi^{(1)}$ and $\chi^{(3)}$ are

$$\chi^{(1)} \approx P^2/E_1^2,$$

$$\chi^{(3)} \approx P^2/E_1^5,$$

respectively , and by induction the conjecture was made /5/ that

$$\chi^{(2n-1)} \approx P^2/E_1^{3n-1},$$

where P is an average transition dipole moment matrix element.

In three-dimensional systems, besides the E_o and E_1 contributions derived above, $\chi^{(2n-1)}$ will also contain a contribution from a critical surface at energy E_2, and in general $E_2 < E_1 < E_o$, so that for a given electron density the impact of E_o is further reduced. In particular, since ω_{cv} and Ω_{cv} vary much more rapidly as we move from E_o to E_2, the corresponding intraband terms also become more dominant. Again using simple considerations for the model density of states, one obtains /7/

$$\chi^{(1)} \approx P^2/E^{5/2}$$

$$\chi^{(3)} \approx P^2/E_2^{9/2}$$

and by induction /5/

$$\chi^{(2n-1)} \approx P^2/E_2^{2n+1/2}.$$

Clearly these scaling laws and the ones related to the E_1 contribution are subject to large uncertainties, as they were derived from a very approximate model for the density of states.

The important point we wish to stress here is that the contributions from the different critical regions are *additive* ; furthermore the energies E_o, E_1 and E_2 can be determined experimentally. Hence one has a very convenient way of evaluating $\chi^{(2n-1)}$ for a given compound or comparing different compounds without resorting to complicated quantum mechanical calculations. In particular, one can follow the implications of any changes of the valence electron density distribution of chemical or other origin.

Clearly to the extent that one wishes crystals with a large transparency region the most favorable case is the one with only the E_o contribution ; in other words, one-dimensional systems, other things being equal, possess the highest odd-n order nonlinearity per valence electron in one direction but clearly not in the other directions ; on the other hand $\chi^{(1)}$ does not show the same enhancement as $\chi^{(3)}$, neither does $\chi^{(2)}$. Thus the striking differences between the inorganic semiconductors like Ge and GaAs on one hand and the polydiacetylenes on the other find a simple and unified explanation quantitatively as well as qualitatively.

Clearly the scaling laws given above were derived within the one-electron approach where electron correlation is altogether neglected ; when the latter is taken into account the expressions of Ω_{cv} and $\chi^{(n)}$ become too complex to allow any analytical or even numerical treatment. However, we may remark that the many-body effects become relevant mostly when the light frequency ω approaches or coincides with one of the critical point energies, E_0, E_1 or E_2 ; well below all these energies these effects can be taken into account by renormalizing the one-electron parameters but the exponents should not be severely affected.

4. ZERO-DIMENSIONALITY SYSTEMS

These are isolated molecular systems like finite polymer chains, metal or semiconductor microcrystallites of size in all directions much smaller than the optical wavelength λ. In comparing such systems with the corresponding infinite ones previously discussed one has to take into account two major effects,the *quantum confinement* effect,which results in a spectral redistribution of the quantum states, and the *dielectric confinement* effect, which takes into account the local field amplitude redistribution because of the interface between the finite size system and its surrounding medium. Clearly both effects crucially depend on the boundary conditions and this greatly complicates the extraction of meaningful scaling laws for the nonlinear polarizabilities of the zero-dimensionality systems.

The dielectric confinement effect, to a good approximation, can be taken into account by the effective medium approach for composites, e.g. /19/ or /20/. The quantum confinement , on the other hand, needs careful consideration for each case ; certain trends however can be extracted by simple models.

Thus for linear conjugated carbon chains without bond alternation one obtains /3/, using a procedure similar to the one used to study the Stark ladder level configuration,

$$\alpha^{(2n-1)} \approx L^{2n+1}\, e^{2n}/a\beta^{2n-1} \tag{4.1}$$

valid for large $N \equiv L/a$, where L is the half-length of the chain, a is the interatomic distance and β is the resonance (hopping) energy between neighboring atoms. In particular one has for n=1 and n=2 respectively

$$\alpha \approx L^3 e^2/a\beta\,, \tag{4.2}$$

$$\gamma \approx L^5 e^4/a\beta^3\,. \tag{4.3}$$

Expression (4.2) was initially derived by DAVIES /21/ and (4.3) by RUSTAGI and DUCUING /10/ using the electron model to describe the electronic states /22/. These scaling laws should in principle apply for cyanine dyes ; however, it is difficult to experimentally test the fifth-power dependence of γ on N as predicted from (4.3). Indeed, as N increases, so does the wavelength of the first absorption peak and the frequency dependence of γ cannot be overlooked ; it may mask any N dependence /23/.

For conjugated carbon chains with bond alternation the scaling laws can only be extracted through extensive numerical calculations of the quantum mechanical expressions of the polarizabilities. If we denote by β_1 and β_2 the two resonance energies there and introduce the delocalization parameter

$$N_d = \frac{\beta_1 + \beta_2}{\beta_1 - \beta_2}, \qquad (4.4)$$

one can show that within the Hückel approximation the analogue of the tight binding approximation used for infinite one-dimensional systems, for a chain of 2N atoms and such that $N_d < N$,

$$\gamma_{2N} \approx N_d^6, \qquad (4.5)$$

which is the same (compare 3.5) as the one derived previously for an infinite one-dimensional semiconductor (e.g. bond alternated infinite chain). This is a very important result since it indicates the validity range of the scaling laws (3.5) and (3.3) derived for one-dimensional systems when defects are present along the chain, extrinsic (impurities) or intrinsic (conjugation defects), and interrupt the periodicity ; as long as the average distance between two neighboring defects is larger than the delocalizaiton length L_d defined by (3.1) and N_d defined by (3.6) or (4.4) the scaling law (3.3) does not break down.

Another class of zero-dimensionality systems that are presently attracting a lot of attention for fundamental as well as technological reasons are the metal or semiconductor microcrystallites embedded in a transparent dielectric like glass or water. Their nonlinear optical properties are being extensively studied /12-14,24-26/. In these systems one actually exploits the resonances in the spectrum of these microcrystallites, like the surface plasma resonance in the metallic particles or the band to band transitions in the semiconductor particles. Because of the presence of resonances the expressions of the polarizabilities contain parameters specific to the resonances and this greatly reduces the scope of the scaling laws ; the quantity of interest here is $\chi^{(3)}(\omega,-\omega,\omega)$ since one may multiresonantly enhance this nonlinearity, which is related to the optical Kerr effect.

There have been attempts to obtain scaling laws for the third-order susceptibility of semiconductor microcrystallites with pronounced quantum confinement effects but the situation there is still not clear either experimentally or theoretically /27,28/ . The case of the metallic particles has also been studied /13,26/ and it was found that scaling laws played a crucial role there in deciding which of several mechanisms (three) determined the optical Kerr effect.

5. CONCLUSIONS

In the preceding discussion we outlined an approach to derive scaling laws for the nonlinear optical susceptibilities and summarized the main results. These laws pertain to the nonresonant valence electron contribution to the nonlinear susceptibilities ; this contribution is the dominant one in the transparency region of the nonlinear materials, the one extending below the lowest electronic transitions and above the vibrational resonances. The derivation of these laws was based on an analysis of the critical points in the joint density of states and an exploitation of the topological features of these critical points. Thus it was shown that these are of three types, denoted E_0, E_1 and E_2 contributions, coming from a point, a line and a surface in the joint density of states space respectively with $E_2 < E_1 < E_0$; one-dimensional semiconductors contain only a E_0 contribution, two-dimensional ones a E_0 and a E_1 contribution and three-dimensional ones all three contributions. These contributions are *additive* and are restricted only by some general sum rules : the total density of states is constant and other similar rules.

These contributions reflect topological features of the valence electron density distribution in k-space and the exponents of the scaling laws critically depend on the dimensionality. As a general rule, one-dimensional semiconductors possess the highest odd-order nonlinearities per valence electron.

These scaling laws only apply to the odd-order susceptibilities and not to the even-order susceptibilities ; the latter are sensitive to local properties in real space like charge asymmetry while the critical point analysis concerns properties in k-space.

In addition these laws were derived within the one-electron description and electron correlation and other many-body effects may affect the exponents, but not severely.

Another important feature of the above analysis is that for systems with large odd-order susceptibilities the dominant contribution comes from intraband mechanisms. As the overlap between valence electron clouds increases and the bands become wider, E_2 and, to a lesser extent, E_1 move to lower energies and their relative contribution increases and even dominates the E_0 contribution ; this occurs at the expense of the transparency region of the crystal.

Scaling laws for the nonlinear susceptibilities as for other physical quantities serve many purposes, fundamental as well as practical. They allow one to single out the dominant contributions and to establish relationships between macroscopic quantities which are very useful in predicting the nonlinearities of new materials by simple inspection, they introduce substantial economy in comparing different materials, in formulating guidelines for the search for new ones and also in setting up bounds on the possible variation of the nonlinear susceptibilities. In particular they allow one to predict the modifications that the nonlinear susceptibilities will incur as the dimensionality of the valence electron distribution is reduced an important aspect in modern technology.

6. REFERENCES

1. C. Flytzanis in "Treatise of Quantum Electronics", H. Rabin and C.L. Tang, Eds, vol 1, part A, (Academic Press, New York 1975)
2. C. Sauteret, J.P. Hermann, R. Frey, F. Pradère, J. Ducuing, R.R. Chance and R.H. Baughman, Phys.Rev.Lett. 36 956 (1976)
3. G.A. Agrawal, C. Cojan and C. Flytzanis, Phys.Rev. B17, 776 (1978)
4. A. Garito and K. Singer, Laser Focus, February 1982 p.59
5. C. Flytzanis, in "Nonlinear Optical Properties of Organic Molecules and Crystals", D. Chemla and J. Zyss, Eds, vol 2 p.121 (Academic Press, New York 1987)
6. W. Ashcroft and N.D. Mermin, "Solid State Physics" (Holt Saunders, Tokyo 1981)
7. M. Cardona and F.H. Pollack, in "Optoelectronic Materials", G.A. Albers, ed. (Plenum, New York 1971)
8. J.A. van Vechten, M. Cardona, D.A. Aspnes and R.M. Martin, Proc.Int. Conf.Semicond. 1970, p. 82
9. G.P. Agrawal and C. Flytzanis, Chem.Phys.Lett. 44, 366 (1976)
10. K. Rustagi and J. Ducuing, Opt.Comm. 10, 258 (1972)
11. K. Rustagi and C. Flytzanis, Opt.Lett. 9, 344 (1985)
12. D. Ricard, P. Roussignol and C. Flytzanis, Opt.Lett. 10, 511 (1985)
13. F. Hache, D. Ricard, C. Flytzanis, J.Opt.Soc.Am. B3, 1647 (1986)
14. S. Schmitt-Rink, D.A.B. Miller and D.S. Chemla, Phys.Rev. 35, 8113 (1987)
15. R.C. Miller, Appl.Phys.Lett. 50, 17 (1964)
16. C. Flytzanis and J. Ducuing, Phys.Rev. 178, 1218 (1969)

17. V.N. Genkin and P.M. Mednis, Sov.Phys.JETP (Engl.Transl.) 27, 609 (1968) ; Eksp.Teor.Fiz. 54, 1137 (1967)
18. D. Penn, Phys.Rev. 128, 2093 (1962)
19. J.C. Maxwell-Garnett, Philos.Trans.R.Soc. (London) 203, 385 (1904) ; 205, 237 (1906)
20. D.A.G. Bruggeman, Ann.Phys. Leipzig 24, 636 (1935)
21. P.L. Davies, Trans.Far.Soc. 48, 789 (1952)
22. H. Kuhn, J.Chem.Phys. 16, 840 (1948)
23. J.P. Hermann and J. Ducuing, J.Appl.Phys. 45, 5100 (1974)
24. P. Roussignol, D. Ricard, J. Lukasik, C. Flytzanis, J.Opt.Soc.Am. B4 5 (1987)
25 P. Roussignol, D. Ricard, C. Flytzanis, and N. Neuroth, Phys.Rev.Lett. 62, 312 (1989)
26. F. Hache, D. Ricard, C. Flytzanis and U. Kreibig, to be published
27. T Takagahara and H. Hanamura, Phys.Rev.Lett. 56, 2533 (1986)
28. T. Takagahara, Phys.Rev. B36, 9293 (1987)

Fundamental Nonlinear Limits and Generation of Quantum Light States

R.E. Slusher, S.L. McCall, A.Mysyrowicz, S. Schmitt-Rink, and K. Tai*

AT&T Bell Laboratories, Murray Hill, NJ 07974, USA
*Permanent address: GPS, Ecole Normale Superieur, Paris, France

I. Introduction

Nonlinear optical interactions have recently been shown to produce a new class of light called squeezed light.[1] The squeezed state is characterized by noise reduction in one of its quadrature-phase field amplitudes below that of ideal laser light (a coherent state) or the normal vacuum state. The noise in the other phase quadrature is correspondingly increased in order to satisfy the quantum uncertainty relation. This new class of light can only be described in quantum terms in contrast to the classical description of light fields which serves well for almost all previous light sources and nonlinear optical interactions. These new quantum states of light have been shown to allow measurement precision below the shot-noise limit for interferometric measurements.[2,3] It is possible to envision optical networks or computers that operate in the quantum noise limit where these new states of light could minimize the fundamental limits on noise and energy consumption. In this paper we will describe the nonlinear optical parameters required to optimize generation of squeezed light. It is clear that the parameters of nonlinear optical materials presently developed are near the required limits. If new materials with larger nonlinearities and lower loss than those presently available are possible, these new nonclassical light effects may become commonplace in optical switching, laser stabilization, interferometric measurement and optical communications.

Squeezed light can be generated by a parametric amplifier. A general schematic of a parametric amplifier is shown in Fig. 1(a). An intense pump beam drives a polarization wave in the nonlinear medium of length L that couples the pump wave to signal and idler waves. Either $\chi^{(2)}$ or $\chi^{(3)}$ processes can be used for the nonlinear coupling. In order to obtain large quantum noise reduction the nonlinear phase shift

$$\Delta\phi_{NL} = \frac{2\pi L}{\lambda}\,\Delta n\,, \tag{1.1}$$

must be of order π where λ is the vacuum optical wavelength of the signal and Δn is the effective nonlinear change in refractive index due to the pump. For example, for $\chi^{(3)}$ processes

$$\Delta n = n_2 I\,, \tag{1.2}$$

where I is the pump intensity in the nonlinear interaction region and n_2 is the nonlinear refractive index. The π nonlinear phase shift converts a large fraction of any incident idler beam into the signal beam. This is just the condition for optical switching. If the incident signal and idler beams are vacuum states, they are converted to a quantum correlated signal and idler "pair of photons" which can be used to create a number of interesting quantum light states. Recombining the signal and idler beams forms a squeezed state. This recombination can be accomplished by an optical cavity, forward phase matching, polarization selection and a number of other techniques.[1]

The quantum coherence of the signal and idler beam in a parametric amplifier makes a high quality paramp a nearly ideal element for many optical operations. In the same parameter range that quantum light states are generated, the paramp can serve as a "noiseless" amplifier, a

Springer Proceedings in Physics, Vol. 36
Nonlinear Optics of Organics and Semiconductors
Editor: T. Kobayashi

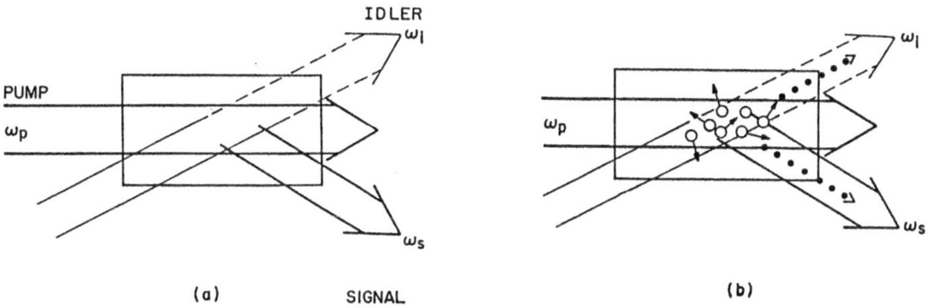

Fig. 1 - A schematic drawing of the ideal parametric amplifier/switch is shown in (a) and a parametric amplifier using real carriers is shown in (b). Diffusion of the carriers can lead to the scattering of pump light into the signal and idler modes as shown by the dotted lines.

tunable oscillator, a beam splitter or a fast low-power switch. Although large pump intensities are required to obtain $\Delta\phi_{NL} \sim \pi$, the pump is not strongly absorbed. In fact, extrapolation to the picosecond switching regime with attojoule switching energies is realistic. Optical Fredkin gate switches have even been proposed[4] where no energy is dissipated in the switching process. These switches are reversible, i.e. they do not destroy information.

It is clear that the quantum coherence of the signal and idler beams created by the parametric process can be destroyed by random deletion of one component of the pair due to linear losses. Thus an essential requirement for generation of quantum light states is that the linear losses be small, i.e.,

$$\alpha L \ll 1, \tag{1.3}$$

where α is the linear absorption coefficient. In general a dissipative loss attenuates the quantum light signal *and* introduces extra noise from the vacuum fluctuations always present in each mode of the light field in its minimum energy state. If a fraction $(1-\eta)$ of the signal and idler beams are lost the noise reduction ratio is reduced by a factor of η and extra noise from the vacuum is added by a factor of $(1-\eta)$. Losses act as a beam splitter with reflectivity $(1-\eta)$ which attenuates the quantum light state and adds in a beam of vacuum fluctuation noise.

The requirements of low loss, large $\chi^{(2)}$ or $\chi^{(3)}$, and no scattering of the pump beam into signal and idler modes (see section II) leads one to search for near resonant nonlinear mechanisms where no real carriers are generated. These requirements typically reduce the nonlinear interactions and necessitate large pump intensities in order to obtain $\Delta\phi_{NL} \sim \pi$. On the other hand, by detuning from a resonance, virtual excitation processes are very fast and are not limited by real carrier lifetimes. For example, a 10 meV detuning from a band edge means response times near 0.1 ps.

In this discussion we will concentrate on semiconductor nonlinearities. However, $\chi^{(2)}$ or $\chi^{(3)}$ processes in organic materials are also promising. First, a simple model is described which shows that pump scattering from real carriers in $\chi^{(3)}$ processes can result in unacceptable randomly scattered photons into the signal and idler modes. Limits to present squeezing experiments are described and a criteria for evaluating nonlinear materials is proposed. Next a model of $\chi^{(3)}$ nonlinearities near an exciton resonance is used to show that the crucial $\chi^{(3)}/\alpha$ ratio is independent of dimensional scaling. A promising new enhancement of $\chi^{(3)}$ and $\chi^{(2)}$ at the midgap energies in semiconductor quantum multiwells is reviewed. Finally initial results of two experiments to measure $\chi^{(3)}$ for GaAs in the lossless regime are presented.

II. A Scattering Limit

Consider a nonlinear interaction caused by excitation of real carriers in a semiconductor. Examples of the interaction include nonparabolic band effects,[5] free carrier (Drude) refraction,[6] and band filling.[7] We will consider a simple model where each electron changes the susceptibility of the material by an amount δX in a small volume δV resulting in scattering of pump light into the signal and idler fields as shown in Fig. 1(b). Models using extended Bloch states for the electrons and holes give similar results unless the electron density is high enough that degeneracy effects become important. Also assume that the electrons are mobile and experience a primarily diffusive motion characterized by a diffusion constant D. We assume that the electrons have a lifetime τ and that a $\chi^{(3)}$ parametric process is used, e.g. near forward four-wave-mixing. Each electron has a scattering cross-section

$$\sigma=\frac{4\pi}{3}k^4\left[(\delta X\delta V)-\frac{e^2}{m^*\omega^2}\right]^2, \tag{2.1}$$

where k is the vacuum wavevector of the pump light. The factor $e^2/m^*\omega^2$ accounts for free carrier Thomson scattering from the electron with effective mass m^*. The number of excited electrons in the interaction volume with cross-sectional area A is

$$N_e = A\dot{N}\alpha L\tau, \tag{2.2}$$

where $\dot{N}$ is the incident pump photon flux. It is assumed that each absorbed photon results in an electron-hole pair. The diffusive motion of the electron will determine the spectrum of scattered light for sufficiently small momentum transfer

$$\Delta k = |\mathbf{k}_p - \mathbf{k}_{s,i}|, \tag{2.3}$$

where $\mathbf{k}_p$, $\mathbf{k}_s$ and $\mathbf{k}_i$ are the wavevectors of the pump, signal and idler beams. Diffusion and carrier lifetime result in a spectrum of scattered radiation with a Lorenzian frequency profile and a frequency half-width

$$\frac{1}{\tau_g}=\frac{1}{\tau}+D\Delta k^2. \tag{2.4}$$

This frequency spread is the inverse of the time constant of the grating of carriers formed by the interference of the pump and signal beams. Faster diffusion results in a broader bandwidth for the scattered spectrum. Consider the light scattered into a mode solid angle $\frac{\lambda^2}{2\pi\varepsilon A}$ where ε is the dielectric constant. Then the number of scattered photons per second per unit bandwidth into the mode is

$$S = N_e\dot{N}\sigma\frac{\lambda^2}{2\pi\varepsilon A}\tau_g. \tag{2.5}$$

This scattering is a consequence of the *random* spatial distribution of carriers in the interaction region. A similar result is obtained for randomly distributed bound carriers, e.g. excitons.

The key point is the relation of this scattering to the nonlinear phase shift. These two effects are fundamentally related since the same polarization is responsible for both. In other words the forward scattering is just the nonlinear change in the refractive index

$$\Delta n = \frac{2\pi(\delta X\delta V)N_e}{LnA} \tag{2.6}$$

and the associated phase shift

$$\Delta\phi_{NL} = \frac{2\pi L \Delta n}{\lambda}, \tag{2.7}$$

which must equal π for large squeezing. Faster diffusion reduces Δn and requires a larger electron density for a large phase shift. This increases the scattering rate according to Eq. (2.4). Combining Eqs. (2.2), (2.4), (2.6) and (2.7) the scattering rate per unit bandwidth is

$$S = \frac{\pi \Delta\phi_{NL}}{\alpha L} \frac{\tau}{\tau_g}. \tag{2.8}$$

This is clearly bad news for reaching the regime for quantum noise reduction. Since $\alpha L \ll 1$ and the last factor on the right of Eq. (2.8) is greater than unity, the number of scattered photons in the mode is much larger than unity. This must be compared with quantum noise where one can think of the vacuum fluctuation level as being "1/2 photon" per mode, corresponding to the ground state of the harmonic oscillator. This means that one requires $S \ll 1$ for good squeezing. If the spectrum of the scattered light is sufficiently broad in frequency relative to the spectral regions where quantum noise reduction is of interest, the scattering is proportionally less and will not be fundamental limit. Unfortunately, typical values of D and Δk may make scattering a major problem. In a waveguide geometry ($\Delta k = 0$) the scattering can be thought of as a time dependent phase shift similar to the GAWBs noise in glass fibers[8] which limits quantum light effects.

This model is not completely general but it does indicate fundamental problems that may arise when trying to generate quantum states of light by exciting carriers in a nonlinear material. Of course scattering from defects, impurities, surfaces, etc. is also a major problem. One is led to concentrate on optimizing the near-resonant nonlinearities where bound electrons are polarized but do not scatter light from the pump because they are spatially ordered in nonrandom arrays. Alternately $\chi^{(2)}$ parametric downconversion may be used so that the spectrum of scattered pump light is easily filtered out.

III. Limits in Present Quantum Light Experiments

Nonlinear media used to generate quantum light states have included atomic Na beams,[1] $MgO{:}LiNbO_3$[1], KTP[9] and glass fibers.[1,8] Except for exceptional crystals of $MgO{:}LiNbO_3$,[1] the nonlinear materials are partially responsible for limiting the degree of squeezing obtained. A recent example is the pulsed squeezing experiment[9] in progress in our laboratory using KTP for parametric downconversion. A $\Delta\phi_{NL} \sim 0.6$ is obtained for 100 ps pump pulses with peak intensities near 10 MW/cm^2. We use a repetitive pulse train to avoid Fourier components of the pulsed homodyne detection currents from obscuring the quantum noise reduction. In this experiment the noise reduction obtained at present is 1.3 dB below the shot noise limit, i.e. we have 1.3 dB of squeezing. This is limited by (1) available pump power, (2) heating in the KTP crystal due to absorption at the pump frequency (probably due to impurities), (3) losses in optical components and crystal coating and (4) imperfections in the spatial mode patterns and wavefronts.

The crystal length used in these experiments is ~1cm. This length limits the degree of focusing for the pump beam and thus the $\Delta\phi_{NL}$ that can be obtained. A waveguide structure can strongly increase the effective intensity-length product leading to much larger $\Delta\phi_{NL}$. The penalty when using waveguides is the difficulty and losses incurred in coupling into and out of the waveguide structure. The method typically used at present to enhance the pump coupling is an optical cavity where either or both the pump and signal/idler modes are resonant. In this case the penalty is a loss in bandwidth, mirror losses, and the difficulty of stabilizing the pump frequency to the cavity frequency. For KTP in a cavity, thermal instabilities in the crystal (due to absorption of the pump) limit the squeezing output.

The switching energy in each pulse is 20 nJ for the present experiments in KTP. A very small fraction of this energy is actually converted to the signal and idler beam. The pump light could be recycled many times. In principle, the energy used for squeezing or parametric switching at the quantum limit (i.e. a small number of photons in the signal beam) is only the gain (hopefully 10-100) times the photon energy, i.e. only attojoules of energy.

It is clear that one needs a factor of ten or greater improvement in the ratio of $\Delta\phi_{NL}/\alpha L$ in order to approach the regime of ideal parametric processes. A reasonable criteria for evaluating nonlinear mechanisms is the evaluation factor

$$\xi = \frac{2\pi\Delta n_{max}}{\alpha\lambda} \gg 1\,, \tag{3.1}$$

where for $\chi^{(3)}$ processes

$$\Delta n_{max} = n_2 I_s\,. \tag{3.2}$$

The saturation intensity I_s typically limits the maximum nonlinear index change (if there is no damage or heating problem). This evaluation factor represents the maximum value for $\Delta\phi_{NL}$. It must be at least 10 for significant generation of quantum light states or nearly ideal parametric switching.

For two examples, consider saturation of a two-level atom and the exciton in GaAs. In the atomic case the absorption decreases as Δ^{-2} for the Lorenzian resonance line shape, where Δ is the number of natural line widths for detuning of the pump. Saturation intensity scales inversely with the square of the dipole moment and directly with Δ^2. The nonlinear index n_2 scales as Δ^{-3}. For Na atoms with a density of $10^{12}cm^{-3}$

$$\xi \sim 0.1\Delta\,, \tag{3.3}$$

so that reasonable values of $\Delta\phi_{NL}$ are obtained for $\Delta \gtrsim 10^2$. As Δ is increased one is eventually limited by the available pump intensity.

For GaAs the absorption below the exciton resonance can in principle decrease exponentially (Urbach tail) as one detunes the pump below the exciton feature. For virtual excitations, the nonlinear index decreases as a power law, $\Delta^{-1.5}$ for detuning much larger than the exciton binding energy E_b. The saturation intensity should increase as a power law. A shaky estimate is $\xi \sim 30$ at 50 meV below the exciton at room temperature. Impurities and heating may result in excessive real carrier excitation as one approaches the exciton resonance from below. In any case the ξ parameter is not extremely encouraging.

In summary, presently available nonlinear media marginally meet the required parameters for ideal quantum light generation. It is clear from the above examples that Δ is crucial in determining the useful nonlinear parameter range. *It is important when discussing any optical nonlinearity to give a value for ξ and to specify Δ.* We now turn to the question of "How large can ξ be?" It seems clear that it can be at least an order of magnitude or two larger than available at present. However, at some point this crucial ratio reaches a fundamental limit. What is that fundamental limit?

IV. Dimension Scaling of $\chi^{(3)}/\alpha$

While extremely nonresonant nonlinearities *cannot* depend on dimensionality, scaling of exciton effects to lower dimensions has been shown to strongly increase the effective oscillator strength and the corresponding nonlinear response *near* the exciton resonance. However, for a simple model[10,11] where saturation or "phase space filling" dominates the nonlinear response, both $\chi^{(3)}$ *and* α increase in the same fashion as the dimensionality is decreased. Near the lowest 1s exciton energy, the linear susceptibility can be written as

$$\chi^{(1)} = 2e^2 r_{cv}^2 \, |\phi_{1s}(r=0)|^2 \frac{1}{E_{1s} - \hbar\omega - i\Gamma}. \tag{4.1}$$

The unit cell valence-to-conduction band dipole strength is $e^2 r_{cv}^2$. The 1s state energy is E_{1s} and Γ is the exciton energy width. The probability of the bound electron/hole excited state being in the unit cell is the square of the 1s wavefunction ϕ_{1s} evaluated at the relative coordinate $r=0$. This probability factor scales as the reciprocal of the volume of the exciton, approximately equal to the Bohr radius cubed, a_B^3. As the dimension L is decreased below a_B the probability factor and corresponding effective oscillator strength is increased by $(\frac{a_B}{L})^{3-D}$ where D is the dimensionality (0D, 1D, 2D, and 3D). The linear absorption ($\mathrm{Im}\chi^{(1)}$) correspondingly increases assuming a high density of spatially uniform, lower dimension structures can be maintained in the interaction volume. For example, for quantum dots, there need to be many quantum dots in a sphere with the Bohr radius. This may be difficult to achieve in many systems since interactions between the dots will be an important factor.

The nonlinear response for the simple model considered here is predominantly caused by what is often called phase space filling, i.e. when the density of virtually excited excitons begins to result in overlap of the exciton wavefunctions. In this same regime the effective exciton resonance frequency as measured by a test beam is shifting due to the AC stark effect. For a detuning $\Delta \gg E_b$ the nonlinear response is nearly identical to the atomic case except that the higher lying band states also make a significant contribution. For moderate and small values of Δ, the 1s exciton will make the leading contribution and phase space filling, equivalent to saturation of an atomic two level system, will dominate. The relative change in susceptibility with pump intensity is then[11]

$$\frac{E^2\chi^{(3)}}{\chi^{(1)}} \sim \frac{\sum_k \{[1-fe(k)-f_h(k)]\phi_{1s}(k)-\phi_{1s}(k)\}\phi_{1s}^*(r=0)}{|\phi_{1s}(r=0)|^2}$$

$$= \frac{-\sum_k [f_e(k)+f_h(k)]\phi_{1s}(k)}{\phi_{1s}(r=0)} = -\frac{N}{N_s}, \tag{4.2}$$

where $f_e(k)$ and $f_h(k)$ are the Fermi distributions for the electron and hole as a function of momentum, $\phi_{1s}(k)$ is the 1s wavefunction in k space ($|\phi_{1s}(k)|^2 \sim |Vol_{1s}|$, the volume of the 1s wavefunction), N is the number of virtually excited excitons and N_s is the saturation density for excitons in the 1s state, equal to $1/|Vol_{1s}|$. The pump intensity is directly related to the square of the electric field E. For real excitons,

$$N \sim \frac{\tau\alpha_o}{\hbar\omega}\left(\frac{\Gamma}{\Delta}\right)^2\left(1-\frac{N}{N_s}\right)I, \tag{4.3}$$

where α_o is the linear absorption at the peak of the exciton resonance, I is the incident pump intensity and

$$\Delta = (E_{1s} - \hbar\omega). \tag{4.4}$$

For virtual excitons, τ is replaced by $1/2\Gamma$. The effective nonlinearity is larger at lower dimension because α_o increases as dimension decreases as described above. However the important point is that the ratio $\chi^{(3)}/\alpha$ is constant as dimensionality decreases. Thus, the increased nonlinear response achieved at lower dimensionality is not useful in quantum light generation and many other applications.

V. Electric Field Enhanced Nonlinearities

It has recently been suggested that electric fields applied to quantum well structures give rise to optical rectification[12] and enhance the $\chi^{(3)}$ and $\chi^{(2)}$ susceptibilities.[13–15] The applied field distorts the quantum well potential resulting in a mixing of the various exciton states. This mixing results in a "giant" intermediate state dipole matrix element in the calculation of both $\chi^{(3)}$ and $\chi^{(2)}$. Recent experiments suggest that the decrease in optical matrix element which results from the applied field may outweigh the electric field enhancement factors for $\chi^{(3)}$ near the exciton resonance.[16] We have started an investigation of the electric field enhanced $\chi^{(3)}$ processes at midgap. At midgap the linear absorption is very low so that the $\chi^{(3)}/\alpha$ values should be very large.

Experimentally the $\chi^{(3)}$ process at midgap is being studied by measuring the two-photon absorption coefficient (Im $\chi^{(3)}$). Quantum multiwells of GaAs (40Å) and barriers of $Al_xGa_{1-x}As$ (100Å) are pumped by a pulsed F-center laser directed in the plane of the wells in a planar waveguide configuration. The two-photon luminescence or the current across a p-i-n structure is measured which is proportional to the two-photon absorption. Initial results at zero applied field are shown in Fig. 2. Two photon absorption begins at a frequency corresponding to the two-photon transition from ground state, via the 1s state, to the 2p exciton level. In this case the 1s to 2p portion of the transition has a "giant" dipole moment comparable to the one predicted to enhance the $\chi^{(3)}$ with an applied electric field. The initial estimates indicate that $\chi^{(3)}$ is not as large as the more optimistic calculated estimates. $\chi^{(3)}$ appears to be in the range below 10^{-10}esu at zero field. The electric field enhanced value of $\chi^{(3)}$ extrapolated from the present experimental results will not be much larger than 10^{-10}esu. Two-photon absorption will also limit the pump intensity that can be used so that the ξ parameter (Eq. (3.1)) may not be as large as required for the ideal paramp regime.

Electric field enhancement of the $\chi^{(2)}$ susceptibility in GaAs is predicted to be a factor 20 larger than the bulk GaAs $\chi^{(2)}$ without electric field (or 400 times the $\chi^{(2)}$ in $LiNbO_3$). This means a reasonably large value of $\xi > 10$ since the electric field can be spatially varied to achieve phase matching. However, absorption of the pump, which is near the exciton absorption may limit the useful nonlinear phase shifts that can be achieved.

This large predicted value for $\chi^{(2)}$ due to electric field distortion of the quantum wells may approach a fundamental limit for optical nonlinearities in semiconductor materials. A large dipole is produced corresponding to the separation of the electron and hole at opposite sides of the quantum well with a width near 100Å. Separations larger than this result in a decrease in optical matrix element because the overlap of the wavefunctions for the electron and hole decreases.

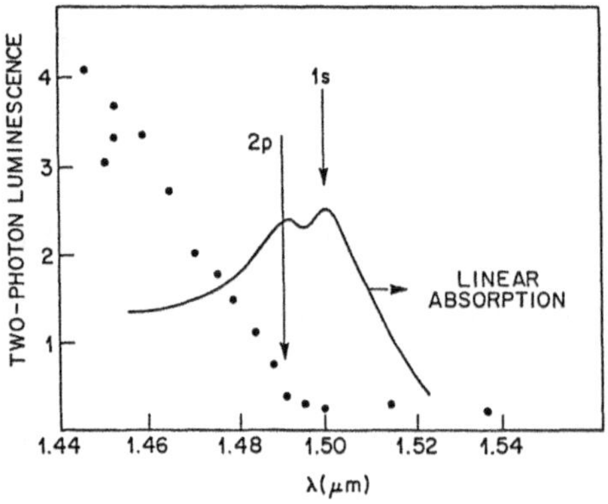

Fig. 2 - Two-photon luminescence is shown as a function of wavelength for 40Å GaAs wells and 100Å $Al_xGa_{1-x}As$ barriers at 12 K. The applied electric field is zero. The solid line is the linear absorption spectrum of the same sample plotted at one-half its measured wavelength. There is no two-photon luminescence (absorption) at the position corresponding to the 1s exciton final state as predicted for zero applied field. The two-photon luminescence begins as expected near the 2p final state level.

VI. Polarization Interferometer Measurements of $\chi^{(3)}$

It is important to measure $\chi^{(3)}$ over extended frequency ranges with small enough pump intensities so that only the basic nonlinearities are observed (e.g. no heating, carrier generation, etc.). Also many of the interesting nonlinear systems may have relatively small interaction lengths so that high measurement sensitivity is required. We have used a polarization interferometer to measure $\chi^{(3)}$ by the AC Kerr effect. By using a frequency shifted, modulated pump the shot-noise limited sensitivity can be achieved. Heating is not a problem with this technique, since only a *rotation* of polarization is measured which is not influenced by thermal changes in refraction index.

The experimental apparatus is shown in Fig. 3. A mode-locked Nd:YAG laser operating at a wavelength of 1.06 μm is used as both the pump, to induce a birefringent index change in GaAs, and as the excitation beam for a simple polarization interferometer to detect the induced birefringence. The pump beam is modulated and frequency shifted by two acousto-optic modulators so that the resulting signal at 50 MHz is in a "quiet" region of the detected photocurrent spectrum where only shot-noise limits the sensitivity.

A measurement of the AC Kerr effect due to $\chi^{(3)}$ is shown in Fig. 4. The signal to noise ratio is 34 dB for a 400 μm interaction length in a GaAs crystal at room temperature and a pump intensity of 0.5 MW/cm^2. The measured value of n_2 is 4×10^{-10} cm^2/kW corresponding to a $\chi^{(3)}$ of 1.6×10^{-10} esu. These values are in good agreement with estimates from third harmonic generation.[17] The sensitivity achieved here should make possible accurate $\chi^{(3)}$ measurements over broad spectral ranges for interaction lengths as small as 1 μm.

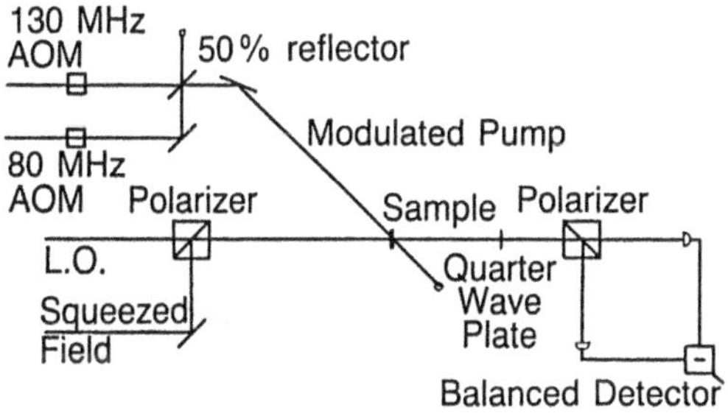

Fig. 3 - Apparatus for an AC Kerr effect measurement of $\chi^{(3)}$ in GaAs at a wavelength of 1.06 μm. All inputs are obtained from a mode-locked Nd:YAG laser. The local oscillator L.O. beam has an average power of 1 mW and the modulated pump beam average power is 1W. The polarization of the L.O. light propagating through the polarization interferometer, formed by the two polarizers, is rotated by the pump induced birefringence of the GaAs sample. The polarization rotation is at a frequency of 50 MHz (the difference between the acousto-optic modulator AOM frequencies) which is detected using a balanced pair of InGaAs photodiodes. Squeezed light can be introduced at the dark port of the interferometer to further increase the sensitivity of the apparatus.[3]

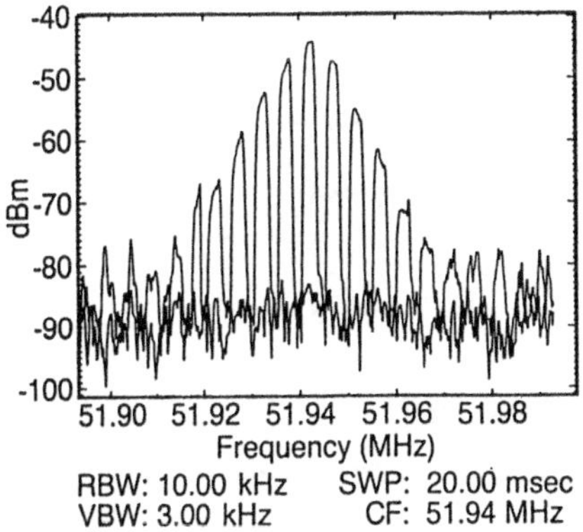

Fig. 4 - An AC Kerr effect measurement of the $\chi^{(3)}$ nonlinearity in GaAs at 1.06 μm is shown as obtained from the apparatus in Fig. 3. The detected photocurrent at the output of the balanced detector is shown on a logarithmic scale as a function of frequency. The basic noise at -90 dBm is due to the photodetector amplifier. Chopping the L.O. beam as frequency is scanned shows an increase in noise to the -82 dBm level due to light shot-noise when L.O. is on (at frequencies well separated from the pump modulation frequency, 51.94 MHz). At the modulation frequency a large 34 dBm signal/noise ratio is obtained due to polarization rotation induced by the pump beam in the GaAs sample.

VII. Conclusions

Semiconductor optical nonlinearities are promising for generation of quantum light states. However, larger values of the ratio of nonlinear phase shift to absorption are needed to achieve the nearly ideal parametric interactions required for quantum limited results. Real carrier nonlinearities can result in light scattering and time dependent phase shifts which degrade quantum limited parametric amplifier performance. Dimensional scaling increases both linear loss and nonlinearity so that no improvement in parametric amplifier performance is promised. Electric field enhanced nonlinearities are presently being studied to achieve large nonlinear phase shifts and low loss. Enhancement of the local fields near small structures due to discontinuities in the dielectric constant can enhance the nonlinear phase shift to linear loss ratio.[10] This regime is another promising direction for study. Yet another possibility are optic phonon mediated nonlinearities[18] for which $\chi^{(3)}/\alpha$ can be optimized by tuning the inverse Raman resonance slightly below the exciton line. It seems clear that the evaluation parameter we suggest to estimate the nonlinear phase shift to loss ratio can in principle be increased by several orders of magnitude before fundamental limits to the nonlinear mechanisms are reached.

REFERENCES

1. "Squeezed States of the Electromagnetic Field," J. Opt. Soc. Am. B*4*, 1450-1741 (1988).
2. Min Xiao, Ling-An Wu, and H. J. Kimble, Phys. Rev. Lett. *59*, 278 (1987).
3. P. Grangier, R. E. Slusher, B. Yurke, and A. La Porta, Phys. Rev. Lett. *59*, 2153 (1987).
4. Y. Yamamoto, private communication.
5. C. K. N. Patel, R. E. Slusher, and P. A. Fleury, Phys. Rev. Lett. *17*, 1011 (1966).
6. I. Ç. Khoo and R. Normandin, Appl. Phys. Lett. *52*, 525 (1988).
7. D. A. B. Miller, C. T. Seaton, M. E. Prise, and S. D. Smith, Phys. Rev. Lett. *47*, 197 (1981).
8. R. M. Shelby, M. D. Levenson, and P. W. Bayer, Phys. Rev. B *31*, 5244 (1985). R. M. Shelby, M. D. Levenson, S. H. Perlmutter, R. G. DeVoe and D. F. Walls, Phys. Rev. Lett. *57*, 691 (1986).
9. R. E. Slusher, P. Grangier, A. La Porta, B. Yurke and M. J. Potasek, Phys. Rev. Lett. *59*, 2566 (1987).
10. S. Schmitt-Rink, D. A. B. Miller and D. S. Chemla, Phys. Rev. B *35*, 8113 (1987).
11. For recent review, see S. Schmitt-Rink, D. S. Chemla and D. A. B. Miller, Adv. Phys., to be published.
12. D. S. Chemla, D. A. B. Miller, and S. Schmitt-Rink, Phys. Rev. Lett. *59*, 1018 (1987).
13. A. Shimizu, Phys. Rev. Lett. *61*, 613 (1988); Phys. Rev. B *37*, 8527 (1988).
14. L. Tsang, D. Ahn, and S. L. Chuang, Appl. Phys. Lett. *52*, 697 (1988).
15. M. Yamanishi, Phys. Rev. Lett. *59*, 1014 (1987).
16. W. H. Knox, J. B. Stark, D. S. Chemla, D. A. B. Miller and S. Schmitt-Rink, to be published.
17. W. K. Burns and N. Bloembergen, Phys. Rev. B *4*, 3437 (1971).
18. B. I. Greene, J. F. Müller, J. Orenstein, D. H. Rapkine, S. Schmitt-Rink and M. Thakur, Phys. Rev. Lett. *61*, 325 (1988).

A Simple Theory for Excitonic Optical Nonlinearity in Semiconductors

T. Hiroshima

Opto-Electronics Research Laboratories, NEC Corporation,
4-1-1 Miyazaki, Miyamae-ku, Kawasaki 213, Japan

Recently, excitonic optical nonlinearities in direct-gap-semiconductors have attracted much attention and have been studied extensively [1]. Of particular interest are the *nonresonant* excitonic nonlinearities for their potential applications to ultrafast all-optical devices. The nonlinear optical properties of exciton systems result, in general, from the deviation of excitons from non-interacting ideal bosons. Not only *the mutual interaction between excitons*, but also *the anharmonic interaction of excitons with the radiation field*, contribute to the excitonic optical nonlinearity. In this paper we develop a simple theory for nonresonant excitonic optical nonlinearity, treating the above mentioned two kinds of anharmonicity on an equal basis. The expression for the nonresonant third-order optical susceptibility $\chi^{(3)}(\omega;-\omega,\omega,-\omega)$ is given.

First of all, we construct the effective Hamiltonian in terms of Bose operators for a semiconductor in the presence of a classical monochromatic laser field $E^{*}exp(i\omega t)+Eexp(-i\omega t)$ in the transparency region below the lowest exciton resonance. This is accomplished by applying Usui's transformation [2] to the relevant Hamiltonian described in the electron-hole picture. This procedure was originally developed by Hanamura [3,4] for the boson formalism of the many exciton theory. The effective boson Hamiltonian in the rotating frame then takes the form:

$$H_{eff}=\sum_{\sigma}(\varepsilon_{x}-\hbar\omega)C_{\sigma}^{\dagger}C_{\sigma}+V_{x}\sum_{\sigma}C_{\sigma}^{\dagger}C_{\sigma}^{\dagger}C_{\sigma}C_{\sigma}-\left[\mu_{x}^{*}E\sum_{\sigma}\left(1-\gamma[C_{\sigma},C_{\sigma}^{\dagger}]_{+}\right)C_{\sigma}^{\dagger}+h.c.\right]. \quad (1)$$

In this Hamiltonian $C^{\dagger}_{\sigma}(C_{\sigma})$ is the creation (anihilation) operator for an exciton with electron spin σ and energy ε_{x}. This operator obeys boson commutation relations. The wave vectors of incident photons are negligibly small, so that the relevant translational wave vectors of excitons are taken to be zero. Furthermore, only the lowest (1*s*) exciton state is retained in Eq. (1). This is justified as long as the off-resonance energy, $\Delta=\varepsilon_{x}-\hbar\omega$, is smaller than or of the same order as the 1*s*-exciton binding energy. The second term on the right-hand side (rhs) of Eq. (1) represents the repulsive exciton-exciton interaction. This term stems from Pauli exclusion principle, and V_{x} is given by

$$V_{x}=-\frac{2}{\Omega^{3}}\sum_{k,k'}V(k-k')\left[\phi^{2}(k)\phi^{2}(k')-\phi^{3}(k)\phi(k')\right], \quad (2)$$

where Ω is the quantization volume of the system, $\phi(k)$ is the Fourier transform of the exciton envelope function, and $V(q)$ is the Coulomb potential. Note that V_{x} is inversely proportional to Ω. The first term on the rhs of Eq. (2) is an exchange self-energy due to fermion exchange (electron-electron and hole-hole exchange) and the second term is a vertex correction which describes the reduction of the Coulomb interaction between an electron and a hole due to Pauli exclusion principle. The third term on rhs of Eq. (1) represents the anharmonic interaction of excitons with the radiation field. This term comes from Usui's transform of the electron-radiation field interaction:

Springer Proceedings in Physics, Vol. 36
Nonlinear Optics of Organics and Semiconductors
Editor: T. Kobayashi

$$H' = -\mu^*_{cv} E \sum_{\mathbf{k},\sigma} a^\dagger_{\mathbf{k},\sigma}\, d^\dagger_{-\mathbf{k},-\sigma} - h.c. \ ,$$

where $a^\dagger_{\mathbf{k},\sigma}$ ($d^\dagger_{\mathbf{k},\sigma}$) is a creation operator for an electron (hole) with wave vector **k** and spin σ, and μ_{cv} is the transition dipole moment between conduction and valence bands. The transition dipole moment between an exciton and the ground state of the crystal is denoted by μ_x, and is given by

$$\mu_x = \Omega^{-1/2} \mu_{cv} \sum_{\mathbf{k}} \phi(\mathbf{k}) \ ,$$

and γ in Eq. (1) is defined as follows:

$$\gamma = \Omega^{-1} \sum_{\mathbf{k}} \phi^3(\mathbf{k}) / \sum_{\mathbf{k}} \phi(\mathbf{k}) \ .$$

The third term on rhs of Eq. (1) is also rewritten in the mean-field approximation as follows:

$$-\left[\mu^*_x E \sum_{\sigma} (1 - \Omega\gamma n_{exc}) C^\dagger_\sigma + h.c.\right] ,$$

where n_{exc} is the density of excitons created virtually by the laser field. The transition dipole moment is reduced by the factor $(1-\Omega\gamma n_{exc})$. This is nothing but the saturation effect due to phase-space filling [5]. Therefore, the factor $\Omega\gamma$ gives the inverse of the saturation density of excitons due to phase-space filling.

On the basis of the Hamiltonian [Eq. (1)], we can calculate the nonlinear polarization driven by the laser field, or the nonlinear optical susceptibility. Here and after it is assumed that E, μ_x, and γ are real without loss of generality. The ground state of the system of coherently pumped excitons described by the Hamiltonian [Eq. (1)] is written by $U|0\rangle$, where $|0\rangle$ is the vacuum of the exciton system, and U is a unitary operator. We assume that U has the following form [6,7]:

$$U = exp\left[\sum_\sigma x_\sigma (C^\dagger_\sigma - C_\sigma)\right] ,$$

where x_σ is the variational parameter (c number) determined by the requirement that $E = \langle 0|U^{-1}H_{eff}U|0\rangle$ should be minimized. From $\partial E/\partial x_\sigma = 0$, the following equation is obtained:

$$\Delta x + 2V_x x^3 - \mu_x(1-3\gamma)E + 6\mu_x E\gamma x^2 = 0 \ , \tag{3}$$

where we put $x = x_\sigma = x_{-\sigma}$ without loss of generality. Equation (3) can be solved by the iteration in E, and the result is

$$x = \frac{\mu_x(1-3\gamma)E}{\Delta}\left\{1 - \frac{2\mu_x^2(1-3\gamma)^2E^2}{\Delta^2}\left[\frac{(1-3\gamma)V_x}{\Delta} + 3\gamma\right] + \cdots\right\} . \tag{4}$$

Polarization density P driven by the laser field is given by the expectation value of the polarization operator:

$$P = \Omega^{-1}\langle 0|U^{-1}\mu_x \sum_\sigma C_\sigma \left(1 - \gamma[C_\sigma, C^\dagger_\sigma]_+\right) U|0\rangle \tag{5a}$$

$$= 2\Omega^{-1}\mu_x x(1 - 3\gamma - 2\gamma x^2) \ . \tag{5b}$$

From Eqs. (4) and (5b), P is calculated as a power series of E, from which the third-order optical susceptibility $\chi^{(3)}(\omega;-\omega,\omega,-\omega)$ for degenerate four-wave mixing is obtained as follows:

$$\chi^{(3)}(\omega;-\omega,\omega,-\omega) = -\frac{4\mu_x^4}{\Omega\Delta^4}(V_x + 4\Delta\gamma) \ , \qquad (6)$$

where we put Ω to be infinity. Note that $\chi^{(3)}$ [Eq. (6)] is independent of Ω. The effect of damping is neglected in Eq. (6), however this is justified as long as the off-resonance energy is much larger than the linewidth due to the transverse relaxation of the exciton. Equation (6) shows that $\chi^{(3)}$ is composed of two parts; the effect of the mutual interaction between excitons (V_x) and that of the phase-space filling ($4\Delta\gamma$). The ratio $4\Delta\gamma/V_x$ is a simple function of Δ and E_B, the binding energy of the bulk exciton. In two-dimensions (2D) $V_x^{(2D)} = 4\pi E_B a_B^2(1-315\pi^2/2^{12})/\Omega$ and $\gamma^{(2D)} = 2\pi a_B^2/(7\Omega)$ so that $4\Delta\gamma^{(2D)}/V_x^{(2D)} = 1.19\Delta/E_B$, where a_B is the effective Bohr radius of the bulk exciton. In three-dimensions (3D) $4\Delta\gamma^{(3D)}/V_x^{(3D)} = 3.23\Delta/E_B$. These results show that the saturation effect due to phase-space filling on the excitonic optical nonlinearity works more effectively in the 3D case than in the 2D case. However, it should be noted that the effect of phase-space filling of excitons is not so pronounced for materials with large E_B. When we use the numerical value for CuCl: $E_B = 200$ meV, and choose the off-resonance energy $\Delta = 10$ meV, $4\Delta\gamma^{(3D)}/V_x^{(3D)} = 0.16$ and $4\Delta\gamma^{(2D)}/V_x^{(2D)} = 0.06$. If we choose the situation of GaAs: $E_B = 5$ meV, and $\Delta = 10$ meV, $4\Delta\gamma^{(3D)}/V_x^{(3D)} = 6.46$ and $4\Delta\gamma^{(2D)}/V_x^{(2D)} = 2.38$. For materials with small E_B, such as GaAs, the effect of phase-space filling on nonresnant excitonic optical nonlinearity is of crucial importance for off-resonance energies of an order of 10 meV.

In conclusion, we have formulated the nonresonant excitonic optical nonlinearity, treating both the mutual interaction between excitons and the anharmonic interaction of excitons with the radiation field on an equal basis. It has been pointed out that the latter effect is nothing but the phase-space filling of excitons and that this works more effectively in three-dimensions than in two-dimensions. The saturation effect due to phase-space filling on nonresonant excitonic optical nonlinearity is quite small for materials with large exciton binding energy, such as CuCl. However, this effect is not negligible even for the off-resonance energy of a few meV, and dominates over the effect of the exciton-exciton interaction for larger off-resonance energies in GaAs and other III-V semiconductors. These results are of great importance for microscopic understanding of nonresonant excitonic optical nonlinearity, such as dispersive optical bistability in GaAs and other III-V semiconductors.

The author would like to thank Professor E. Hanamura (University of Tokyo) and Professor M. Yamanishi (Hiroshima University) for informative discussions. Thanks are also due to Dr. H. Yokoyama and Dr. R. Lang for their helpful comments.

REFERENCES

1. A special issue on excitonic optical nonlinearity, J. Opt. Soc. Am. B **2**, No.7 (1985).
2. T. Usui, Prog. Theor. Phys. **23**, 787 (1960).
3. E. Hanamura, J. Phys. Soc. Jpn. **29**, 50 (1970); **37**, 1545 (1974); **37**, 1553 (1974).
4. E. Hanamura, Solid State Commun. 11, 485 (1972); see also E. Hanamura and H. Haug, Phys. Rep. **33 C**, 209 (1977).
5. S. Schmitt-Rink, D.S. Chemla, and D.A.B. Miller, Phys. Rev. B **32**, 6601 (1985).
6. A. Shimizu, Phys. Rev. B **37**, 8527 (1988).
7. T. Hiroshima, E. Hanamura, and M. Yamanishi, Phys. Rev. B **38**, 1241 (1988).

Part II

Semiconductors

Ultrafast Dynamical Nonlinearities in Epitaxial III-V Structures

J.L. Oudar

Centre National d'Etudes des Télécommunications,
196, Avenue Henri Ravera, F-92220 Bagneux, France

Direct-gap semiconductors, such as GaAs, present strong, resonantly enhanced, dynamic optical nonlinearities. The saturation of their absorption caused by the photoexcitation of electron-hole pairs and its associated nonlinear refractive index have attracted particular interest since both effects allow the operation of optical gates and bistable devices. Since the response time of these nonlinear optical effects is governed by the relaxation dynamics of the excited states, it is important to improve our understanding of the relaxation mechanisms which influence the behaviour of electron-hole pairs.

The recent development of tunable femtosecond light pulses made it possible to study directly in the time domain the carrier-carrier collisions, and the resulting *momentum relaxation* of the electrons and holes within their bands. This is performed by creating a highly nonequilibrium electron-hole distribution with an intense ultrashort light pulse. Such a nonequilibrium situation actually results from the fact that selection rules normally restrict the number of optically-coupled states to a relatively small k-space domain. As a result, the dynamics of absorption saturation depends directly upon the carrier scattering rates out of (and back into) these optically-coupled states, through carrier-phonon and carrier-carrier collisions.

On the other hand, the response time of nonlinear optical devices based on nonlinear refraction, such as Fabry-Perot bistable devices, depends primarily on the recombination dynamics of the photoexcited carriers. Our studies on semiconductor Multiple Quantum Well (MQW) structures have shown that the carrier lifetime can be drastically reduced by stimulated recombination when the MQW acts as a waveguide for the luminescence light. This represents another ultrafast process, potentially useful for increasing the speed of semiconductor nonlinear optical devices.

In the following these two topics will be discussed, in the light of recent experiments performed on the dynamics of absorption saturation in GaAs-based epitaxial structures.

I. Optical studies of carrier momentum relaxation

Femtosecond pump-and-probe experiments allow one to time-resolve the orientational relaxation of an initially anisotropic momentum distribution of electron-hole (e-h) pairs [1] through the technique of anisotropic absorption saturation. This effect occurs when e-h pairs are excited by linearly polarized light, since this polarization imposes a preferential direction in the k-space distribution of electrons and holes. With intense ultrashort pulses the generation rate can become larger than the collision rate, making it possible to saturate preferentially the e-h pairs whose electron (or hole) wavevector points in a given direction. This anisotropic momentum distribution can be revealed by a corresponding anisotropy in the transmission coefficients of a weak probe pulse polarized parallel or perpendicular to the pump pulse polarization. With a mixed probe polarization (e.g. at 45° to the pump), this differential attenuation induces a rotation of the probe polarization, which can be measured as a function of probe delay, as shown in Fig. 1. From the decay of this polarization signal, one can directly infer the orientational relaxation time of photoexcited e-h pairs.

This effect has some relation with the observation of linear polarization in the hot luminescence of GaAs, as discussed by Zakharchenya *et al.* [2], and an even closer relation with the orientational gratings responsible for the nonlinear coupling between orthogonally polarized beams in degenerate four-wave mixing [3,4]. In these early works however, the orientational relaxation could not be directly observed in the time domain, due to the lack of temporal resolution.

Springer Proceedings in Physics, Vol. 36
Nonlinear Optics of Organics and Semiconductors
Editor: T. Kobayashi

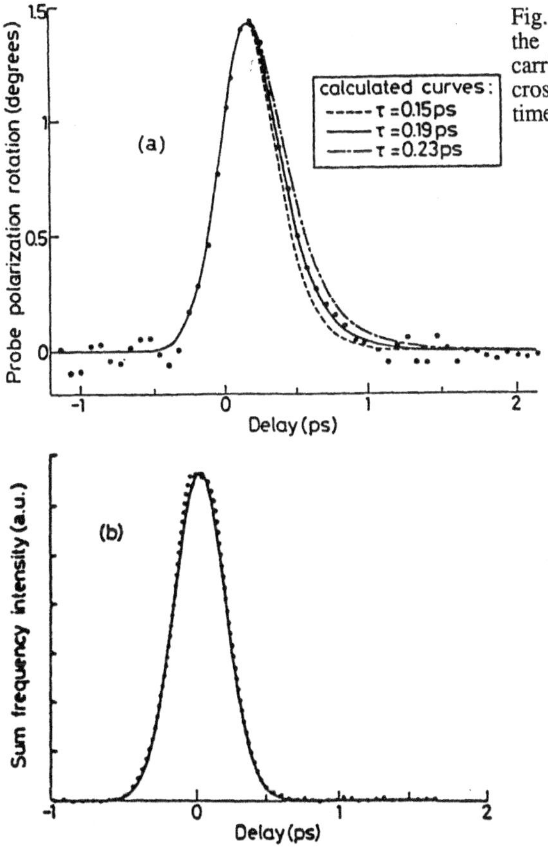

Fig. 1: (a) Polarization signal corresponding to the orientational relaxation of photoexcited carriers in GaAs; (b) pump-probe intensity cross-correlation curve giving the instrumental time resolution

Another example is the observation of a dynamical spectral hole-burning feature centered at the pump frequency in pump-probe experiments with femtosecond pulses [5]. This effect also comes from a preferential saturation of the optically-coupled states, but this time it is observed in the spectral domain with broadband femtosecond pulses. Provided that the pump frequency spectrum is small enough - close to the Fourier-transform limit - a preferential bleaching occurs in the immediate neighbourhood of the pump frequency. This feature is characteristic of nonthermal carrier distributions, and provides information on the primary steps of carrier relaxation, i.e. the first scattering events out of the optically-coupled states. In bulk GaAs and in GaAs MQWs the effect is more easily observed when the photon excess energy above the band gap is smaller than the energy of an optical phonon. This prevents the electron relaxation by optical phonon emission, and leaves the carrier-carrier interactions as the main scattering processes. A lineshape analysis allows one to separate the ultrafast contribution of nonthermal carriers from the thermalized distribution of carriers generated at earlier times, and permits the determination of the e-h pair dephasing time T_2. With bulk GaAs at 15 K [5], T_2 is found equal to 300 fs at 19 meV above the band gap. Calculations of the plasma dynamics in the femtosecond regime [6] are found in quite good agreement with the experimental curves. These calculations show that the carrier dephasing time first increases with the plasma density, but then stabilizes or even decreases at densities higher than 10^{17} cm^{-3}, due to the plasma degeneracy and the screening of the Coulomb interaction.

A similar effect is observed in GaAs MQWs [7]. Our experimental results [8] obtained on 114 Å thick quantum wells under the same excitation conditions as in our previous experiment are displayed in Fig. 2. The lineshape analysis allowed us to determine $T_2 = 150$ fs in the MQW sample, which is significantly shorter than for bulk GaAs. This indicates larger carrier-carrier scattering rates in quantum wells, which is mainly attributed to the reduced screening efficiency of the e-h plasma, a characteristic feature of two-dimensional structures.

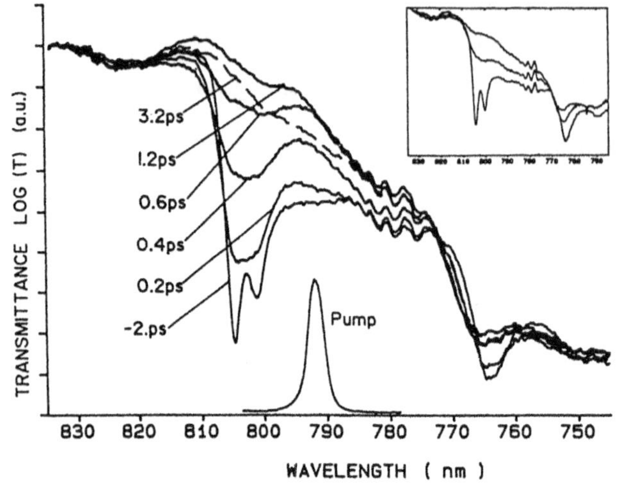

Fig. 2: Time-resolved transmission spectra at various probe delays after the beginning of the pump pulse, under narrow-band excitation at 792 nm. Insert: transmission spectra under broad-band excitation; the wiggles around 780 nm are instrumental (from ref.8).

II. Nonlinear effects governed by the carrier lifetime

While semiconductor optical gates based on carrier-induced nonlinear refraction or absorption can be switched on very rapidly by ultrashort (picosecond or femtosecond) light pulses, their switch-off time is limited by the carrier lifetime (usually in the nanosecond range). An example of such a switching behaviour is displayed in Fig. 3, which represents the operating characteristics of a monolithic intrinsic bistable device developed in our laboratory [9]. The lower right oscilloscope trace represents the light pulse incident on the device at 880 nm wavelength, while the upper right trace is the reflected output. The left-hand trace is the input-output curve showing a hysteresis loop characteristic of the bistable regime. The spikes at the beginning and at the end of the output pulses correspond to the switch-on and switch-off respectively, and occur on the time scale of about 10 ns.

In order to increase the speed of this type of device, one possible approach is to deliberately increase the nonradiative recombination rate, e.g. by ion implantation or using surface recombination. The drawback of this approach is that a lot of heat is generated through these nonradiative transitions. This difficulty could be overcome if the stored energy could be rapidly drained off the device material in the switch-off process. This is precisely what occurs when the carriers recombine through stimulated emission, so it would be very interesting to obtain efficient stimulated emission with reasonable pulse energies.

Recent studies of the transient behaviour of absorption saturation in low temperature MQW samples have revealed a surprisingly fast recovery of absorption (within 10 ps at 15 K) after excitation by intense ultrashort light pulses [10]. This has been attributed to the effect of the large stimulated recombination rate caused by amplified spontaneous luminescence guided along the MQW structure, as confirmed by streak camera observations of the edge-emitted luminescence [10], and by measurements of time-resolved luminescence spectra with subpicosecond resolution [11].

In order to obtain more detailed information about these processes, pump-and-probe transmission experiments [12] were performed with a dual-wavelength set-up of tunable infrared picosecond parametric generators. The sample used consisted of 60 GaAs wells of 75Å width separated by $Al_{0.3}Ga_{0.7}As$ barriers of 85Å, sandwiched between two $Al_{0.3}Ga_{0.7}As$ layers of 1 μm thickness.

The temperature variation of the absorption recovery time in the spectral region above the first subband excitons (about 20 meV above the band gap) was found to be particularly strong between 80 K and 140 K, with an 1/e relaxation time varying from less than 20 ps to more than 250 ps

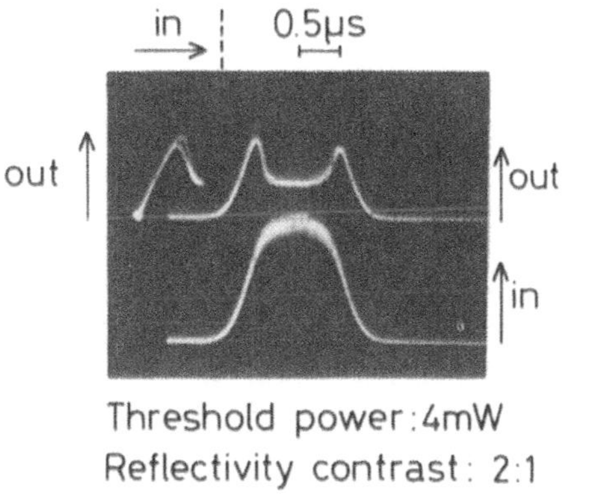

Fig. 3: Nonlinear response of a monolithic GaAs/AlAs optical bistable device (from ref. 5)

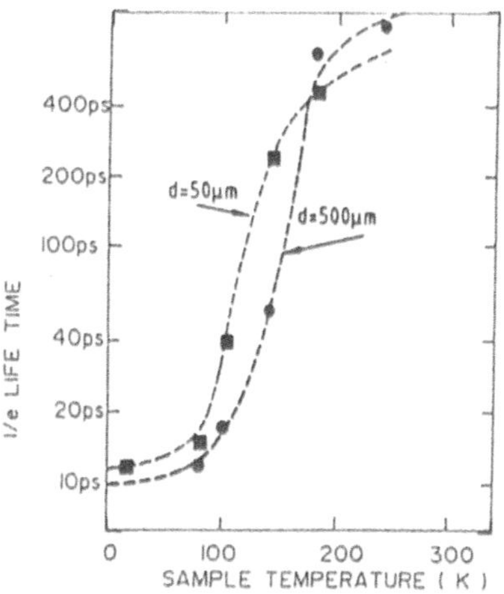

Fig. 4: Temperature dependence of the absorption recovery time for two values of the excited region diameter d.

respectively (see Fig. 4). This indicates a strong temperature dependence of the gain that accompanies band-filling in the continuum absorption region above the exciton lines.

At a given temperature in the range 100-160 K, a faster recovery is observed for larger spot diameters but similar power densities (see Fig. 4 for spot diameters 50 and 500 μm). This is expected since the amplification factor increases rapidly with the spot diameter d. However,the steepness of the temperature dependence is unexpected from conventional theories of gain in quantum wells (see e.g. Ref.13). More specifically our data indicate that at electron-hole densities around 3×10^{11} cm^{-2} the intrinsic gain coefficient g is smaller than 180 cm^{-1} at 180 K, but larger than 1800 cm^{-1} at 100 K. Dutta has predicted a much smaller dependence of gain with temperature [13] since an intrinsic gain of 200 cm^{-1} at 200 K and 550 cm^{-1} at 100 K was calculated for a given current density. This discrepancy is attributed to the excitonic effects neglected in conventional theories, which cause a large gain enhancement due to strong electron-hole correlations at low temperatures. Calculations that include the effect of electron-hole correlations on the gain of QWs [14] predict a strong deviation from the free-particle approximation, with an increasing excitonic enhancement at lower temperatures.

In several cases, especially at high pump intensity, two distinct relaxation regimes were clearly observed, with different characteristic time scales. This is explained by considering that when the gain-diameter product is much larger than unity a fast stimulated recombination occurs, until the gain g is stabilized to an effective threshold value $g_0(d)$. Then the recombination proceeds much more slowly, essentially through spontaneous decay channels. At the beginning of this second stage of relaxation the gain coefficient is not zero but a value close to $g_0(d)$. This is confirmed experimentally by the fact that the stabilized transmission value at a typical photon energy $h\nu = E_g + 20$ meV is higher for smaller pump spot diameters d (since $g_0 \propto 1/d$, the larger stabilized gain at smaller spot diameters is accompanied by more band-filling). The observation of two distinct relaxation regimes provides an opportunity to study the optical absorption spectra of MQWs under excitation conditions in which the gain coefficient is known to be a few hundred cm^{-1}. This is interesting because such gain values are typical of quantum well lasers. This range of gain, however, is quite difficult to measure directly in pump-probe experiments since a 100 cm^{-1} net gain results in a 1% transmission increase over a 1 μm thick MQW sample, which is hardly distinguishable from a corresponding decrease of Urbach's tail absorption.

The technique was thus to excite the material so that a fast stimulated recombination occurred, until the gain coefficient was stabilized to the effective threshold value. One then probed the sample transmission spectrum at an appropriate delay after the pump pulse. This was done systematically as a function of temperature for pump and probe spot diameters of 300 and 150 μm respectively, and probe delays in the range 50-100 ps. In most cases the transmission spectra were effectively stationary over this delay range, and were relatively independent of the pump intensity above some minimum value. Such stabilized spectra, which, from the above considerations, correspond to $g = 250 \pm 50$ cm^{-1}, are represented in Fig. 5 for selected values of the sample temperature. In all cases the excitonic lines are

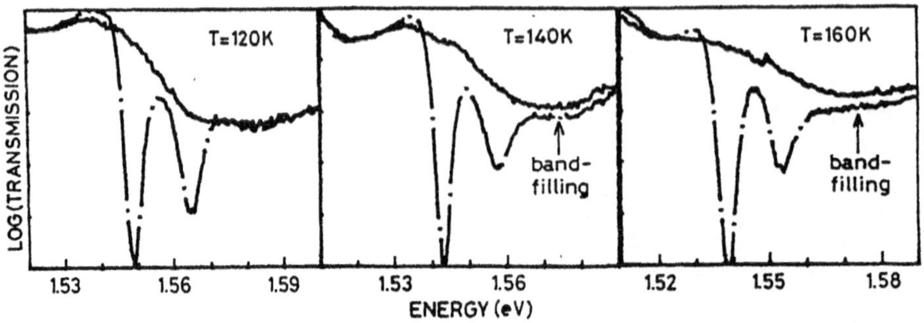

Fig. 5: Transmission spectra at temperatures T; dash-dotted lines: unexcited sample; full lines: quasi-stationary regime of small but nonzero gain.

fully saturated, and the striking result is that, while a substantial band-filling appears at temperatures higher than 120 K, this band-filling is absent at lower temperatures. This observation of gain without band-filling is also attributed to the excitonic effects in quantum wells. From the low value of the pump energy density needed to reach the fast recombination regime at 120 K (a few $\mu J/cm^2$ at 1.59 eV) we estimate that the e-h pair density required for stimulated emission is only about 10^{11} cm^{-2}, which is comparable to the saturation density of excitons [15]. This indicates that phase-space filling of quantum-well excitons may lead not only to absorption saturation but also to a significant gain at low temperatures. The low temperature spectra of Fig. 5, with only the excitons being bleached while some gain is present, may suggest that stimulated recombination is directly due to the excitons, although complementary results would be needed to confirm this interpretation. In any case, even in the plasma regime, the strong electron-hole correlations in 2 dimensions are expected to lead to a strong excitonic enhancement of the gain. As a final remark, we note that kT at 120 K is approximately equal to the exciton binding energy E_b in the investigated structure. These effects should scale as kT/E_b and thus occur at higher temperatures in microstructures with more tightly bound excitons such as quantum wires and quantum dots of III-V compounds, or in quantum wells of II-VI semiconductors.

III. Conclusion

In conclusion, we have briefly reviewed some experimental studies of ultrafast processes governing the dynamic optical nonlinearities of epitaxial III-V structures. The fastest processes involve the momentum relaxation of the photoexcited carriers, the time scale of which lies in the femtosecond range. They involve very complex phenomena, since the photoexcited carriers are in a highly nonequilibrium stage on this time scale. Carrier recombination occurs on a somewhat longer time scale, but under favorable conditions such as in the stimulated case discussed here can also be very fast. A proper use of these effects should find applications for increasing the processing rate of all-optical gates.

Acknowledgement

The subjects briefly described here are the results of a close collaboration with my colleagues, both at the Centre National d'Etudes des Télécommunications, Bagneux, and at the Laboratoire d'Optique Appliquée, at Ecole Polytechnique-ENSTA, Palaiseau. I would like to thank them for fruitful discussions and a pleasant collaboration.

REFERENCES

1 J.L. Oudar, A. Migus, D. Hulin, G. Grillon, J. Etchepare and A. Antonetti, Phys. Rev. Lett. 53, 384 (1984).
2 B. P. Zakharchenya, D. N. Mirlin, V. I. Perel' and I. I. Reshina, Sov. Phys. Usp. 25, 143 (1982).
3 A. L. Smirl, T. F. Boggess, B. S. Wherrett, G. P. Perryman and A. Miller, Phys. Rev. Lett. 49, 933 (1982).

4 J. L. Oudar, I. Abram and C. Minot, Appl. Phys. Lett. 44, 689 (1984).
5 J.L. Oudar, D. Hulin, A. Migus, A. Antonetti and F. Alexandre, Phys. Rev. Lett. 55, 2074 (1985).
6 J. Collet, J.L. Oudar and T. Amand, Phys. Rev. B34, 5443 (1986).
7 W.H. Knox, C. Hirlimann, D.A.B. Miller, J. Shah, D.S. Chemla and C.V. Shank, Phys. Rev. Lett. 56, 1191 (1986).
8 J.L. Oudar, J. Dubard, F. Alexandre, D. Hulin, A. Migus and A. Antonetti, J. de Physique (Paris), Colloque C5, 511 (1987).
9 R. Kuszelewicz, J. L. Oudar, J. C. Michel and R. Azoulay, Appl. Phys. Lett. 53, 2138 (1988).
10 J. Dubard, J.L. Oudar, F. Alexandre, D. Hulin and A. Orszag, Appl. Phys. Lett., 50, 821 (1987); erratum *ibid.* 50, 1696 (1987).
11 D. Hulin, M. Joffre, A. Migus, J.L. Oudar, J. Dubard and F. Alexandre, J. de Physique (Paris) Colloque C5, 267 (1987).
12 J.L. Oudar and J.A. Levenson, paper ThB7, XVIth Internat. Conf. on Quantum Electron., Tokyo (July 1988).
13 N. K. Dutta, Electron. Lett. 18, 451 (1982).
14 S. Schmitt-Rink, C. Ell and H. Haug, Phys. Rev. B33, 1183 (1986).
15 S. Schmitt-Rink, D.S. Chemla and D.A.B. Miller, Phys. Rev. B32, 6601 (1985).

Room-Temperature Optical Nonlinearities in Multiple-Quantum-Well GaAs/AlGaAs and Bulk GaAs

A. Chavez-Pirson, H.M. Gibbs, and S. Koch

Optical Sciences Center, University of Arizona, Tucson, AZ 85721, USA

Interest in understanding the character and microscopic origin of the nonlinear optical properties of semiconductors has been increasing, both because of the underlying physics and the potential application to all-optical and opto-electronic devices. With the advent of crystal-growth techniques such as molecular beam epitaxy (MBE) and metal-organic chemical vapor deposition (MOCVD), growth of high-quality bulk and multiple-quantum-well (MQW) crystals of precise layer thickness is now possible. Interestingly, the optical properties of MQW GaAs/AlGaAs are qualitatively different from those of bulk GaAs. The quasi-two-dimensional nature of MQWs changes the shape of the density of states at the fundamental absorption edge. In addition, the physical confinement of electron-hole pairs in MQW structures enhances excitonic effects relative to bulk. The increased binding energy of the exciton makes possible the observation of the exciton resonance even at room temperature [1]. In this paper, we study experimentally the room-temperature, frequency-dependent optical absorptive and dispersive nonlinearites of bulk GaAs and MQW GaAs/AlGaAs. We also explicitly compare the magnitude of the MQW nonlinear refractive index changes as a function of well size.

In studying the optical nonlinearities of bulk and MQW materials, there are several relevant temperature and time regimes to consider. At low temperature (4 K), the exciton feature dominates the absorption spectrum in both bulk and MQW crystals. At this temperature, excitonic effects are the major contribution to the nonlinearity [2,3]. As the temperature increases, scattering with LO phonons broadens the exciton resonance. At room temperature, the bulk-exciton feature is only just resolvable in the spectrum. The time regimes associated with the nonlinearity are as follows. In both bulk and MQWs, excitons at room temperature are thermally ionized in less than a picosecond. Within a picosecond, a photo-generated exciton gas decays into an electron-hole plasma [4]. Because of fast (< 1 ps) intraband relaxation processes, the electron-hole plasma quickly reaches a quasi-equilibrium distribution within the band. As carriers recombine, radiatively or nonradiatively, the index and absorption will return to their values prior to optical excitation. The carrier lifetimes are on the order of nanoseconds in our samples. In our experiments, we study the quasi-steady-

Springer Proceedings in Physics, Vol. 36
Nonlinear Optics of Organics and Semiconductors
Editor: T. Kobayashi

state regime in which the electron-hole distribution has equilibrated in the band. We note that the relevant parameter to describe the nonlinear effect is N, the electron-hole density. This originates from the experimental fact that resonant or nonresonant excitation produces the same nonlinear effects, as long as the total carrier density is the same. In our experiments, the steady-state carrier density is determined by balancing the pumping rate of electrons and holes with the carrier decay rate.

We study the nonlinear effects using a pump-and-probe technique, and an optical multichannel analyzer (OMA). The pump beam originates from a cw tunable dye laser (Styrl 9) pumped by an argon ion laser. The infrared pump laser wavelength is tuned well inside the bandgap of each sample (600 Å) to efficiently generate electrons and holes. A broad-band probe beam (700 Å) is derived from the photoluminescence of a MQW sample pumped by 100 mW from the argon laser. Depending on the MQW sample to be studied, a separate MQW platelet is selected whose photoluminescence best covers the spectrum of interest. The pump and probe beams are synchronously modulated by acousto-optic modulators to produce 1.0-μs and 0.8-μs rectangular pulses, respectively, at a 20-kHz rate. These pulse widths are much longer than the carrier lifetime, assuring a quasi-steady state throughout the duration of the pulse. Under these pumping conditions, thermal effects are observed to be negligible. The transmitted probe pulses are collected and integrated by the OMA. The pump and probe beams are focused to a 15-μm-diameter spot size, but because of carrier diffusion, the effective pump spot is 20 μm; in this way, the probe spot area is uniformly excited.

Four different MBE-grown samples were used in this work: 76-, 152-, 299-Å GaAs/AlGaAs MQWs, and bulk GaAs. These samples were grown in the same MBE chamber in order to minimize the variability of sample quality. The first MQW contains 63 periods, each period consisting of a 76-Å GaAs layer followed by 81 Å of $Al_{0.37}Ga_{0.63}As$, for a total GaAs thickness of 0.45 μm. The second sample contains 100 periods, each period consisting of a 152-Å GaAs layer followed by 104 Å of $Al_{0.33}Ga_{0.67}As$, for a total of 1.52 μm. The third sample contains 61 periods, each period consisting of a 299-Å GaAs layer followed by 98 Å of $Al_{0.36}Ga_{0.64}As$, for a total GaAs thickness of 1.8 μm. The bulk sample consisted of a 2.05-μm layer capped on both sides with $Al_{0.30}Ga_{0.70}As$. All the samples were anti-reflection coated (< 2%) with a quarterwave layer of silicon nitride to minimize the Fabry-Perot effects from the front and back surfaces of the sample.

The nonlinear absorption spectrum of bulk GaAs as a function of carrier density is shown in Fig. 1. The room-temperature microscopic origins of the

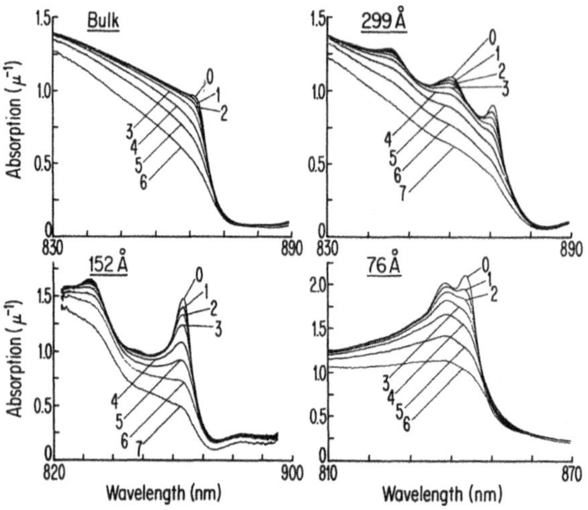

Figure 1. Nonlinear absorption spectra for MQW and bulk GaAs. Curve 0 in each set represents the unpumped (linear) absorption spectrum. The highest numbered curve in each set corresponds to a carrier density N of 4.5×10^{18} cm^{-3} for bulk, 3.8×10^{18} cm^{-3} for the 299-Å MQW, 2.8×10^{18} cm^{-3} for the 152-Å MQW, and 9.6×10^{17} cm^{-3} for the 76-Å MQW. Each lower numbered curve within a set represents a 40% reduction in N from the next higher numbered curve.

electronic nonlinearity in bulk GaAs are explained by many-body effects, which include plasma screening of the long-range Coulomb interaction, bandgap renormalization, and band filling [5-7]. Plasma screening of the Coulomb interaction reduces the exciton binding energy. When this screening reaches a critical level (the Mott density), the screened Coulomb potential is not strong enough to support any bound states. The screening also suppresses the Coulomb enhancement factor of the continuum states. This appears as an overall reduction in the absorption coefficient near and above the bandgap. The bandgap renormalization is a consequence of the fact that the single-particle energies of electrons and holes change with increasing plasma density. The energy gap is reduced because of the self-avoiding correlation between equally charged particles and because of the screened exchange interaction. Band filling results from the occupation of states in the conduction and valence bands. As electrons and holes fill the individual states in the bottom of the conduction band and the top of the valence band, those states are no longer available for other carriers to occupy. As a result of the Pauli exclusion principle, the band-filling effect reduces the valence-band to conduction-band transition probability.

The nonlinear absorption spectra of the MQWs as a function of carrier density are shown in Fig. 1. As a result of the confinement in one dimension, the MQW band edge is shifted to shorter wavelengths relative to bulk, and the exciton resonance is clearly resolved. The 76-Å sample even shows the splitting of the light- and heavy-hole transitions. For MQWs, the dominant nonlinearities are phase-space filling, screening, and exchange effects [8,9]. These effects combine to saturate the exciton resonance. It is important to note that the influence of screening on the absorption changes in two dimensions is somewhat reduced in comparison to the three-dimensional case. Phase-space filling reduces the number of available electron and hole states from which excitons can be formed, and effectively weakens the net attraction between electrons and holes. Changes in the effective exchange interaction also reduce the electron-hole attraction, resulting in a decrease of the exciton binding energy. The resulting increase of the exciton diameter reduces the exciton oscillator strength. At the highest carrier densities, the exciton disappears, leaving the step-like density-of-state feature that is characteristic of a two-dimensional system.

The nonlinear refractive index changes associated with each sample (see Fig. 2) are obtained by a Kramers-Kronig transformation of the corresponding absorption changes:

$$\Delta n(E) = \frac{\hbar c}{\pi} P \int_{E_1}^{E_2} \frac{\Delta\alpha(E')dE'}{E'^2 - E^2}, \tag{1}$$

where $\Delta n(E)$ and $\Delta\alpha(E)$ are the refractive index and absorption changes as a function of photon energy, respectively. The applicability of this analytic relationship to these experimental conditions was verified by a direct interferometric technique [10]. The spectral shape of the nonlinear refractive

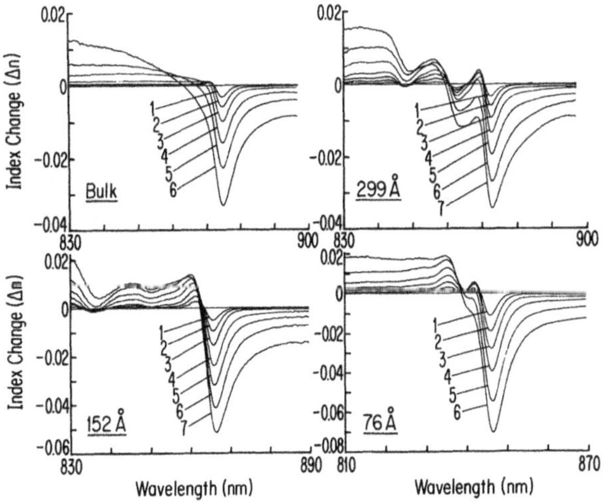

Figure 2. Nonlinear refractive index changes for MQW and bulk GaAs. The curve numbers in each set correspond to the same carrier densities described in Fig. 1.

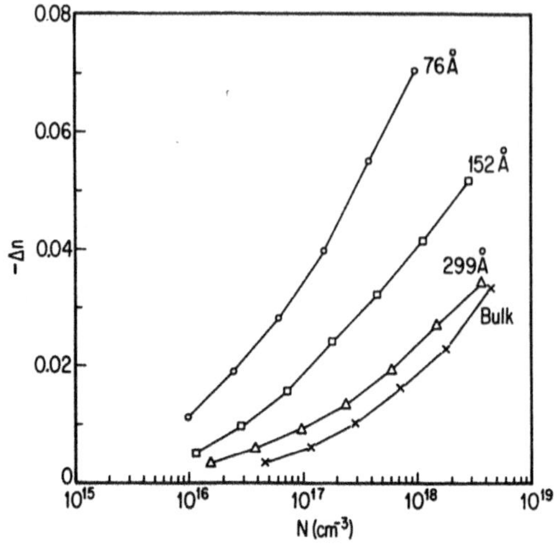

Figure 3. The maximum refractive index change, Δn, as a function of carrier density, N, for MQW and bulk GaAs.

index changes for all the samples is characterized by a bandgap resonant dip at which the index change is at a maximum, followed by a slowly decreasing fall-off for longer wavelengths. We pick the maximum index change for each sample and plot it as function of carrier density (see Fig. 3). This bandgap resonant nonlinearity is primarily attributable to plasma screening and band filling in bulk GaAs and excitonic saturation in MQWs. From these curves, we see that larger index changes are associated with the narrower wells.

The electron-hole carrier concentration for the different pumping levels is determined through a rate-equation approximation:

$$\frac{dN}{dt} = -\frac{N}{\tau} + \frac{\alpha(\omega)I}{\omega}, \tag{2}$$

where τ is the electron-hole lifetime, I is the incident intensity, ω is the pump radial frequency, and $\alpha(\omega)$ is the absorption coefficient measured at the pump frequency. In steady state ($dN/dt = 0$), the carrier density N can be calculated. All the parameters in the equation are directly measured from the experiment except for the carrier lifetime. To determine the lifetime for each sample, the experimental setup was modified. A subnanosecond pump source (0.5 ns) from a nitrogen-pumped dye laser generates the carriers. A weak probe laser with a pulse length of 4 μs is tuned to the wavelength which has the maximum absorption change. A fast risetime (100 ps) avalanche photodiode and fast oscilloscope (1 GHz) monitor the transmitted probe pulse. Immediately after the pump pulse hits the sample, the probe transmission increases, indicating a decrease in the absorption coefficent. As the carriers recombine, the absorption recovers to its original level. A typical absorption-recovery trace for the bulk sample is shown in Fig. 4.

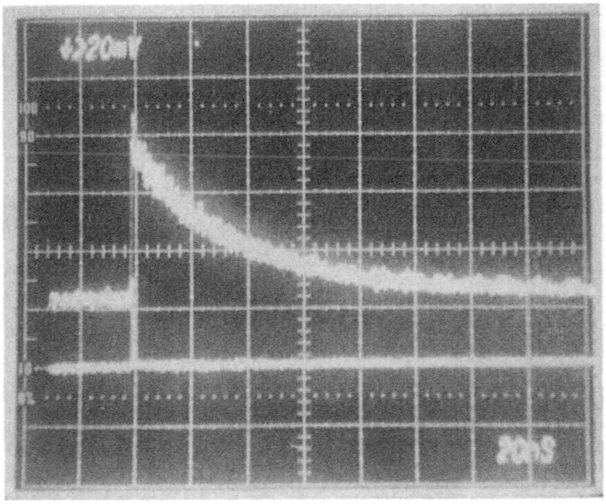

Figure 4. Time evolution of the absorption recovery for bulk GaAs. The top trace shows the probe transmission through the sample before and after the pump pulse. The bottom trace shows the residual pump pulse transmission through the sample with no probe pulse incident.

Similar traces were obtained for all the samples. Table 1 presents the single-exponential decay times associated with each sample. There is a systematic reduction in the carrier lifetime with decreasing well size, which is probably the result of nonradiative recombination at the barrier interfaces. This carrier lifetime is used to calculate the carrier density.

Table 1. Carrier lifetime in bulk and MQW GaAs

Sample	Lifetime
Bulk (2.05 μm)	32 ns
299 Å MQW	24 ns
152 Å MQW	18 ns
76 Å MQW	3 ns

To compare the nonlinearities of the different samples, the peak nonlinear refractive index change per carrier was calculated as a function of carrier density. These curves are plotted in Fig. 5, showing that MQWs with narrow wells have a higher nonlinearity than those with thick wells at all carrier densities. The graph also shows that each MQW has a higher nonlinearity than bulk. The 76-Å sample has three times the nonlinearity of bulk at high carrier densities, and ten times that of bulk at low carrier densities. These results indicate that MQWs with thin wells have larger nonlinearities than the other samples. PARK et al. [11] obtained similar $\Delta n/N$ results in a pump-probe experiment using nanosecond pulses. However, they did not explore the low-carrier-density

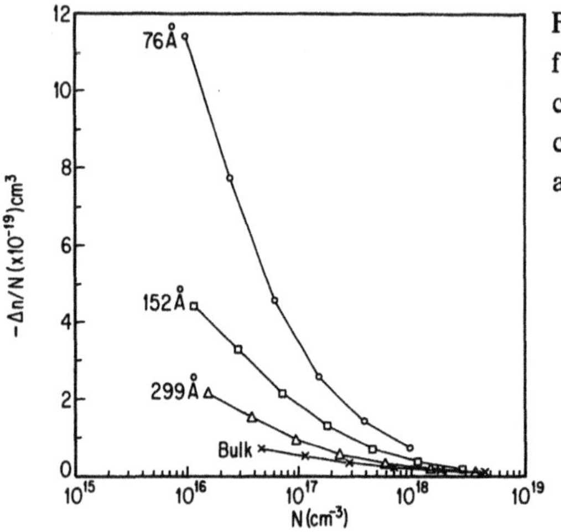

Figure 5. The maximum refractive index change per carrier, $\Delta n/N$, as a function of carrier density, N, for MQW and bulk GaAs.

regime where the difference in $\Delta n/N$ is largest, nor did they explicitly take into account the variation of carrier lifetime with well size. The results of the lifetime measurements have an important consequence for quasi-steady-state operation in these samples. Although the peak value of $\Delta n/N$ is higher for the MQWs, the shorter lifetime of MQWs demands a stronger pumping rate (higher intensity) to generate the same carrier density as that in the longer-lifetime bulk sample. Also the larger peak absorption for the MQWs means that the detuning must be greater in the MQW for the same intracavity loss and carrier generation absorption. These factors may explain the experimental result that the threshold power for optical bistability is nearly constant for both MQWs and bulk.

In conclusion, we have studied the room-temperature, frequency-dependent optical nonlinearities in bulk GaAs and GaAs/AlGaAs MQWs. From the measured nonlinear absorption changes, we deduce the dispersive changes by a Kramers-Kronig transformation. The magnitude of the nonlinear response for each sample is determined as a function of the carrier density. The ratio of the maximum nonlinear refractive index to the carrier density increased for MQWs with smaller well sizes, with the 76-Å MQW exhibiting a nonlinearity that was an order of magnitude larger than bulk at low excitation levels.

Acknowledgments

The authors acknowledge the support of the National Science Foundation (Lightwave Technology). We thank A. C. Gossard and W. Wiegmann of AT&T Bell Laboratories for the MBE samples. We also thank J. Salik for antireflection coatings of the samples and B. McGinnis for the use of his dye laser for lifetime measurements.

References

1. D. A. B. Miller, D. S. Chemla, D. J. Eilenberger, P. W. Smith, A. C. Gossard, W. T. Tsang: Appl. Phys. Lett. **41** (1982).
2. J. Shah, R. F. Leheny: Phys. Rev. B **16**, 1577 (1977).
3. G. W. Fehrenbach, W. Schafer, J. Treusch, R. G. Ulbrich: Phys. Rev. Lett. **49**, 1281 (1982).
4. D. S. Chemla, D. A. B. Miller, P. W. Smith, A. C. Gossard, W. Wiegmann: IEEE J. Quantum Electron. **20**, 265 (1984).
5. Y. H. Lee, A. Chavez-Pirson, S. W. Koch, H. M. Gibbs, S. H. Park, J. F. Morhange, A. Jeffery, N. Peyghambarian, L. Banyai, A. C. Gossard, W. Wiegmann: Phys. Rev. Lett. **57**, 2446 (1986).
6. H. Haug, S. Schmitt-Rink: J. Opt. Soc. Am. B **2**, 1135 (1985).
7. L. Banyai, S. W. Koch: Z. Phys. B **63**, 283 (1986).
8. S. Schmitt-Rink, D. S Chemla, D. A. B. Miller: Phys. Rev. B **32**, 6601 (1985).
9. D. S. Chemla, S. Schmitt-Rink, D. A. B. Miller: In *Nonlinear Optical Properties of Semiconductors*, ed. by H. Haug (Academic, New York, 1988).
10. Y. H. Lee, A. Chavez-Pirson, B. K. Rhee, H. M. Gibbs, A. C. Gossard, W. Wiegmann: Appl. Phys. Lett. **49**, 1505 (1986).
11. S. H. Park, J. F. Morhange, A. D. Jeffery, R. A. Morgan, A. Chavez-Pirson, H. M. Gibbs, S. W. Koch, N. Peyghambarian, M. Derstine, A. C. Gossard, J. H. English, W. Wiegmann: Appl. Phys. Lett. **52**, 1201 (1988).

Optical Nonlinearity Due to Resonant Intersubband Transitions in Doped Quantum Wells

D.J. Newson and A. Kurobe

Toshiba R&D Center, Saiwai Ku, Kawasaki, Kanagawa 210, Japan

1 Introduction

Novel growth techniques have enabled the fabrication of abrupt semiconductor heterointerfaces, allowing electrons to be confined in potential wells in which they form a two-dimensional electron gas. Most of the work on optical bistability in semiconductors has been at frequencies close to the band-gap/1/. However, the switching times of such effects are usually limited by slow electron-hole recombination. We have considered an alternative source of optical non-linearity (ONL), namely due to intersubband transitions within the conduction band of doped semiconductor quantum wells (QW). The wavelength of such transitions is in the 10 micron range. Since their relaxation time is about 10ps, fast switching is expected. Below, we outline our theoretical investigations concerned with such intersubband transitions.

With suitable, polarized far infra-red radiation (FIR) electrons in such QW can be excited from the ground subband to higher subbands. There can be a high dipole moment for such transitions, causing strong absorption/2/, which has been used to make FIR detectors/3/ and modulators/4/, and has been proposed for novel ONL materials/5,6,7/. As the subbands are parallel at small wave-vectors the absorption spectrum is peaked. Such peaks persist up to 300K because of the large intersubband energy, E_{10}. In the following, an intersubband transition is one in which the electron changes subbands, while an electron remains in the same subband after an intrasubband event.

2 Comparison Between Interband and Intersubband ONL

There are several mechanisms which can cause ONL in intersubband transitions. There are similarities to those in semiconductor interband transitions, thus it is instructive to compare the mechanisms which cause ONL in the two cases.

2.a Excitons

Excitons are one of the major causes of band-edge ONL. However, due to the band structure shape, the intersubband equivalent of an exciton is energetically unstable. Thus excitonic-like effects will not occur in intersubband transitions.

2.b Saturation Effects

Under high interband excitation the effective band-gap increases due to state filling (the Burstein-Moss effect). The intersubband equivalent is saturation of the absorption peak. The peak absorption falls because of phase-space filling effects. By analogy this effect can be called subband filling.

Springer Proceedings in Physics, Vol. 36
Nonlinear Optics of Organics and Semiconductors
Editor: T. Kobayashi

2.c Many-Body Effects

In the presence of low charge densities the band-gap decreases because of many-body effects. The intersubband equivalent will cause a shift of the absorption peak, which can cause ONL.

We first discuss the shift of absorption peak due to many-body effects. Later, we consider the saturation characteristics of the intersubband transition, obtained from a Monte Carlo simulation of the carrier scattering processes.

3 Peak Shift Due To Many-Body Effects

To be useful for optical switching devices the peak shift must be comparable to the absorption peak width (about 10meV). We performed quantum mechanical calculations including many-body effects to estimate the size of the peak shift under intersubband resonant excitation.

Calculation of the many-body interaction energy of an electron gas comprises two parts: the Hartree and exchange-correlation (xc) terms. The former ensures that the wave functions obey Poisson's equation, and is thus proportional to the carrier density. The exchange and correlation terms attempt to account for the avoidance of electrons arising from the Pauli principle and Coulomb repulsion, respectively. Electrons behave as though surrounded by unoccupied holes of radius the Fermi wavelength and the Thomas-Fermi wavelength, respectively. The exchange (correlation) energy is then roughly proportional to the Fermi (Thomas-Fermi) wave-vector. The Hartree (xc) term raises (lowers) the electron energy levels. In the local-density functional approach/8/, it has been shown that the many-body problem can be treated as a single electron feeling the potential of its own unoccupied holes (the value of which is simplified to be a function of the local charge density) in addition to the Hartree and other external potentials. This method has been applied to semiconductor devices/9/. We used the expression of Hedin and Lundqvist/8/ to model the xc term.

We used the effective mass approximation, considering the finite well height, and giving the effective mass and dielectric constant the values 0.067 and 12.0, respectively. Wells were n-type, modulation doped, containing electrons of sheet density N_e. The calculations were made for zero temperature when only the ground subband was occupied without radiation.

Since the intersubband energy E_{10} depends on the subband occupations due to many-body interactions it will be changed by resonant intersubband excitation. Change in E_{10} alters the relative positions of the pump beam and the absorption peak maximum, leading to optical nonlinearity for energies close to the resonance peak.

In calculating the size of this effect we used the same formalism to describe the system when subjected to intense excitation as in the absence of excitation; we must justify this assumption. The Hartree energy depends on the charge distribution rather than the $\underline{K}$-space distributions in the well plane, so the Hartree approximation should be valid even under intense excitation. Our justification for the xc term lies in the small value of the electron scattering time in a non-equilibrium distribution compared to the intersubband relaxation time (of the order of 10ps.), thus the electron distribution will thermalize within each subband.

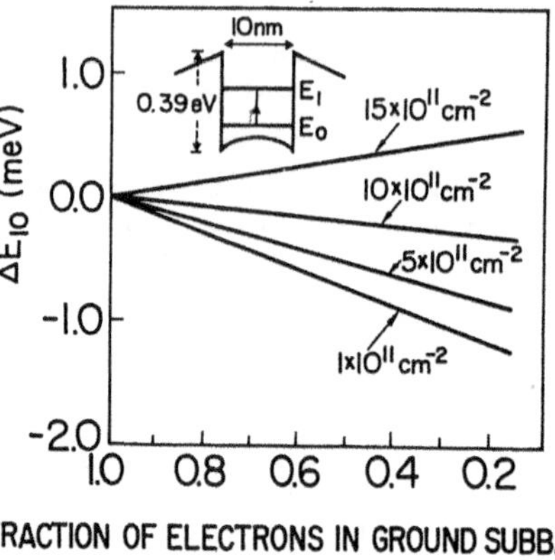

Fig.(1) Calculated change in energy between the lowest two subbands (E_{10}) against the fraction of electrons in the ground subband. The structure is a 10nm modulation doped GaAs quantum well clad by $Al_{0.5}Ga_{0.5}As$. The electron concentrations are 1,5,10 and 15×10^{11} cm^{-2}, for which the calculated values of E_{10} are 111.1, 107.4, 102.5 and 97.6 meV, respectively, in absence of excitation. The intersubband system is very closely related to a two-level system. It will thus be impossible to directly pump more than half the electrons to the higher subband, except on time-scales shorter than the Rabi time.

The effect of excitation on E_{10} depends on whether the xc or Hartree term dominates. Fig.(1) shows the calculated results for a 10nm wide GaAs QW, modulation-doped to various levels of N_e. The Hartree term is proportional to N_e, whereas the xc term, which is approximately linear in the Fermi wave-vector, has a sublinear dependence on N_e. Thus, while both terms increase with N_e, the Hartree term increases faster. For large N_e ($>5\times10^{11}cm^{-2}$) the Hartree term dominates the xc term, giving us confidence in our results. The maximum change in E_{10} is smaller than 1meV for practical values of N_e, since changes in the Hartree and xc terms cancel to a large extent. This value is too small for practical use, so we try to enhance it, proposing a novel QW structure in the next section.

4 Proposal Of A Stepped Quantum Well Structure

The Hartree potential between two parallel charged sheets is proportional to their separation, and thus the Hartree energy will be larger in wide QW than in a narrow QW. This suggests that the change in E_{10} can be enhanced by using a stepped well, such as is shown in the inset of Fig.(2). A QW of this type should be engineered so that the ground state is confined in the narrow well, while the first excited state lies above the top of the narrow well. Physically, a much larger change in Coulomb energy will occur in such a structure compared to a square QW because the charge can spread out further on excitation.

Fig.(2) shows results for the case of a 20nm QW containing a 4nm well at its centre, the step height (0.2eV) being chosen to make E_{10} close to the 10.6 micron line of the CO_2 laser. A peak shift of up to 7meV seems to be possible on intense illumination. Such a value is promising for optical non-linear devices. The dipole moment is only slightly larger than for the narrower 10nm QW (2.0nm compared to

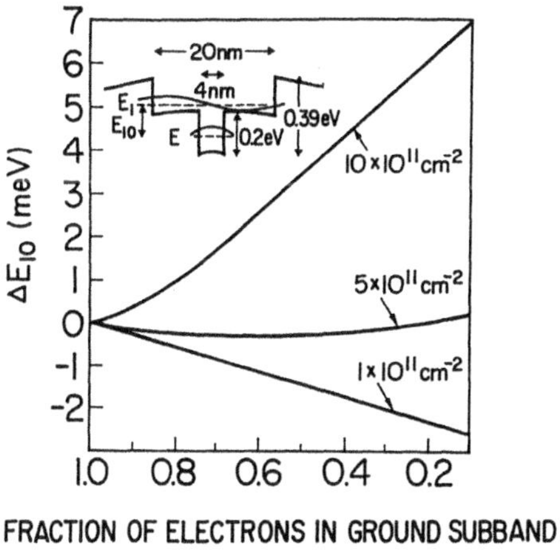

Fig.(2) Change in E_{10} against fraction of electrons in the ground subband, for the stepped quantum well illustrated in the inset. The narrow well is 4nm of GaAs, the wider being $Al_{0.26}Ga_{0.74}As$; the total width is 20nm. Cladding is $Al_{0.5}Ga_{0.5}As$. The electron concentrations are 1,5 and 10×10^{11} cm^{-2}, for which the zero excitation values of E_{10} are 143.5, 133.0 and 116.7 meV, respectively. A change in E_{10} of 7meV would occur if all the electrons occupied the higher subband.

1.8nm); however, the effective nonlinearity is much larger than for the square QW.

We now estimate the size of the optical nonlinearity. The number of excited electrons (n_s) is given by $n_s = aIt/E$, where a is the absorption coefficient per well, I the incident intensity, and t the shorter of subband relaxation time and pulse length. a is about 0.001-0.01 (i.e. 0.1-1% of incident light absorbed per well) for light incident at Brewster's angle. Assuming $n_s = 2\times10^{11}cm^{-2}$ for a useful change in E_{10}, we estimate an intensity of $200kWcm^{-2}$ for switching. For propagation in a single mode waveguide, this leads to a switching power of about 0.1W, switching times of several picoseconds being possible.

5 Effect Of Conduction Band Non-Parabolicity

We have also considered the effect of conduction band non-parabolicity (NP) on the properties of the absorption peak. NP reduces the intersubband energy separation at large **K**. Such high **K**-vectors states cannot contribute to the FIR absorption, reducing the maximum absorption and broadening the peak; both of which are disadvantageous for intersubband mechanisms/13/. Here we assume an infinitely deep QW for simplicity. For electrons in a QW the energy is given as a function of the in-plane dispersions (K_x, K_y) by the equation/14/

$$E(\underline{K}) = \alpha_0 K_z^4 + [\hbar^2/2m^* + (2\alpha_0 + \beta_0)(K_x^2 + K_y^2)]K_z^2 + (\hbar^2/2m^*)(K_x^2 + K_y^2) + (2\alpha_0 + \beta_0)K_x^2 K_y^2 + \alpha_0(K_x^4 + K_y^4) \quad (1)$$

where α_0 and β_0 are the non-parabolicity parameters. m^* is the band-edge effective mass. In an infinitely deep QW of width L, the confining wave-vectors (K_z) take the values n/L for the nth. subband denoted by $e_n(\underline{K})$. This gives the wave-vector dependent energy intersubband separation to be

$$E_{10}(\underline{K}) = e_1(\underline{K}) - e_0(\underline{K}) = E_{10} + (2\alpha_0 + \beta_0)\Delta K_z^2 \, \underline{K}^2 . \qquad (2)$$

The optical absorption is obtained from integrating

$$\alpha = \frac{e^2 \mu c \, \langle \phi_f | z | \phi_i \rangle^2}{2 n_r \hbar^2 \omega L} \frac{\Gamma}{2} \iint d\underline{K} \, \frac{E_{10}^2(\underline{K}) \, [f(\underline{K}) - g(\underline{K})]}{(E_{10}(\underline{K}) - \hbar\omega)^2 + (\Gamma/2)^2} \qquad (3)$$

where $\hbar\omega$ is the photon energy, Γ is the absorption peak width, n_r the refractive index, μ the permeability and c the speed of light. The importance of this effect obviously increases with K_f and hence with doping level. At zero temperature Eqn.(3) can be integrated exactly, giving a peak shift of $(2\alpha_0 + \beta_0)\Delta K_z^2 K_f^2/2$. At finite temperature Eqn.(3) must be integrated numerically/13/. For example in a 10nm InGaAs QW, modulation doped to $1\times10^{12}\,cm^{-2}$, at 300K NP shifts the absorption peak by about 2meV to a lower energy, and reduces the peak height by about 25% compared to the non-parabolic case. The effects are small for GaAs wells, and should not be a major problem in practical applications.

6 Simulation Of Intersubband Relaxation

We have also performed a Monte Carlo simulation of intersubband relaxation in modulation doped QW. This simulation was done, in part, to explain the experiments of Seilmeier et al./10/, who measured the time resolved absorption spectra in modulation doped quantum wells which had been subjected to intense FIR. They found the absorption to relax with a time constant of about 10ps, even though the intersubband LO phonon time is about 1ps /12/. In order to explain this discrepancy we have considered:

(1) conduction-band non-parabolicity, this causes electrons to initially relax to states of high momentum from which they take a long to recover.

(2) screening of the LO phonon modes via a Lindhard dielectric function. Screening reduces the net scattering rate.

(3) 'hot phonon' effects. When a large number of electrons relax by emitting LO phonons the phonon density rises above the thermal value given by Bose-Einstein statistics. Such phonons can be reabsorbed, exciting electrons back up to the higher subband. This can dramatically reduce the net scattering rate.

(4) LO phonons decay to acoustic phonons because of crystal anharmonicity. We used the bulk GaAs decay time of 7ps.

(5) intrasubband scattering included Landau damping and plasmon emission effects.

Fig. (3) shows the calculated relaxation of the intersubband system in a 8nm GaAs QW after being excited by a square pulse of duration 2ps. The two curves are for high and low excitation intensities: the low intensity pulse excites about 10% of the electrons into the higher subband, while the high intensity pulse is ten times stronger. Once the pulse is finished, the optical absorption (obtained from integrating Eqn. (3)) relaxes exponentially. The relaxation rate is higher at large excitation because phonon stimulated excitation to the higher subband is hindered by the Pauli exclusion principle. We point out that these results are sensitive to the value of the LO-phonon decay time: in a system which has a short LO-phonon decay time, e.g.

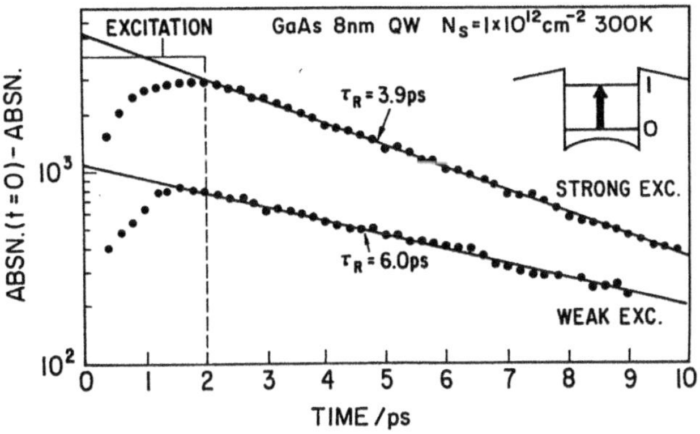

Fig.(3) Relaxation of intersubband absorption against time after FIR excitation between 0 and 2ps. 8nm GaAs QW, $N_s = 1\times10^{12}cm^{-2}$. A screened LO phonon time of 2ps was used.

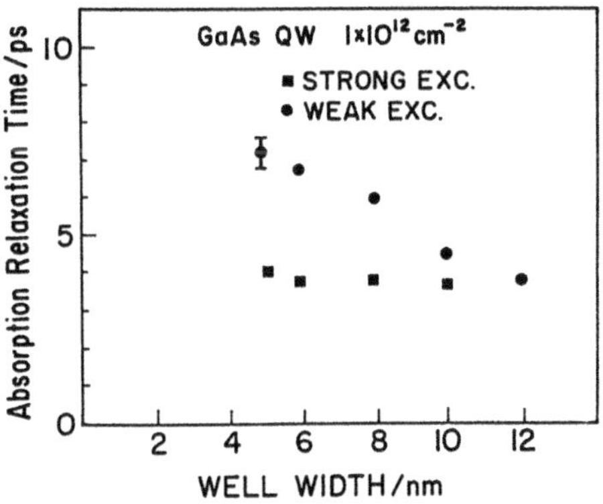

Fig.(4) Effect of well width on relaxation rate, for high (circles) and low (squares) excitation.

alloy semiconductors such as InGaAs, intersubband relaxation will be faster as phonon absorption effects will be less important.

Fig. (4) plots the QW width dependence of the relaxation rate. The high excitation data all show a very similar relaxation time of about 4ps, while the low excitation results show a relaxation time which rises with decreasing width, up to 7.3ps at 5nm.

It is difficult to compare these results with experiment/10,11/, as the number of excited carriers depends on the the absorption coefficient and on the actual intensity reaching the sample, which is dependent on waveguiding conditions and the angle of incidence. We estimate that the conditions of the experiment are intermediate between our high and low intensity limits. The data has a longer relaxation time than our results in narrow wells, indicating that the LO-phonon interaction is weaker than used in our simulation. This is understandable in narrow wells as the overlap between the wave-functions of ground and excited subbands decreases since the higher subband is only weakly bound. In narrow wells transfer to L-valleys may become important.

7 Simulation Of Intersubband Saturation

The same computer program can give the saturation characteristics of the intersubband system under cw radiation. In a homogeneously broadened system the absorption depends on intensity (I) as α =

$\alpha_0/(1+I/I_s)$. I_s is known as the saturation intensity. However, in a system in which there is a range of resonant frequencies, the absorption saturates in a much slower manner, as $\alpha = \alpha_0/(1+I/I_s)^{1/2}$. To predict which is a more accurate description, we have used the program but with the system subject to radiation for 10ps. Fig. (5) shows the time dependence of absorption under cw radiation: it initially falls rapidly, but soon converges to the cw value. We have taken the value of absorption after 10ps to be the cw value. In Fig. (6) we plot the resultant data in the form α_0/α vs. I, in order to compare with homogeneous broadening. A good fit is seen, implying that the system is much closer to being homogeneously than inhomogeneously broadened. In highly non-parabolic semiconductors, the saturation will be closer to inhomogeneous broadening.

We estimate I_s for the n-MQW system to be about 700 $kWcm^{-2}$, although this depends on the exact value of the absorption coefficient and on the incident properties of the radiation. This value is much lower than that in p-Ge (4 $MWcm^{-2}$), whose absorption saturates inhomogeneously. Thus, the n-MQW system may be useful for passive mode-locking of 10 micron range lasers. The first measurements of saturation of intersubband absorption have recently been made/16/. Their data was close to being homogeneously broadened with I_{sat} about 800$kWcm^{-2}$.

We note that intersubband saturation has been considered from a density matrix viewpoint /7/. A linear dependence of saturation on intensity was obtained, as only the first term in the expansion of absorption was used. Intersubband saturation in p-MQW samples has recently been considered /17/. Absorption was predicted to saturate inhomogeneously; however, in that work absorption peak broadening was assumed to only come from carrier lifetime effects, which may give an optimistic value of I_s.

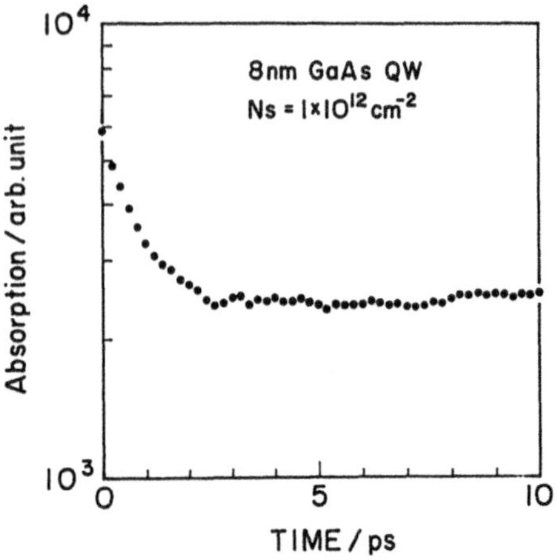

Fig.(5) Time resolved intersubband absorption under cw FIR radiation.

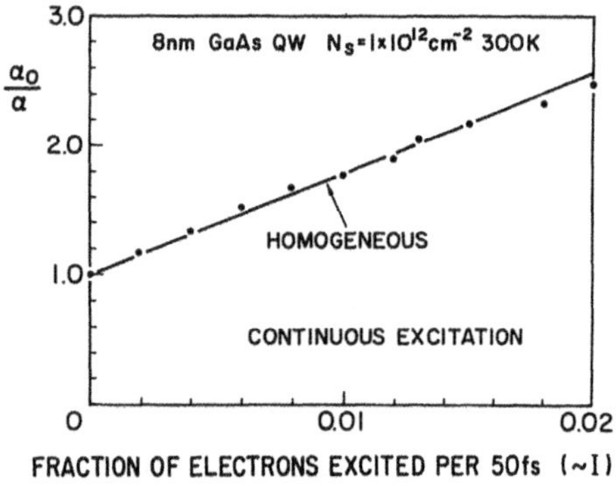

Fig.(6) Saturation of the intersubband absorption in 8nm GaAs QW. Plot of α_0/α against fraction of electrons excited per 50fs(at time = 0) (proportional to intensity). The fitted line is for homogeneous broadening.

8 Conclusions

We have considered the optical non-linearities present in the intersubband transition in n-type modulation doped quantum wells, namely a peak shift due to many-body effects and a reduction in absorption due to subband-filling effects. We predict that a large non-linearity, due to the change in many-body effects on excitation, should be possible in a stepped quantum well. We predict a switching energy of about 0.1W. The saturation intensity of the intersubband transition is expected to be around 700 $kWcm^{-2}$; in GaAs QW the absorption should saturate homogeneously.

A Monte Carlo simulation of relaxation after pulsed excitation gave relaxation times is in reasonable agreement with recent experiments.

References

1. D.S. Chemla, D.A.B Miller J.Opt.Soc.Am.B 2,1155 (1985)
2. L.C. West, S.J. Eglash Appl. Phys. Lett. 46,1156 (1985)
3. B.F. Levine, K.K. Choi, C.G. Bethea, J. Walker, R.J. Malik Appl. Phys. Letts. 50,1092 (1987)
4. A. Harwit, T.S. Harris Jr. Appl. Phys. Lett. 50,685 (1987)
5. S.Y. Yuen Appl. Phys. Letts. 43,813 (1983)
6. D.J. Newson, A. Kurobe Appl. Phys. Letts. 51,1672 (1987)
7. D. Ahn, S.L. Chuang Jour. Appl. Phys. 62,3052 (1987)
8. L. Hedin, B.I. Lundqvist J. Phys. C 4,2064 (1971)
9. T. Ando, A.B. Fowler, F. Stern Rev. Mod. Phys. 54,437 (1982)
10. A. Seilmeier, H.J. Hubner, G. Abstreiter, G. Weimann, W. Schlapp Phys. Rev. Lett. 59,1345 (1987)
11. A. Seilmeier, M. Worner, H.J. Hubner, G. Abstreiter Paper ThB-8, International Conference On Quantum Electronics Tokyo (1988)
12. B.K. Ridley J. Phys. C. 15,5899 (1985)
13. D.J. Newson, A. Kurobe, Semicond. Science and Tech. (in press)
14. U. Ekenberg Phys. Rev. B 36,6152 (1987)
15. R.B. James, D.L. Smith Phys. Rev. Lett. 21,3502 (1979)
16. F.H.Julien, J.M. Lourtioz, N. Herschkorn, D. Delacourt, J.P. Pocholle, M. Papuchon, R.Planel, G. le Roux Appl. Phys. Letts. 53, 116 (1988)
17. Y.C. Chang, R.B. James SPIE 792,214 (1987)

Relaxation Processes and Optical Nonlinearity in Semiconductor Microcrystals Embedded in Glasses

M. Tomita[1] *and M. Matsuoka*[2]

[1]Department of Physics, Faculty of Science, Shizuoka University, 836 Ohya, Shizuoka 422, Japan
[2]Institute for Solid State Physics, University of Tokyo, Roppongi, Minato-ku, Tokyo 106, Japan

1. Introduction

Semiconductor microcrystals have attracted increasing attention to their large nonlinear optical susceptibilities due to quantum confinement effects [1,2]. These microcrystals are often embedded in other crystals, polymers or glass hosts [3-5]. Microcrystals have large boundary surface area. Therefore information on this surface is also very important for understanding the nonlinear optical properties of these microcrystals besides the quantum confinement effect. In this report we discuss the relaxation processes and optical nonlinearity of semiconductor-doped glass on the basis of a four-level system which includes surface trapping state.

2. Nonlinear Photoluminescence

We have examined commercially available colored glass filters. For example, the photoluminescence of a Toshiba R63 color filter is shown in Fig.1. Samples were immersed in bubble-free liquid nitrogen to avoid the well-known photo-darkening effect which complicates the luminescence and nonlinear optical response of these materials. The excitation light source was second harmonic light of a Nd^{3+} YAG

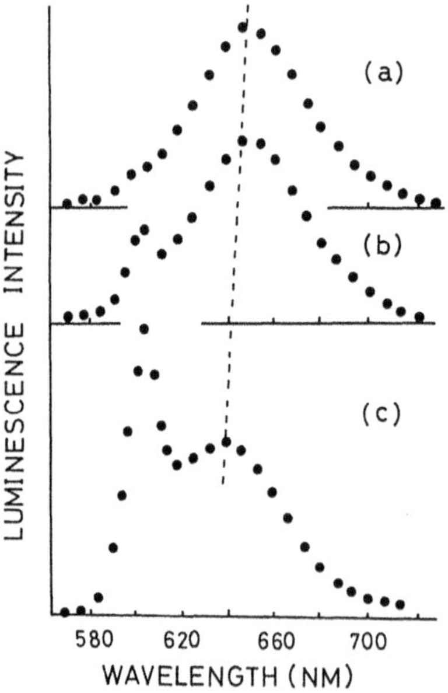

Fig.1 Photoluminescence spectra of a R63 filter under the excitation intensity of (a) 20 mW/mm^2, (b) 100 mW/mm^2, (c) 230 mW/mm^2.

Springer Proceedings in Physics, Vol. 36
Nonlinear Optics of Organics and Semiconductors
Editor: T. Kobayashi

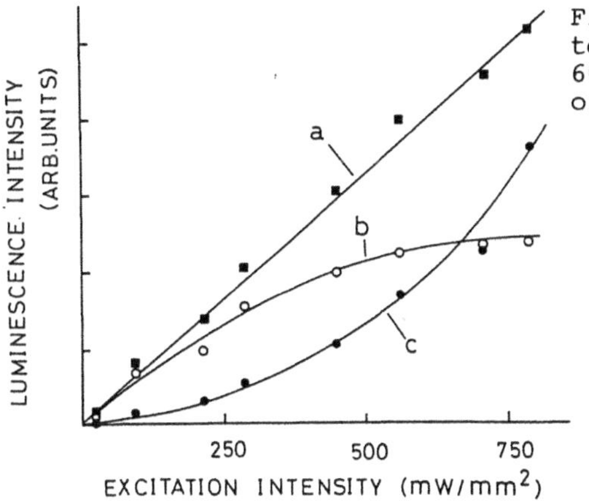

Fig.2 Intensities of (a) the total (b) the 650 nm, (c) the 605 nm luminescence as a function of the excitation intensity.

laser. For a weak excitation intensity (a) the photoluminescence spectrum consists of a low energy broad band peak at 650 nm. Since this luminescence shows a nonexponential decay with time constants spread from 2 μsec to more than 20 μsec, it is attributed to that from the surface trapping state. For increased intensities (b and c), a sharp peak at 605 nm appears. The quantitative behavior of the photoluminescence is shown in Fig.2 as a function of the excitation power. The total luminescence increases linearly. The low-energy luminescence increases sub-linearly and saturates at a certain power. Fig.2 suggests that in these glasses photo-excited carriers are firstly captured into the surface trapping state and the 605 nm luminescence appears after the population saturation of this low-lying trapping state.

3. Relaxation and Optical Nonlinearity in a Four-Level System

We consider a four-level system as shown in Fig.3. In the figure, states |C> and |V> represent the conduction and valence bands, |a> and |b> trapping states, γ_{CV} and γ_{ab} radiative decay rates, and γ_{Ca} and γ_{bV} nonradiative decay rates. Under the relatively weak excitation, the state |a> acts as a recombination center that emits the 650 nm luminescence. For the increased excitation intensity, population saturation occurs in this state, and the number of the trapping states available is reduced. The decay rate γ_{Ca} is therefore expressed as

$$\gamma_{Ca}(N_a) = K\cdot(N_0 - N_a) , \qquad (1)$$

where N_0 is the number of trapping states, N_a is the population of the state |a>, and K is the transition probability of an electron

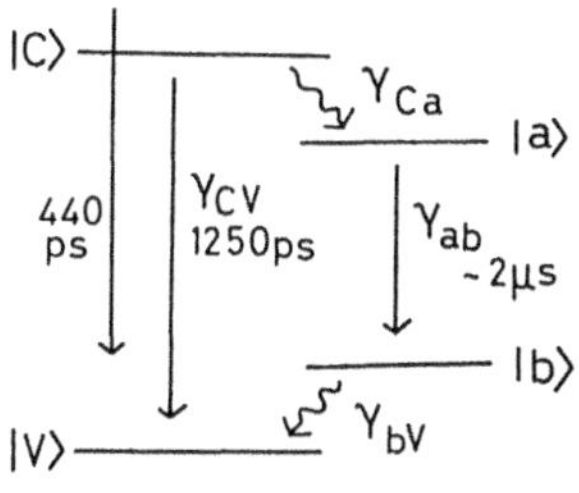

Fig.3 Four-level model of the semiconductor-doped glass.

from the conduction band to a trapping center. In the weak excitation limit we assume that almost all of the photo-excited carriers relax into the trapping state $|a\rangle$, so that

$$\gamma_{CV} << \gamma_{Ca}(0) = K \cdot N_0 \ . \qquad (2)$$

When the excitation intensity is increased, the radiative lifetime of the high energy luminescence becomes longer as the competition with γ_{Ca} lessens. This behavior of the radiative lifetime is shown in Fig.4 with the result of the pump-probe experiment. The same decay times, 440 psec and 1250 psec in (b) and (c), mean that our pump-probe experiment was done at the level of the most intense excitation powers of the photoluminescence experiments.

We can estimate the trapping rate γ_{Ca} in two ways consistently. Firstly, it is estimated as less than 20 psec from the radiative lifetime of the 605 nm luminescence under the weak excitation limit. Secondly, it is estimated as

$$\frac{\gamma_{Ca}(0)^{-1}}{\gamma_{CV}^{-1}} = \frac{dN_C/d\phi \mid_{\phi\to 0}}{dN_C/d\phi \mid_{\phi\to\infty}} \ . \qquad (3)$$

where ϕ is the pumping rate and N_C is the population of the state $|C\rangle$. The trapping rate γ_{Ca} is estimated as $<$ 11psec^{-1} from the slopes of the curve in Fig.2.

The population grating is responsible for nonlinear optical responses, phase conjugation, and transient gratings. The above nonlinear behavior of the photoluminescence in these glasses means that the fast response grating is constructed only after the saturation of the trapping state. This mechanism explains [6] some complicated behavior of the nonlinear optical response in these glasses already reported [7-10]. In Fig.5 the steady-state

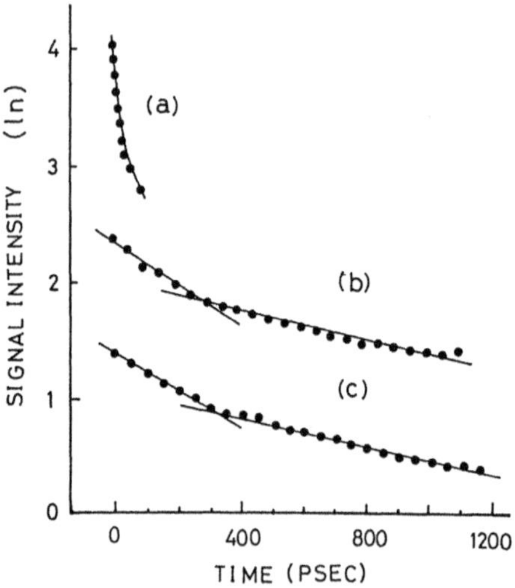

Fig.4 Semilog plots of time dependences of 605 nm luminescence under the excitation intensity of (a) 5 mW/mm^2, (b) 500 mW/mm^2. (c) is the decay of pump-probe signal.

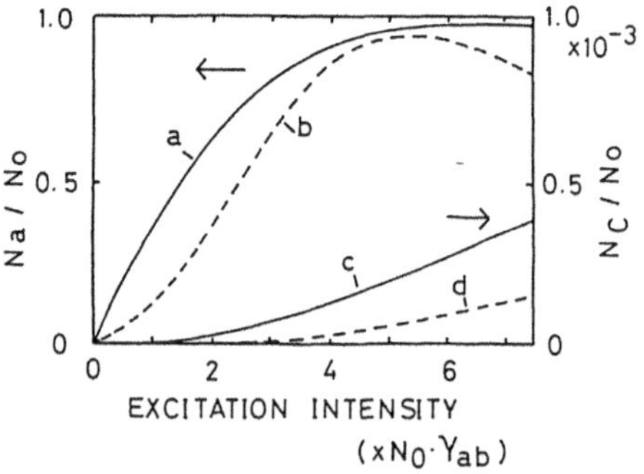

Fig.5 Calculated curves. (a), (c) Normalized population in the states $|a\rangle$ and $|C\rangle$, respectively. (b), (d) Normalized square of the first-order Fourier component of the population gratings in the states $|a\rangle$ and $|C\rangle$, respectively.

population in the states |C> and |a> and the square of the first-order Fourier components of the population gratings in these states are calculated as a function of the normalized excitation intensity. All parameters used here are obtained from the experimental results. The square of the first-order Fourier component of the population grating in |a> that corresponds to the slow response diffraction signal decreases absolutely under increased intensity due to the saturation effect.

4. External Control of the Optical Properties Through the Trapping States

We propose here a new application of these surface states. If we use an additional optical pulse that saturates the population of the trapping state independently, we can externally control the optical properties of these glasses. To confirm this idea, we performed two experiments.

The first one was an experiment with two superposed mode-locked pulse trains, one for a strong pre-excitation and the other for a weak probe, delayed with respect to the other. It turns out that, even though the weak probe pulses cannot excite the 605 nm luminescence by themselves (Fig.1 a), they can excite it with the pre-excitation pulses which enter within the lifetime of the trapping state [6,11]. This result suggests that it is possible to use the fast and large nonlinear response of the inter-band transition of the microcrystals selectively, if the pre-excitation pulses saturate the trapping state beforehand.

In the second experiment, the effect of the pre-excitation pulses on the transient grating was examined in the experimental configuration shown in Fig.6. The signal excitation beams k_1, k_2, k_3 were from a cw mode-locked Rh6G dye laser, the pulse width was about 3 psec and the repetition rate was 80 MHz. The signal beam emitted in the direction of $k_3+k_1-k_2$ was detected by a photo-multiplier. The pre-excitation beam was from the second harmonic of the cw YAG laser that pumped the dye laser. Fig.7 shows the diffraction signal as a function of the delay time of the k_3 beam. It turns out that the slow response component of the diffraction signal corresponding to the trapping state is decreased by the pre-excitation pulses[6,11].

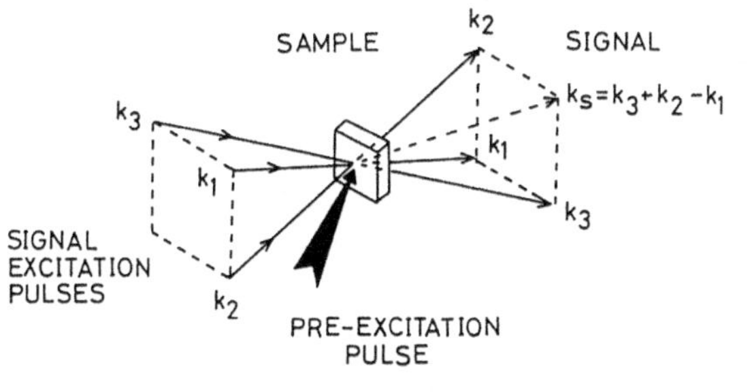

Fig.6 Schematic diagram of the four-wave mixing experiment.

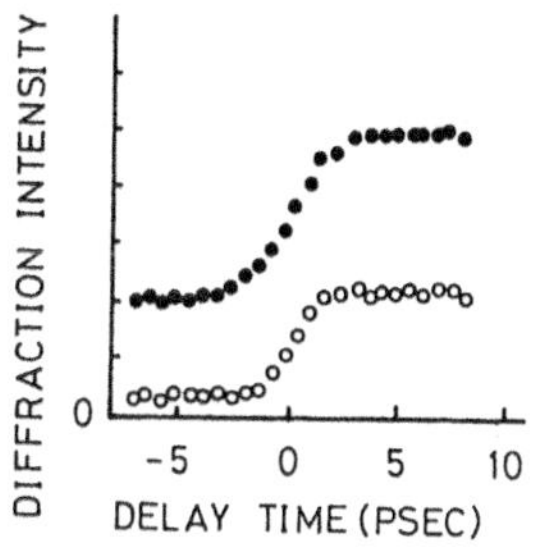

Fig.7 Diffraction signals as a function of the delay time of the k_3 pulses without (filled circles) and with (open circles) the pre-excitation pulses.

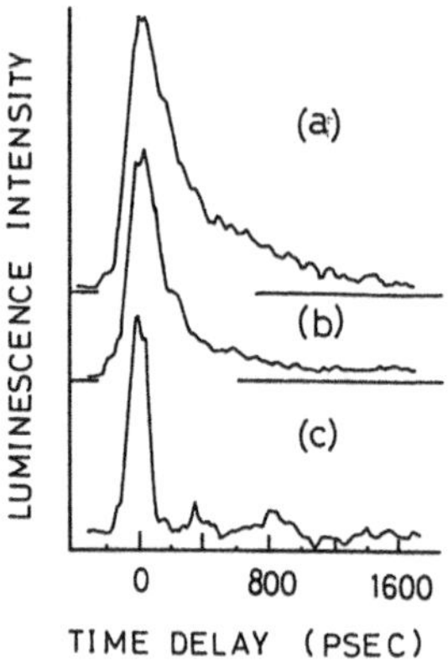

Fig.8 Time dependences of the 605 nm luminescence of a R63 filter at room temperature; (a) in fresh condition, (b), (c) after irradiation of 1 W/mm^2, 532 nm light, for 0.2 sec and 20 sec, respectively.

5. Photo-darkening Effect

At room temperature, photo-darkening occurs easily [12]. After it occurs, the photoluminescence intensity decreases significantly and the lifetime of the 605 nm luminescence becomes shorter (Fig.8). We can consider a possible model of photo-darkening effect on the basis of the above four-level model [13]. Photo-induced irreversible changes of optical properties are intrinsic features in amorphous materials. In semiconductor-doped glasses, there exist a lot of nonequiliburium dangling bonds in the glass host. The photo-darkening effect could be explained in terms of the interaction between the microcrystalline surface and these bonds. If the effective number of the trapping centers increases after irradiation of light (probably due to activation of a nonradiative decay channel between the states $|a\rangle$ and $|b\rangle$ through these bonds, or effective decrease of the carriers due to trapping by these bonds), the radiative lifetime of the 605 nm luminescence becomes shorter since the saturation intensity increases. This model explains not only the decrease of the total luminescence intensity but also, in a certain condition, the relative decrease of the 650 nm luminescence intensity after the photo-darkening so far reported [12].

6. Conclusion

We have clarified the mechanism of the intensity-dependent carrier recombination process in the semiconductor-doped glass. Photo-excited carriers are firstly captured into the surface trapping states with a very fast nonradiative decay. This interpretation of the intensity-dependent photoluminescence is consistent with recent experimental results in the femtosecond range [14]. This mechanism plays an important role in the intensity-dependent optical nonlinear properties as well as in the intensity-dependent photoluminescence. Recently much interest has been directed toward the crystalline size dependence of the optical nonlinearity in semiconductor-doped glasses. In the research in this direction, one should pay attention to the large variation between samples of the number of trapping states. Quantum confinement effects [15.16] in these glasses should be investigated on the basis of the above four-level system. We proposed a new method which may be used to analyze the size effect in these glasses. This pre-excitation technique may also be used to operate an optical switch by a weak driving power or control switching characteristics externally by using optical pulses.

References

1. T. Takagahara, E. Hanamura: Phys. Rev. Lett. 56, 2533 (1986)
2. E. Hanamura: Phys. Rev. B37, 1273 (1988)
3. T. Itoh: Phys. Stat. Solidi (b) 145, 567 (1988)
4. Y. Wang, A. Suna, W. Mahler, R. Kasowski: J. Chem. Phys. 87, 7315 (1987)
5. N.F. Borrelli, D.W. Hall, H.J. Holland, D.W. Smith: J. Appl. Phys. 61, 5399 (1987)
6. M. Tomita, T. Matsumoto, M. Matsuoka: J. Opt. Soc. Am. B6, to be published.
7. R.K. Jain, R.C. Lind: J. Opt. Soc. Am. 73, 647 (1983);
8. P. Roussignol, D. Ricard, K.C. Rustagi, C. Flytzanis: Opt. Comm. 55, 143 (1985);
9. D. Cotter: In Ultrafast Phenomena V, 274 (Springer 1986);
10. J.T. Remillard, D.G. Steel: Opt. Lett. 13, 30 (1988)
11. M. Tomita, T. Matsumoto, M. Matsuoka: In Ultrafast Phenomena VI (Springer 1988)
12. P. Roussigonol, D. Ricard, J. Lukasik, C. Flytzanis: J. Opt. Soc. Am. B4, 5 (1987)
13. M.Tomita, M.Matsuoka: submitted.
14. M.C. Nuss, W. Zinth, W. Kaiser : Appl. Phys. Lett. 49, 1717 (1986)
15. J. Warnock, D.D. Awshalom: Appl. Phys. Lett. 48, 425, (1986)
16. E. Hanamura: Phys. Rev. B38, (1988) to be published

Dephasing Time of CdS Microcrystallites Measured by a Simple Method Using Incoherent Light

K. Misawa [1], *T. Hattori* [1], *T. Kobayashi* [1], *Y. Ohashi* [2], *and H. Itoh* [2]

[1] Department of Physics, University of Tokyo, 7-3-1 Hongo, Bunkyo-ku, Tokyo 113, Japan

[2] Central Research Institute, Mitsui Toatsu Chemicals Inc., Kasama-cho 1190, Sakae-ku, Yokohama 247, Japan

1. Introduction

Semiconductor microcrystallites are believed to be among promising materials used in optical nonlinear devices because of their large nonlinearity due to excitons confined in the particles. After JAIN and LIND [1] reported a nonlinear susceptibility $\chi^{(3)}$ as large as 10^{-8} esu, many groups have been studying physical properties of microcrystallites using commercial semiconductor-doped glass filters [2-5]. YAO et al. [2] measured the decay time of the photoluminescence due to carrier recombination ranging from 80 ps to less than 16 ps. They also obtained the decay time constant of a population grating by degenerate four-wave mixing (DFWM) to be 30 ps. However ROUSSIGNOL et al. [3] reported much longer decay time constants of tens of nanoseconds. They explained the difference in the carrier recombination times by several orders of magnitude in terms of a darkening effect: darkened samples have shorter decay times due to efficient free-carrier recombination, while non-darkened ones have longer decay times due to recombination of carriers trapped in impurities or surface defects. The darkening is a photochemical process due to carriers which fill trapping states permanently. The carrier recombination time even in the same sample depends on the laser intensity or it decreases during the irradiation of laser light [4].

Phase decay in microcrystallites has also been studied. TOKIZAKI et al. investigated the dephasing process in a commercial filter [5]. The relaxation process relevant to trapping states depends on samples, because such trapping states are of extrinsic nature. These commercial filters are easily available as samples containing microcrystallites of CdS_xSe_{1-x}. However, since an excessive amount of zinc is used in the manufacturing process, microcrystallites with different stoichiometries such as ZnS and ZnSe are contained. Therefore commercial filters are not appropriate for the investigation of the basic properties of microcrystallites. In order to discuss physical properties of microcrystallites, it is necessary to study with a well-defined stoichiometry [6]. Thus we prepared pure CdS microcrystallites doped in a polymer film by "precipitation-in-gel" (PIG) method [7]. In the present paper we report measurement of the dephasing time (T_2) in CdS microcrystallites using temporally incoherent light. In Chap. 2 a signal-selective method to measure T_2 which we recently proposed is described [8]. Experimental results on CdS microcrystallites by the new method are discussed in Chap. 3.

2. Method

Recently a new spectroscopic technique using incoherent light has been presented [9] and verified experimentally [10]. Measurement with incoherent light has a time resolution which is determined by the correlation time of the light (τ_c), not by the temporal pulse duration. We have applied this technique to measure the dephasing time in several materials by 2-beam forward DFWM. For example, we resolved the electronic T_2 in a film of polydiacetylenes, poly(4,6-decadiyne-1,10-diol bis((n-butoxycarbonyl)-methyl) urethane) (poly-3BCMU) [11]. After the finite

Springer Proceedings in Physics, Vol. 36
Nonlinear Optics of Organics and Semiconductors
Editor: T. Kobayashi

inhomogeneous broadening had been taken into account, the estimated T_2 was 50-60 fs at a wavelength of 648 nm, and was 140-160 fs at 582 nm [12]. The incoherent light source was a home-made broad-band dye laser pumped by a nanosecond N_2 laser. Although it is a very simple apparatus, we could resolve dephasing times in fs regime. In addition, taking advantage of easy tunability of the laser used in this method, we could discuss the dependence of dephasing process on the conjugation length by the wavelength dependence experiment.

We proposed a simple and signal-selective method to measure T_2 using incoherent light. It is performed in the Kerr-shutter configuration (Fig. 1) [8]. The scheme of this method is similar to heterodyne-detected accumulated photon echo experiments [13,14]. A different point in our method from the latter is that the signal intensity is proportional to the squared absolute value of the third-order polarization, and that the signal appears only when the pump beam is present. So far the Kerr-shutter configuration has been applied to investigate the molecular dynamics in various liquids [15], and coherent coupling between pump and probe beams in optical Kerr effect in transparent media was pointed out where dephasing is instantaneous [16]. In systems resonant with the incident light, the dephasing is not instantaneous, which leads to a phase-memory effect on the coherent-coupling contribution to the signal intensity. We calculated the trace of signal intensity as a function of the delay time (τ) of the probe beam with respect to the pump beam for a two-level system and demonstrated that T_2 can be obtained from the trace.

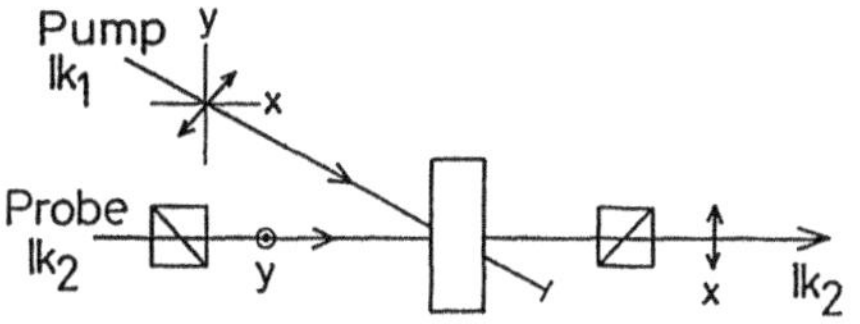

Fig. 1: DFWM in the Kerr-shutter configuration

Here, the following two assumptions were made for the system to be studied,

(i) extremely large inhomogeneous broadening, and

(ii) $T_1 \gg \tau_c$, T_2.

The assumptions correspond to the condition to be satisfied in accumulated photon echoes. Besides them we implicitly assumed that the temporal pulse duration is so long that the population distribution of the excited and ground states does not vary during dephasing process. Using ultrashort pulses, we cannot measure T_2 in the Kerr-shutter configuration, because the population change would be significant.

The electric field of incident light is given as

$$\mathbf{E}(\mathbf{r},t) = \mathbf{e}_1 E(t)\exp[i(\mathbf{k}_1\mathbf{r}-\omega t)] + \mathbf{e}_2 E(t-\tau)\exp[i(\mathbf{k}_2\mathbf{r}-\omega(t-\tau))] \,. \tag{1}$$

On assumption (i), the $\mathbf{k}_2$ component of the third-order polarization is

$$\mathbf{P}^{(3)}(t) \propto \int_{-\infty}^{t} dt' \int_{-\infty}^{t'} dt'' \exp[-(t'-t'')/T_1 - 2(t-t')/T_2] \Big[\bar{\chi}\mathbf{e}_2\mathbf{e}_1\mathbf{e}_1 E(t'-\tau)E(t'')E^*(t''+t'-t) + \bar{\chi}\mathbf{e}_1\mathbf{e}_2\mathbf{e}_1 E(t')E(t''-\tau)E^*(t''+t'-t)\Big] \tag{2}$$

where $\mathbf{e}_i$ is the unit polarization vector of $\mathbf{E}_i$, and $e_{1x} = e_{1y} = 1/\sqrt{2}$; $e_{2x} = 0$; $e_{2y} = 1$. $\bar{\chi}$ is a fourth-rank tensor of the third order optical susceptibility. Its nonzero components are χ_{xxxx}, χ_{xxyy}, χ_{xyyx}, and χ_{xyxy}, and those obtained by permutation among indices x, y, and z. Here we take the definition of χ_{ijkl} as the index k associates with the complex conjugate of the electric field E^* [17].

These tensor components are the products of four mean values of the dipole moment and are assumed to be independent of temporal evolution, that is, the material has only single T_1 and T_2. Since only the light with x-polarization is detected, we take the x-component of the polarization. Therefore,

$$(\bar{\chi} e_2 e_1 e_1)_x = (\chi_{xyxy} + \chi_{xyyx})/2$$
$$(\bar{\chi} e_1 e_2 e_1)_x = (\chi_{xyxy} + \chi_{xxyy})/2 \tag{3}$$

and we define "polarization factor" as

$$g = (\chi_{xyxy} + \chi_{xxyy}) \;/\; (\chi_{xyxy} + \chi_{xyyx}) \;. \tag{4}$$

On assumption (ii), we can rewrite (1) with the electric-field autocorrelation function $G(u) = \langle E(t)E^*(t+u)\rangle$ as

$$P_x^{(3)}(t) \propto \int_{-\infty}^{t} dt' \exp[-2(t-t')/T_2] \left[E(t'-\tau)G(t-t') + g\, E(t')G(t-t'-\tau) \right] . \tag{5}$$

In the case of electrons in semiconductors, g is equal to unity [17]. The signal intensity, which is the statistical average of $|P_x^{(3)}|^2$, is given as

$$\begin{aligned} I^{(3)}(t) \propto \int_{-\infty}^{t} dt' \int_{-\infty}^{t} ds' \exp[-2(2t-t'-s')/T_2] \Big[& G(t'-s')G(s'-t)G(t-t') && \leftarrow \text{background term} \\ & + G(t'-s'-\tau)G(s'-t+\;)G(t-t') && \\ & && \leftarrow \text{cross terms} \\ & + G(t'-s'+\tau)G(t-t'-\tau)G(s'-t) && \\ & + G(t-t'-\tau)G(s'-t+\tau)G(t'-s') \Big] . && \leftarrow \text{decay term} \end{aligned} \tag{6}$$

If τ_c is much shorter than T_2, G(u) can be replaced with a delta function $\delta(u)$. After some calculations, we get the normalized signal intensity as follows:

$$\begin{aligned} I(\tau) &= 1 && (\tau < 0) \\ I(\tau) &= [1 + 2\exp(-2\tau/T_2)]^2 && (\tau > 0) \;. \end{aligned} \tag{7}$$

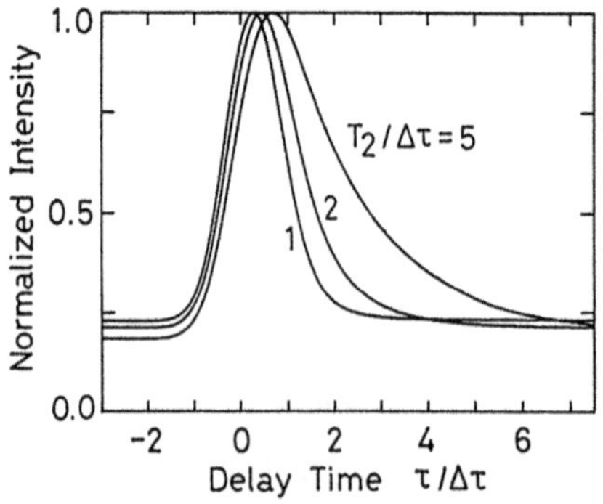

Fig. 2: Calculated curves of the DFWM signal intensity in the Kerr-shutter configuration as a function of the normalized delay $\tau/\Delta\tau$ for several values of $T_2/\Delta\tau$

In another case, we assume that G(u) is expressed as a Gaussian function $G(u) = \exp[-u^2/\Delta\tau^2]$. Here we define $\tau_c = \sqrt{2}\Delta\tau$ to be a temporal width (FWHM) of $|G(u)|^2$. We can experimentally determine the defined τ_c by measuring the intensity of the diffracted signal from thermal grating in dye solution at room temperature, because its delay time dependence represents $|G(u)|^2$. Figure 2 shows calculated signal intensity as a function of the normalized delay $\tau/\Delta\tau$ for several values of $T_2/\Delta\tau$. The decay tail is evident for $T_2 > 2\Delta\tau$, and it offers information about T_2.

3. Experimental

In PIG method, CdS colloids are precipitated in aqueously swollen gels of crosslinked copolymer of N-acryloyl pyrrolidine and urethane acrylate. The particle size is limited by the volume of the cells made of crosslinked-polymer lattice. Therefore it is not necessary to use other chemicals for stabilizing colloids.

The absorption spectrum of the sample in Fig. 3 exhibits a shoulder around 420 nm which corresponds to the exciton absorption. The width of the shoulder is estimated as about 40 nm. If the size of the particles is uniform, exciton absorption should appear as a narrow peak. ALIVISATOS et al. reported homogeneous line width of CdSe microcrystallites to be 17meV (4nm) [18]. Hence the particles in this sample may have a broad distribution of diameters.

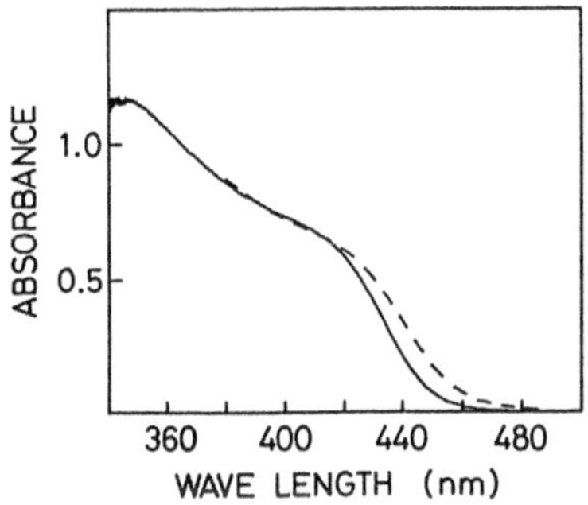

Fig. 3: Absorption spectra of CdS microcrystallites doped in a polymer film;
solid line: at 2K;
dashed line: at room temperature

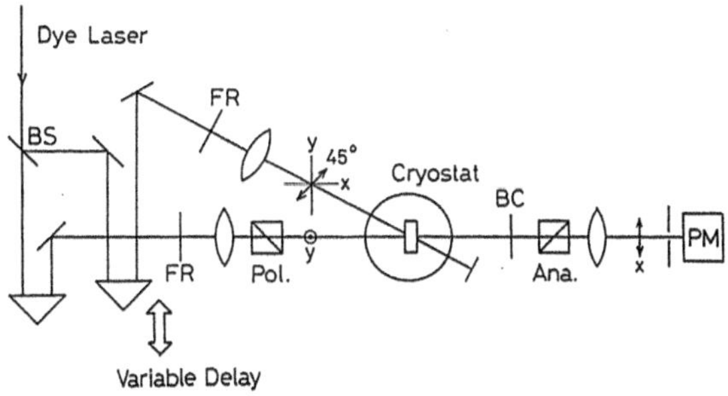

Fig. 4: Experimental setup for T_2 measurement in the Kerr-shutter configuration; PM:Photomultiplier; BS:Beam splitter; FR:Fresnel rhomb; BC:Babinet-Soleil compensator

Figure 4 shows the experimental apparatus. An incoherent light source was a broad-band dye laser pumped by a nanosecond excimer laser. The laser dye was stilbene 420 selected to be nearly resonant on the exciton absorption. The dye laser had no tuning elements [11], and τ_c was about 120 fs. The laser beam which was linearly polarized by a Glan-Thompson prism was divided into two beams with equal intensity. One of them (probe beam) is focused on the sample through a polarizer. Then it is directed to a photomultiplier through an analyzer. A Babinet compensator and a pinhole were used to suppress the background. The other (pump beam) is polarized 45° with respect to the probe beam, variably delayed with a translational stage, and focused at the same point. Samples were immersed directly in superfluid helium. The signal transmitted through a crossed analyzer was detected as a function of the delay of the probe with respect to the pump.

We could successfully detect the signal by the new method using the Kerr-shutter configuration, while we failed in the ordinary 2-beam DFWM configuration. It is because quick and precise alignment can be done in the former method. That

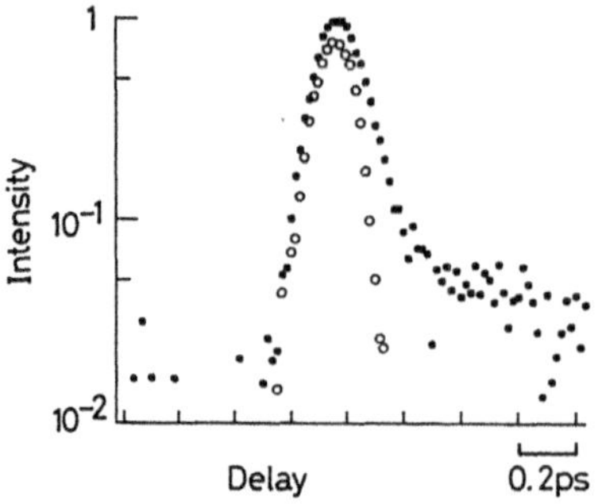

Fig. 5: Semi-logarithmic plots of the delay time dependence of signal intensity at 2K obtained from CdS microcrystallites; signal (•); autocorrelation (o)

is why we called the new method "signal-selective". Two decay components can be seen in the obtained data from CdS microcrystallites with a mean diameter less than 50 Å (Fig. 5). From this trace we estimate the decay-time constant to be 200 fs for the faster component. This value is shorter than that of excitons confined in a GaAs quantum-well which is reported to be a few picoseconds by SCHULTHEIS et al. [19]. Since their sample contained a single quantum-well, they could tune the wavelength of the incident laser beam just resonant on the lowest level of the exciton absorption. However, the exciton absorption of our sample is inhomogeneously broadened. Thus we also excite higher levels of excitons and the dephasing of the higher levels may be faster. On the other hand, the homogeneous line width reported in Ref. [18] represents a dephasing time of 80 fs. Dephasing time obtained indirectly by hole-burning is shorter than that obtained by DFWM. Systematic errors tend to broaden the hole width.This means underestimation of dephasing time.

For more detailed and systematic study, it is necessary to control the particle size and get a sample with narrower distribution of particle diameter.
We are still in the process of experimenting to find the exact resonance with the lowest sub-level.

References

1. R.K. Jain and R.C. Lind: J. Opt. Soc. Am. 73, 647 (1983)
2. S.S. Yao, C. Karaguleff, A. Gabel, R. Fortenbelly, C.T. Seaton, and G.I. Stegeman: Appl. Phys. Lett. 46, 801 (1985)
3. P. Roussignol, D. Ricard, J. Lukasik, and C. Flyzanis: J. Opt. Soc. Am. B 4, 5 (1987)
4. M. Mitsunaga, H. Shinojima, and K. Kubodera: J. Opt. Soc. Am. B 5, 1448 (1988)
5. T. Tokizaki and T. Yajima: Ultrafast Phenomena VI, (Springer, Berlin), to be published.
6. Y. Wang and W. Mahler: Opt. Comm. 61, 233 (1987)
7. Y. Ohashi, H. Ito, T. Hayashi, A. Nitta, H. Matsuda, S. Okada, H. Nakanishi and K. Kato: Nonlinear Optics of Organics and Semiconductors ed. by T. Kobayashi (Springer, Berlin), to be published.
8. K. Misawa, T. Hattori, and T. Kobayashi: submitted to Opt. Lett.
9. N. Morita and T. Yajima: Phys. Rev. A 30, 2525 (1984)
10. S. Asaka, H. Nakatsuka, N. Fujiwara, and M. Matsuoka: Phys. Rev. A 29, 2286 (1984)
11. T. Hattori and T. Kobayashi: Chem. Phys. Lett. 133, 230 (1987)
12. T. Kobayashi, T. Hattori, A. Terasaki, and K. Kurokawa: Revue Phys. Appl. 22, 1773 (1987)
13. W.H. Hesselink and D.A. Wiersma: Phys. Rev. Lett. 43, 1991 (1979)
14. S. Saikan, Y. Kanematsu, R. Shiraishi, T. Nakabayashi, and T. Kushida: J. Lumin. 38, 15 (1987)

15. D. McMorrow, W.T. Lotshaw, and A. Kenney-Wallace: IEEE J. Quantum Electron. QE-24, 443 (1988)
16. J.L. Oudar: IEEE J. Quantum Electron. QE-19, 713 (1983)
17. B.S. Wherrett, A.L. Smirl, and T.F. Boggess: IEEE J. Quantum Electron. QE-19, 680 (1983)
18. A.P. Alivisatos, A.L. Harris, N.J. Levinos, M.L. Steigerwald, and L.E. Brus: J. Chem. Phys. 89, 4001 (1988)
19. L. Schultheis, A. Honold, J. Kuhl, K. Kohler, and C.W. Tu: Phys. Rev. B 34, 9027 (1986)

Very Large $\chi^{(3)}$ Due to Coherency of an Exciton in PbI_2

T. Ishihara and T. Goto

Department of Physics, Faculty of Science, Tohoku University,
Sendai 980, Japan

The third-order optical nonlinear susceptibility $\chi^{(3)}$ is estimated quantitatively from the population oscillation effect in PbI_2. The value of $\chi^{(3)}$ amounts to -10^{-4} esu at the exciton resonance. Such a large $\chi^{(3)}$ can be explained only if the nonlinear interaction takes place over many unit cells, which comes from coherency of the translational motion of an exciton. The coherence volume is considered to be limited by the relaxation of exciton, which is estimated from the exciton absorption width.

1. Introduction

Recently, optical nonlinearity in semiconductors and organic compounds has been studied extensively. One of the aims of these researches is to seek a way to realize large optical nonlinearity. Along this line, many authors have been discussing excitonic optical nonlinearity in low dimensional materials. These studies are attractive in their own right. It seems, however, that we have not yet fully understood what determines $\chi^{(3)}$ due to an exciton state even in bulk semiconductors.

On the other hand, in organic conjugate chain compounds, it is known that $\chi^{(3)}$ depends strongly on the chain length [1]: the longer the chain length is, the larger $\chi^{(3)}$ is. This comes from the electron delocalization. How about semiconductor crystals, which may be regarded as three dimensional infinite chains? It is impossible that $\chi^{(3)}$ is infinitely large. What limits $\chi^{(3)}$?

Such a problem has not been discussed yet, because there are few works which estimate $\chi^{(3)}$ of the bulk exciton [2,3]. In this report we try to estimate the value of $\chi^{(3)}$ experimentally. The material we have chosen is PbI_2, in which the parameters for exciton is well known and of which evaporated films of good quality are easily obtained.

The details of the experiments and the method of analysis have been already reported [4]. In this paper besides the nonlinear measurement, we also measure luminescence spectra of the sample excited resonantly in order to check the origin of the exciton width and discuss the excitonic $\chi^{(3)}$ in relation to relaxation of an exciton.

2. Experimental

Purified PbI_2 was evaporated on a fused quartz substrate resulting in a thickness of 200 nm. The optical nonlinearity was measured by the nondegenerate pump and probe method: Pump and probe beams were obtained from two independent dye lasers pumped by an N_2 pulsed laser. The probe

Springer Proceedings in Physics, Vol. 36
Nonlinear Optics of Organics and Semiconductors
Editor: T. Kobayashi

light was filtered to eliminate broad background emission by the use of a premonochromator. The transmission of the probe beam was measured as a function of the pump photon energy Ω_2. The spectral and pulse widths of both dye lasers were 0.05 meV and 2.5 ns, respectively. The pump and probo beams were of parallel polarization and focused on the sample at an angle of 20 degrees. The size of the focused spots on the sample was about 100 μm. The samples were immersed directly in liquid N_2 or pumped liquid He.

As for the tunable excitation source for emission spectra at 77 K, a dye laser excited by an excimer laser was used. The repetition rate was 100 Hz. The output of the dye laser was converted to quasi-CW light using an optical fiber bundle [5] in order to avoid high density excitation effect.

3. Results

Figure 1 shows the transmitted probe intensity I_1 in the logarithmic scale as a function of photon energy difference between pump and probe beams $\Omega_2 - \Omega_1$. Each spectrum a-h corresponds to that for a probe photon energy Ω_1 shown by a downward arrow in Fig.1(B). The structure at $\Omega_2 - \Omega_1 \fallingdotseq 0$ for each spectrum is due to the population oscillation effect [6]. The shape of the structure changes remarkably as Ω_1 is tuned around Ω_0. These changes can be reproduced fairly well by the simple two-level model as was shown in reference [4]. Note that the intensity change is maximum when Ω_1 is located at the exciton absorption peak. The solid curve in Fig.1(B) is the optical density spectrum of the same sample as is used for Fig.1(A).

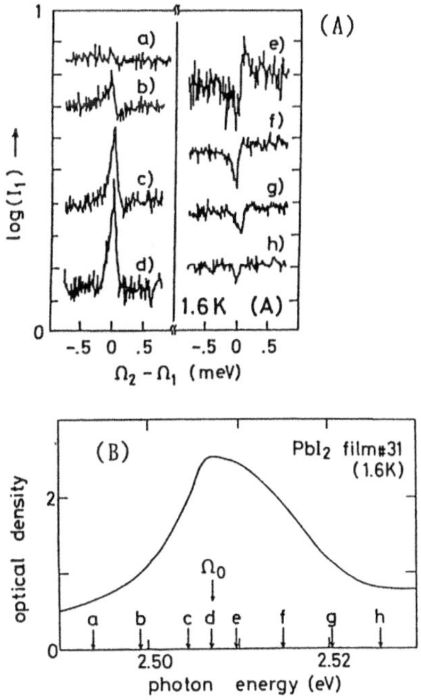

Fig.1
(A) Induced absorption spectra at 1.6K on a logarithmic scale as a function of pump-probe detuning. The spectra a)-h) correspond to different probe energies shown in (B). The pump intensity is 1 MW/cm^2.
(B) Optical density spectrum for the same sample as in (A). The letters a,b,... refer to the values of Ω_1 in (A).

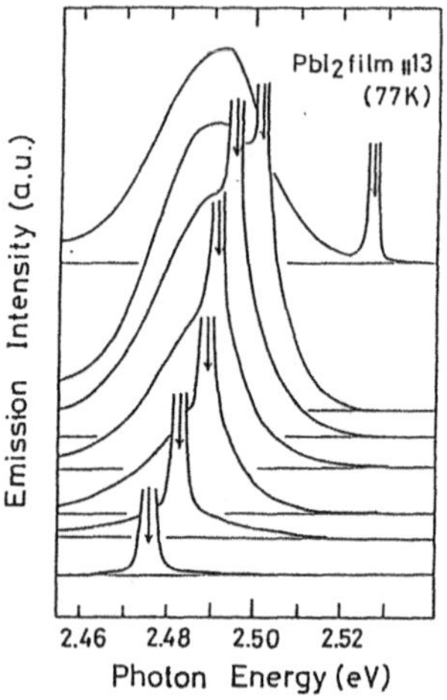

Fig.2
Emission spectra of a PbI_2 film at 77 K. Excitation energies are shown by downward arrows in each spectrum.

Figure 2 shows emission spectra at 77 K excited resonantly in the exciton region. The excitation energies are shown by downward arrows in the figure. The topmost spectrum is excited by the light corresponding to interband transition. Note that the emission width is about 30 meV which is much larger than that of a bulk exciton. In order to clarify the origin of the width, we change the energy of excitation photons. As is shown in the lower spectra in Fig.2, the emission bands are essentially the same, which means that the widths are not inhomogeneous. Thus the situation is quite different from that of excitons in microcrystals [7].

4.Discussion

From the change of transmission in Fig. 1A d), we can estimate the nonlinear absorption coefficient α_{NL} at the exciton resonance to be

$$\alpha_{NL} = \alpha \log(I_1'/I_1)/I_2 \log(e) = -2\times10^5 \text{ cm/MW}, \tag{1}$$

where α is the linear absorption coefficient for the exciton peak. I_1 and I_1' are the intensities of the probe transmission at and beside the Rayleigh peak. The pump beam intensity is denoted by I_2. From this value of α_{NL}, $\chi^{(3)}$ is estimated to be

$$\mathrm{Im}\,\chi^{(3)} \fallingdotseq -10^{-4} \qquad \text{esu}. \tag{2}$$

If we assume that a unit cell in the crystal is treated as a two-level oscillator with its dipole moment μ_0 and that the oscillators are independent, the maximum $\chi^{(3)}$ is

$$\chi^{(3)} = i4N\mu_0^4/\gamma\Gamma^2, \tag{3}$$

where N is the density of unit cells, γ and Γ are longitudinal and transverse relaxation energies, respectively. The relaxation energy γ is estimated from the width of the structure in Fig.1(A) and Γ is estimated from the absorption width of FIg.1(B). The latter is justified from the homogeneity of the width. Using known parameters, the imaginary part of $\chi^{(3)}$ is estimated to be about -10^{-6} esu, which is 2 orders smaller than the experimentally estimated value.

We infer that this discrepancy comes from the assumption of independent oscillators. The exciton state is extended in the crystal coherently and the nonlinear interactions occur all over the coherent region which is limited by interruption of the coherence. Recently FELDMANN et al. [8] and HONOLD et al. [9] discussed a coherent extension of an exciton in relation to its lifetime.

Supposing that an exciton is coherent over N' unit cells, the exciton has a dipole moment of $\mu=\sqrt{N'}\mu_0$. This can be proved easily in the tight binding model. The factor $\sqrt{N'}$ comes from the center of mass motion of an exciton. The expanse of relative motion (Bohr radius) is reflected to μ_0. As the unit volume consists of N/N' coherent regions in this case, $\chi^{(3)}$ should be revised as

$$\chi^{(3)} = i4(N/N')(\sqrt{N'}\mu_0)^4/\gamma\Gamma^2 = i4NN'\mu_0{}^4/\gamma\Gamma^2. \tag{4}$$

Referring to the experimental estimation of $\chi^{(3)}$, we obtain $N'\sim10^2$.

Now we check the value N' reasonable. As is given by FELDMANN et al. homogeneous line width Γ is related to coherent volume as

$$N'\sim 2\pi\hbar^2 z/m\Gamma V_0 = 400, \tag{5}$$

where z is the thickness of PbI_2 layer, M the translational mass and V_0 the volume of a unit cell. In the estimation above we assume that the exciton in PbI_2 is two-dimensional, but it is not essential for order estimation. Thus we consider that the $\chi^{(3)}$ is determined by the exciton coherence in the crystal.

References

1. C.Flytzanis:
   Nonlinear Optical Properties of Organic Molecules and Crystals, ed. D.S.Chemla and J.Zyss (Academic Press, 1987) Vol.2,p.121.
2. G.Mizutani and N.Nagasawa: Opt.Commun.50,31(1984).
3. M.Kuwata: J.Lumin.38,247(1987).
4. T.Ishihara and T.Goto: J.Phys.Soc.Jpn.57,2191(1988).
5. T.Ishihara: J.Phys.Soc.Jpn.57,2573(1988).
6. M.Sergent III :Phys.Report.43,223(1978).
7. T.Itoh,Y.Iwabuchi and T.Kirihara: Phys.Stat.Sol.(b)146,531(1988).
8. J.Feldmann,G.Peter,E.O.Gobel,P.Dawson,K.Moore,C.Faxon and R.J.Elliott: Phys.Rev.Lett.59,2337(1987).
9. A.Honold,L.Schltheis,J.Kuhl and C.W.Tu: Technical Digest of XVI International Conference on Quantum Electronics p.624,(1988).

Excitons and Excitonic Molecules in CuCl Microcrystals

T. Itoh, F. Jin, Y. Iwabuchi, and T. Ikehara*

Department of Physics, Faculty of Science, Tohoku University, Sendai 980, Japan

1 Introduction

Nonlinear optical property of excitons in semiconductor microstructures is one of the current interesting subjects among various quantum size effects on electron-hole systems confined in real space. In CuCl microcrystals the confinement of exciton translational motion has been found to play a dominant role in the quantum size effect on account of the small exciton Bohr radius of 0.68nm.[1,2] This characteristic nature is in contrast with that in GaAs quantum wells where the electron and the hole in excitons are almost individually confined. Since an exciton has a large macroscopic dipole moment on account of its coherent excitation, the rapid radiative decay of excitons, in other words, exciton superradiance may occur in low-dimensional systems due to breakdown of the exciton translational symmetry[3], as far as the excitons keep three-dimensional bulk character in their internal motion but lose their polariton nature. The radiative decay time should depend on the microcrystal size if the size is smaller than the exciton coherence length. The exciton superradiance also brings about the size-dependent optical nonlinearity. On the other hand, since the interaction between two excitons confined in a small microcrystal may be stronger than that in bulk materials, the formation of excitonic molecules is expected to bring about a characteristic optical nonlinearity. In this paper two kinds of experimental results are reported which may show the exciton superradiance and the existence of excitonic molecules in CuCl microcrystals embedded in NaCl single-crystalline matrices.

2 Experimental

CuCl microcrystals of nm in size were grown in NaCl matrices with nominally 1 mol% concentration by the method described in [1]. The matrices were cleaved into platelets or ground into fine powders and directly immersed in liquid nitrogen for the optical measurement. Picosecond time-resolved luminescence was measured by use of the second harmonic of a Pyridine 2 dye laser pumped by a CW-ML YAG laser as the weak excitation source, and of a subtractive double monochromator followed by a synchroscan streak camera as the detection system. For intense ns excitation a B-PBD dye laser pumped by a nitrogen laser was used. The maximum power density at the illuminated surface is about 10kW/cm^2.

3 Exciton Superradiance

The exciton (Ex) luminescence spectrum consists of a single band inhomogeneously broadened on account of the wide size-distribution of microcrystals and the peak energy is shifted from the bulk Z_3 exciton energy toward the higher energy side due to the confinement effect. The amount of blue shift, ΔE, is given by

$$\Delta E(a) = \frac{\hbar^2}{2M_x}\left(\frac{\pi}{a}\right)^2, \tag{1}$$

*On leave of absence from Yanbian University, China

Springer Proceedings in Physics, Vol. 36
Nonlinear Optics of Organics and Semiconductors
Editor: T. Kobayashi

where M_x is the translational mass of excitons and a, the radius of a spherical microcrystal or the thickness of a platelet. The Ex luminescence yield per CuCl molecule is found to be at least one order of magnitude larger for microcrystals than for bulk single crystals at 77K. This high luminescence yield, together with rather high absorption coefficient, successive re-absorption and re-emission processes become important for Ex luminescence spectra in platelet matrices. Therefore, in order to measure the true luminescence decay time, fine powders of the matrices having average diameter of 2μ are used for the time-resolved luminescence measurement. The excitation energy is 3.39eV which is located above the bulk $Z_{1,2}$ exciton absorption peak. Figure 1(a) shows the temporal behavior of Ex luminescence at different energies, the positions of which are indicated on the time-integrated Ex luminescence spectrum shown in (b). The luminescence decay consists of at least two components. The slow one shows the decay time, τ_{slow}, of about 1.5ns almost regardless of the energies of observation, while the decay time for the fast component gradually decreases as the energy of observation approaches the bulk Z_3 exciton energy from the higher energy side. Below the bulk Z_3 exciton energy, the fast component tends to diminish. The average decay time for the first 100ps duration is taken as a transient decay time, τ_{tr}, and is plotted by closed circles in Fig.2 as a function of the energy difference, ΔE. The corresponding crystallite size, a, is obtained from (1) and is shown on the top. The origin of the fast component may be tentatively assigned to the radiative decay because the exciton luminescence yield is very high and the decay due to exciton relaxation toward the lower energy side does not fit with the observed energy-dependent change of the decay time. The exciton radiative decay rate, $1/\tau$, theoretically predicted for the exciton superradiance in spherical microcrystals[3] is given by

$$\frac{1}{\tau(a)} = \frac{64\pi\varepsilon}{3} \left(\frac{a}{\lambda}\right)^3 \frac{\Delta E_{LT}}{\hbar} , \qquad (2)$$

where λ is the wavelength of the luminescence, ΔE_{LT}, the exciton L-T splitting and ε, the static dielectric constant. Together with (1), $1/\tau(a)$ becomes proportional to $(\Delta E)^{3/2}$ and the relation is drawn by a solid line in Fig.2 where ΔE_{LT}=5.65meV, ε=5.6 and M=2.3m_0 (m_0:electron mass in vacuum).[4] Although the

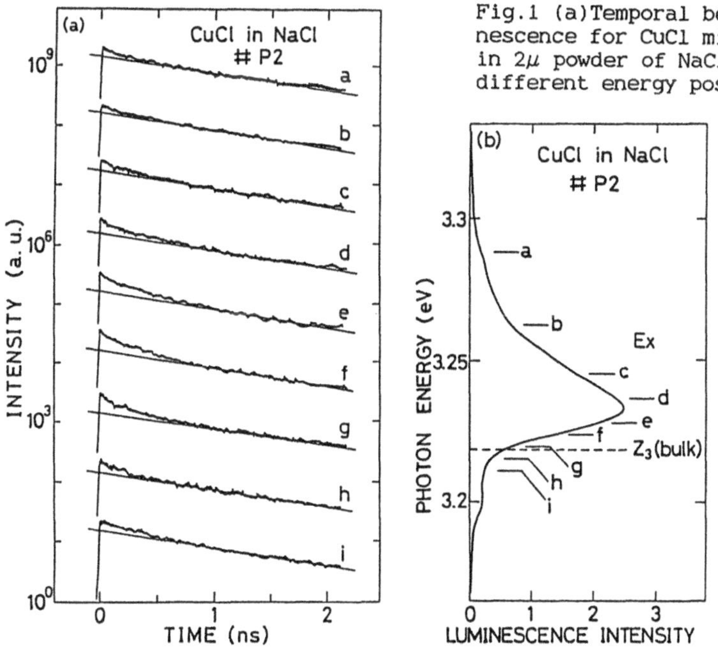

Fig.1 (a)Temporal behaviors of Ex luminescence for CuCl microcrystals embedded in 2μ powder of NaCl matrix observed at different energy positions indicated on the luminescence spectrum shown in (b). Each curve in (a) is drawn with its magnitude shifted by an appropriate amount

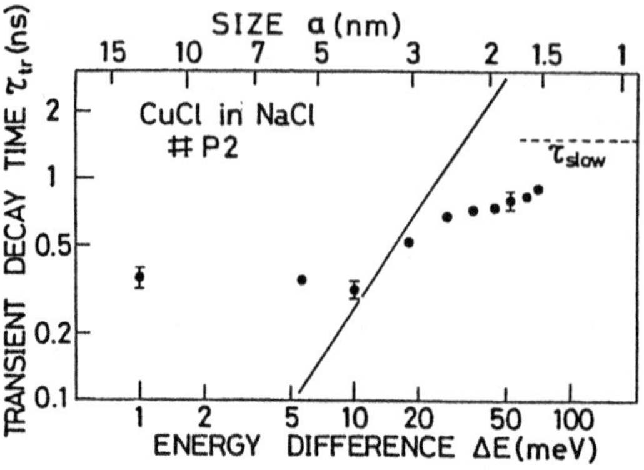

Fig.2 Transient decay time, τ_{tr}, versus the energy shift, ΔE, from the bulk Z_3 exciton. The corresponding crystallite size, a, derived from (1) is indicated at the top. A solid line represents the theoretical values of (2) for Z_3 excitons in CuCl microcrystals

real shape of CuCl microcrystals is known to be somewhat oblate[1], qualitative agreement between the experimental and theoretical values is obtained for a smaller than 4nm. Therefore, if the observed size-dependence of the decay time is really caused by the exciton superradiance, the exciton coherence length is assumed to be at least 4nm which is 6 times as large as the exciton radius.

4 Excitonic Molecules

Under intense laser excitation at 77K a nonlinear luminescence band, M, appears at 3.17eV where the excitonic molecule band exists in a bulk single crystal.[4] With selective excitation at different energies one can observe in Fig.3 peaks 1 to 5 on the Ex band and 1' to 5' on the M band, respectively. Each peak on the Ex band has been assigned to the Ex luminescence in microcrystals of a specific size, in which the excitation energy is resonant with some excited state of a confined Z_3 or $Z_{1,2}$ exciton.[1] The peaks on the M band can be assigned in a similar way. The peaks n and n' are found to form a pair because the peak energy of (n+1)' coincides with that of n' when the peak energy of (n+1) is shifted to that of n with the decrease of the excitation energy by an appropriate amount. This fact means that the M luminescence at n' and the Ex

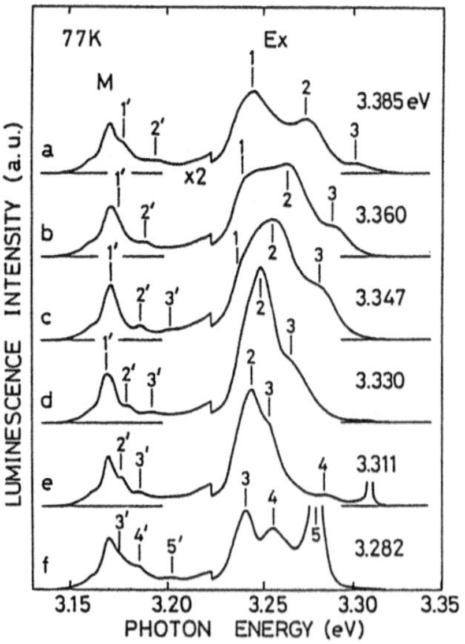

Fig.3 Luminescence spectra of CuCl microcrystals embedded in NaCl matrices of platelet shape under intense excitation at different energies which are indicated on the right hand side. A pair of peaks denoted by n and n' correspond to the Ex and M luminescence, respectively, in microcrystals of the same size. Several peaks on Ex band appear with excitations at 1:$Z_{1,2}$ light exciton, 2 & 3:$Z_{1,2}$ heavy excitons split by k-linear term, 4:Z_3 exciton + optical phonon and 5:Z_3 exciton itself, respectively

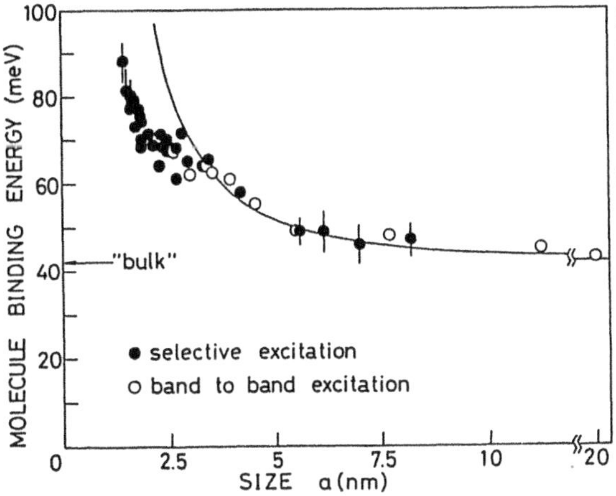

Fig.4. Binding energy of excitonic molecule versus the crystallite size. A solid curve represents the relation derived from (3) on the basis of the molecule confinement in the translational motion

luminescence at n are emitted from the microcrystals of the same size. Therefore, one can obtain the energy relation between the M and Ex for each size. The energy difference between n and n' corresponds to the binding energy of excitonic molecules, E_m^b, and is plotted by closed circles in Fig.4 as a function of crystallite size, a, derived from (1). The values of E_m^b which are obtained from the peak energy difference under non-resonant intense UV laser excitation for different samples with larger crystallite sizes are also plotted in Fig.4 by open circles. For large microcrystals the excitonic molecules are considered to be confined in their translational motion like the excitons[1] and the energy shift of the M band, $\Delta h\nu_m$, from the bulk one is given by

$$\Delta h\nu_m(a) = -\frac{\hbar^2}{2}\left(\frac{1}{M_x} - \frac{1}{M_m}\right)\left(\frac{\pi}{a}\right)^2, \qquad (3)$$

where M_m is the translational mass of an excitonic molecule and is equal to $2.3M_x$ [4]. In Fig.4 E_m^b derived from (3) is drawn by a solid curve. One can see that the curve fits well to the experimental points for a larger than 3nm. This fact indicates that the effective size of the excitonic molecule is about 3nm which is about 4 times larger than that of the exciton. For the size less than 3nm the internal motion of the excitonic molecule gradually changes and E_m^b shows true increase, giving the value almost twice as much as that of the bulk crystals for a equal to 1.5nm. Based on the assumption that two electron-hole pairs are strongly confined in a small microcrystal, E_m^b is calculated to be about 110meV for a equal to 2nm.[5] The discrepancy between the experimental and theoretical values suggests that the individual confinement of an electron and a hole is still not a good picture for the excitonic molecule of CuCl microcrystals even for such a small crystallite size.

5. Conclusion

In CuCl microcrystals the transient decay time of exciton luminescence just after the excitation is found to decrease with increase of crystallite size up to 4nm. This fact is interpreted as the exciton superradiance. The confinement of excitonic molecules is found to occur in their translational motion for microcrystals larger than 3nm in size and their binding energy tends to increase for smaller crystallite size due to the change in their internal motion.

References

1. T. Itoh, Y. Iwabuchi, M. Kataoka: phys. stat. sol. (b)145, 567 (1988); T. Itoh, Y. Iwabuchi, T. Kirihara: ibid. (b)146, 531 (1988)
2. A.I. Ekimov, Al.L. Efros, A.A. Onushchenko: Solid State Commun. 56, 921 (1985)
3. E. Hanamura: Phys. Rev. B38, 1228 (1988)
4. T. Itoh, S. Watanabe, M. Ueta: J. Phys. Soc. Jpn. 48, 542 (1980)
5. H. Takagahara: Proc. 16th Int. Conf. Quantum Electronics, Tokyo (1988)

CdS Particle Doped Polymer Films for Nonlinear Optics

Y. Ohashi[1], *H. Ito*[1], *T. Hayashi*[1], *A. Nitta*[1], *H. Matsuda*[2], *S. Okada*[2], *H. Nakanishi*[2], *and M. Kato*[2]

[1]Central Research Institute, Mitsui Toatsu Chemicals, Inc. 1190, Kasama-cho, Sakae-ku, Yokohama 247, Japan
[2]Research Institute for Polymers and Textiles, 1-1-4 Higashi, Tsukuba, Ibaraki 305, Japan

CdS particle doped polymer films were prepared by "Precipitation-In-Gel"(PIG) method in which CdS particles were precipitated in aqueous swollen polymeric gels of crosslinked N-acryloyl pyrrolidine copolymers by the double decomposition reaction between cadmium acetate and sodium sulfide. Ultra-fine particles of average size from 30 to 60 Å were formed in the polymer films. The absorption edge was blue-shifted with the decrease of particle size. THG was investigated with 1.06 -1.50 µm pulsed laser light. Even by the THG measurements, the doped film shows a resonant-enhancement effect.

1.INTRODUCTION

Recently, materials doped with ultra-fine particles of a semi-conductor (tens to hundreds of Å in diameter) have received attention in the field of third order nonlinear optical research because of their potentiality of large and fast nonlinearity. Polymer composite materials are attractive by reason of their processing flexibility, mechanical stability, etc. A composite material of CdS particles and an ionic polymer was reported by DuPont group /1/, but the particles in ionic polymer had intrinsic defects like sulfur vacancies.

We report here a novel preparative method, namely "PIG" method, of polymer films doped with fine CdS particles, the properties of the doped polymer films and the results of THG measurement with the tuning TH wavelength close to the absorption edge of these doped polymer films.

2.RESULTS and DISCUSSION

Polymer films (about 100 µm thick) were prepared by UV-initiated polymerization of the cast film of N-acryloyl pyrrolidine (N-APR) and urethane acrylate (UA) monomer mixture. UA has two double bonds in the molecule and acts as a crosslinking agent to give a crosslinked polymer network. Resulting polymer films show a unique thermosensitive property, i.e. its hydrophilicity changes with temperature /2/. The polymer films are swellable in water as well as polar organic solvents. Especially in water, swelling volume depends on the temperature: it becomes bigger at lower temperature. This suggests that the size of particles precipitated in the films can be changeable with the swelling volume.

CdS particle doped polymer films were prepared by "PIG" method. In the "PIG" method, at first, the polymer films were swollen in aqueous cadmium acetate solution and then the swollen films were dipped into aqueous sodium sulfide solution. CdS particles were precipitated in aqueous swollen gels by the double decomposition reaction between cadmium acetate and sodium sulfide. The CdS precipitated gel films were dried in vacuum and were subjected to the following experiments.

The cross section of the doped films was measured by transmittance electron microscopy(TEM). Particles were confirmed to be dispersed uniformly in polymer films with no aggregation. The particle size was determined by X-ray diffraction(XRD). Figure 1 shows XRD data of the CdS doped polymer film prepared under the condition: N-APR 75 wt%, UA 25 wt%, cadmium acetate concentration 2.25 mol/l, sodium sulfide concentration 0.65 mol/l, reaction temperature 5°C. Three major XRD peaks observed at 2Θ of 27°,44°and 53° confirm that the precipitated CdS is crystalline and the form is cubic. The particle size was determined as 57 Å for this doped polymer film from the peak width by applying Scherrer's equation.

Size of the particles could be controlled by the polymer compositions and the reaction conditions. Figure 2 shows the average size dependent on the concentrations of aqueous sodium sulfide solution and reaction temperature(Ts) at constant concentration of cadmium acetate solution (2.25 mol/l), and N-APR 75wt%, UA 25wt%. CdS particles of average size from 30 to 60 Å could be obtained reproducibly. Particle size was very sensitive to the temperature, reflecting the thermosensitive property of the film.

The UV-VIS absorption spectra of the doped films were measured to confirm the dependence of the absorption edge shift on particle size. Figure 3 shows the absorption spectra. Absorption edge was blue-shifted with the decrease of particle size. This size-dependent shift of CdS absorption edge is mostly due to the quantum confinement effect (the increase of electron-hole kinetic energy resulting from their spatial confinement) in the case of small semiconductor particles. From these absorption data, this effect becomes apparant when the particle size was smaller than 60 Å for CdS. The shortest wavelength of the absorption edge was 415 nm. The exciton absorption was observed as the marked shoulder in the absorption spectra at room temperature and this shoulder was also blue-shifted with the decrease of particle size. High optical density was also obtained.

The photoluminescence of the doped films was measured to discuss the photogenerated carrier relaxation process in these particles. Figure 4 shows the room temperature luminescence spectra for various sizes. A broad emission peak appeared between 520 and 580 nm, and it was

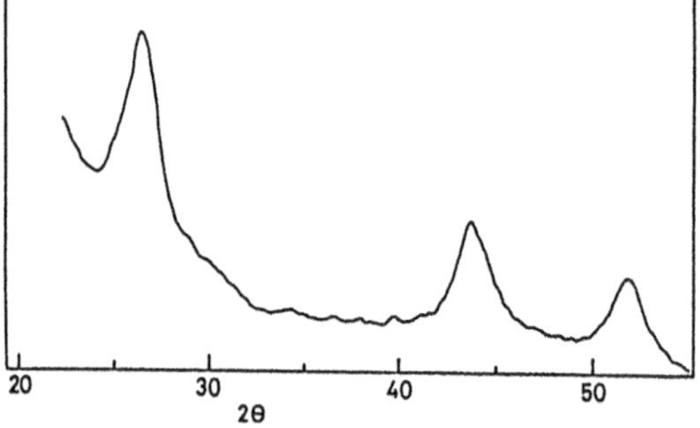

Fig.1 XRD spectra of CdS in N-APR film

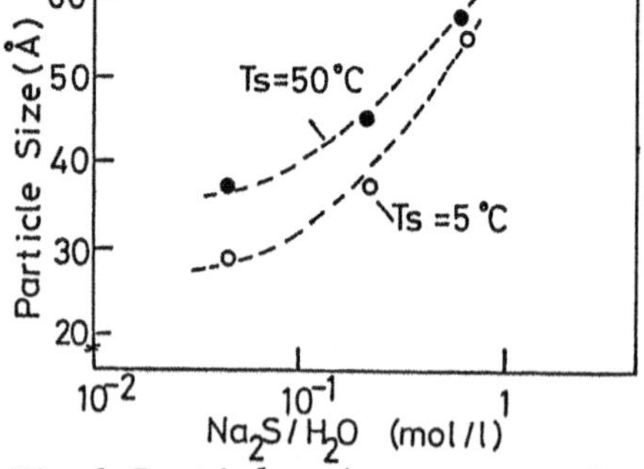

Fig.2 Particle size vs reaction conditions

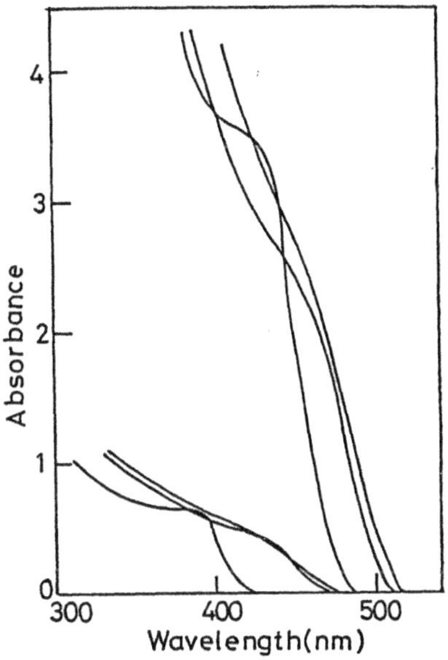

Fig.3 Absorption of CdS in N-APR film

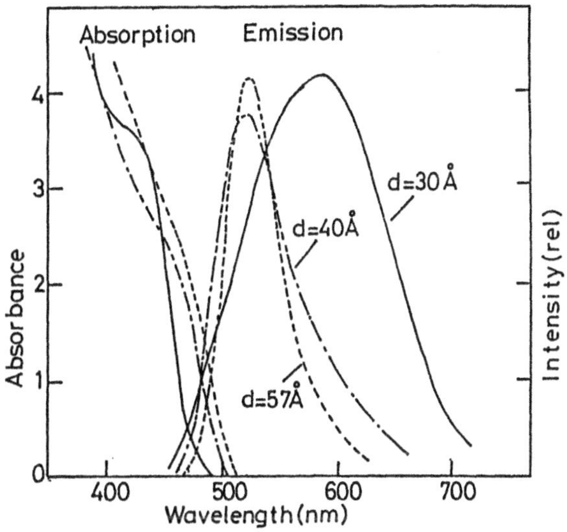

Fig.4 Size dependent absorption and emission of CdS in N-APR film

red-shifted as the particle size decreased. At particle size of 57 Å, which is comparable to the dimension of the bulk-like CdS, a rather sharp emission band peaking near the absorption edge was observed. On the other hand, a broad band peaking at longer wavelength was observed for smaller size. We feel that these emission properties are mostly attributed to the surface recombination of trapped electrons by holes. On the other hand, the emission peak at 690 nm attributed to the intrinsic defect of sulfur vacancy, reported in the case of CdS/Nafion composite /1/, was not observed in our doped films. Furthermore, recent results of T2 measurement on our "sulfur vacancy free" particle-doped films in Kerr-shutter configuration showed the materials to be fast responding /3/.

THG of doped films was investigated with 1.45 - 2.10 µm pulsed laser light(5 - 6ns), which was obtained by wave-mixing of Nd:YAG and Dye lasers. Table 1 shows the relative intensities of THG to that of fused quartz at three different wavelengths, together with refractive indexes measured at 0.6328 µm by ellipsometry. At 1.50 µm, whose TH wave was in nonresonant region, THG of the doped films was very weak, i.e. almost the same intensity as those of matrix polymer or quartz standard. On the contrary, at 1.45 µm whose TH wave was at the edge of resonant region, stronger TH was generated: especially, sample-3 gave a 10 times stronger TH than that of the quartz. At the shorter wavelength of 1.06 µm of Nd:YAG laser, THG became weaker again. This may be partly due to the strong absorption by the particles themselves. Thus, it has been clarified that even by the THG measurements CdS particle doped polymer film shows a resonant-enhancement effect,though it is fairly weak.

As a conclusion,"PIG" method was demonstrated to be a promising method to prepare nonlinear optical polymeric materials doped with CdS semiconductor ultra-fine particles.

Table 1 Relative Intensities of THG

	Fundamental wavelength(μm) 1.06	1.45	1.50	n
Quartz	1	1	1	1.4587
Sample-1	1.6	2.1	1.2	1.4426
Sample-2	1.2	5.7	1.3	1.4741
Sample-3	1.5	11.2	1.4	1.4798

Sample-1 non-doped film
Sample-2 CdS doped film (particle size :47 Å)
Sample-3 CdS doped film (particle size :57 Å)

ACKNOWLEDGMENT

We would like to thank Prof. T. Kobayashi, University of Tokyo, for stimulating discussions and evaluating the T2 of the films.

REFERENCES

1. Y.Wang,and W.Mahler:Opt.Comm,61,3,(1987) 233.
2. T.Takaki,H.Ito, and A.Nitta: Polymer Preprints, Japan,36,3, (1987) 716
3. k.Misawa,T.Hattori,Y.Ohashi,H.Ito,and T.Kobayashi:ICUP Conference, late news, (1988)

Part III

Organics: Calculations

Design of Novel Conjugated Molecules with Large Hyperpolarizabilities

J.O. Morley

Fine Chemicals Research Centre, ICI Colours and Fine Chemicals, Hexagon House, Blackley, Manchester, M9 3DA, UK

1. Introduction

A considerable number of calculations have been reported on the hyperpolarisability of conjugated organic molecules and a number of potentially interesting systems have been discovered which have important applications in nonlinear optics as frequency doublers or phase modulators[1]. While a number of different methods have been used to derive the hyperpolarisability, the results described here have been obtained with a version of the CNDO/S method, coupled with a sum-over-states procedure to obtain the first hyperpolarisability[2].

This method involves an initial configuration interaction treatment of ground (g) and excited state (n and n') wave functions and then evaluation of the hyperpolarisability tensor from the improved wave functions using the expression[3]

$$\beta^{2\omega}_{ijk} + \beta^{2\omega}_{ikj} = (-e^3/4\hbar^2) \times$$

$$\sum_{n'}\sum_{n}\Bigg[\left(r^j_{gn'}r^i_{n'n}r^k_{ng} + r^k_{gn'}r^i_{n'n}r^j_{ng}\right)\left\{\frac{1}{(\omega_{n'g}-\omega)(\omega_{ng}+\omega)} + \frac{1}{(\omega_{n'g}+\omega)(\omega_{ng}-\omega)}\right\}$$

$$+\left(r^i_{gn'}r^j_{n'n}r^k_{ng} + r^i_{gn'}r^k_{n'n}r^j_{ng}\right)\left\{\frac{1}{(\omega_{n'g}+2\omega)(\omega_{ng}+\omega)} + \frac{1}{(\omega_{n'g}-2\omega)(\omega_{ng}-\omega)}\right\}$$

$$+\left(r^j_{gn'}r^k_{n'n}r^i_{ng} + r^k_{gn'}r^j_{n'n}r^i_{ng}\right)\left\{\frac{1}{(\omega_{n'g}-\omega)(\omega_{ng}-2\omega)} + \frac{1}{(\omega_{n'g}+\omega)(\omega_{ng}+2\omega)}\right\}\Bigg]$$

(equation 1)

Here r^i_{gn} is defined as

$$r^i_{gn} = \left\langle \psi_g \middle| \sum_{m=1}^{N} r^i(m) \middle| \psi_n \right\rangle \quad \text{or} \quad \langle g | r^i | n\rangle$$

where $r^i(m)$ is the ith component of the position vector of electron m (of N), $r^i_{n'n} = \langle n' | r^i | n\rangle$, ω_{ng} is the eigenvalue of $\psi_{n'}$ relative to the ground state ψ_g (the electronic transition energy), e is the magnitude of the electronic charge and ω is the frequency of the applied radiation field. The indices n and n' may be restricted to run over excited states in order of increasing energy , provided

$$\langle g | r | g\rangle = 0,$$

Springer Proceedings in Physics, Vol. 36
Nonlinear Optics of Organics and Semiconductors
Editor: T. Kobayashi

which holds only in the electronic charge centroid system as previously discussed[2]. The matrix elements, r_{nn}, are then derived in this procedure as the difference between the ground and excited state dipole moments, $\Delta\mu$, i.e.,

$$-\Delta\mu/e = r_{nn} = \langle n|r|n\rangle - \langle g|r|g\rangle$$

and rapid convergence of the second order sum occurs for representative molecules such as aniline and 4-amino-4'-nitrostilbene[2]. All components of the tensor are calculated, though the most relevant quantity, which can be related to experimental data as in previous studies[2,4-7] is the vector component, β_x defined as

$$\beta_x = \beta_{xxx} + 1/3 \sum_i (\beta_{xii} + 2\beta_{iix})$$

where β_x is aligned to lie along the direction of the dipole moment. Because the absorption frequency (ω_{ng}) of conjugated organic molecules often coincides with the second harmonic (2ω) of the applied field of fundamental frequency, ω, there is resonance enhancement of the calculated value of β_x through terms in the denominator of equation (1). Consequently the results reported here are based on the zero frequency, or static field, value β^o_x as in earlier studies[2,4-7], although (1) is an approximate representation when used in this way[8]. However, the results previously calculated at zero frequency using this expression show the same trends as those produced at applied frequencies of either 1.17 or 0.656 eV (for molecules which show no resonance enhancement), thus demonstrating the usefulness of the static field value as an effective measure of the hyperpolarisability.

The results obtained by this method in conjunction with experimental data from electric field induced second harmonic generation in solution[1] have generally established that the largest hyperpolarisabilities are associated with molecules which contain an electron donor at one end of a suitable conjugation path and an electron acceptor at the other. The value of the hyperpolarisability for any given system is controlled both by the relative efficacy of the substituents and the length and nature of the conjugation path. In some systems such as azulenes and pyrazolines the most important factor appears to be the stabilisation of the excited state properties by appropriate substituents.

The calculations reviewed here were carried out on molecules which were constructed from related examples found in the Cambridge Crystallographic Database using molecular graphics procedures. Where no experimental information was available, the structures of relevant molecules were generated from guessed structures using ab initio molecular orbital optimisations.

2. Effect of Substituents on the Hyperpolarisability

In conjugated systems, the calculated hyperpolarisability is derived mainly from pi-pi* transitions[4], and for monosubstituted benzenes(I) a limited number of excited states contribute to most of its value[6]. It follows that a picture of the change in pi-electron distribution in the aromatic ring on excitation for a given substituent should lead to a greater understanding of the electronic factors which contribute to the overall calculated value. At the present time the substituent constant, σ_R^o, has often been used to measure the relative efficacy of substituents in terms of their ability to attract or donate electrons[9].

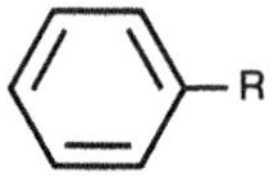

(I)

Furthermore, it has been established that the σ_R^o value of a given substituent is directly proportional to the calculated ground state pi-electron charge (q) in the aromatic ring using the following relation[9] where K is a constant:

$$\sigma_R^o = K\sum q \quad .$$

Thus nitro- and cyano- groups are classified as electron acceptors and amino- and hydroxy- groups are classified as electron donors as far as their ground state properties are concerned. However, the major contribution to the hyperpolarisabilty arises from the excited state properties which have now been calculated for the dominant excited states using the semi-empirical method[6]. A comparison between the calculated ground state pi-charges at this level of approximation and those from STO-3G calculations[10] show a substantial overestimation for electron acceptors such as the aldehyde and nitroso- groups. Because these results are crucial to any scale of substituent constants the corresponding pi-charge distributions have been examined at the 3-21G level. The results obtained show few changes in the pi-charge distribution for +M groups from the STO-3G method, but substantial differences for -M groups. On balance the semi-empirical results show a good correlation with the 3-21G calculations at the ground state level and it seems likely that the excited state results are also reliable. A comparison of the substituent constants derived from the ground state pi-charges with those determined experimentally either using reactivity criteria[11] or 19F N.m.r. studies[12] shows that a reasonable correlation is obtained for most substituents though the effect of electron attracting groups appears to be underestimated. Considerable enhancement of the theoretical σ_R^o values occur on excitation for most substituents if the first five excited states are considered[6]. The results obtained from these studies (Table 1) show that the excited state substituent constants follow the same pattern as the ground state values, i.e.

$$NMe_2 > NH_2 > OMe > OH$$

for electron donors, and

$$NO > NO_2 > CHO > CO_2H > CN$$

for the corresponding electron acceptors.

Table 1. Calculated ground and excited state substituent constants for monosubstituted benzenes

R	Experimental σ_R^o Values: Reactivity	Experimental σ_R^o Values: ^{19}F n.m.r	Calculated σ_R^o Values: Ground state	Calculated σ_R^o Values: Excited states
NH_2	-0.48	-0.48	-0.47	-0.75
NMe_2	-0.52	-0.54	-0.60	-1.25
OH	-0.40	-0.43	-0.43	-0.54
OMe	-0.41	-0.43	-0.48	-0.63
F	-0.35	-0.32	-0.21	-0.25
CO_2H		0.21	0.17	0.24
CHO		0.27	0.22	0.81
CN	0.10	0.21	0.13	0.14
NO_2	0.16	0.19	0.14	1.10
NO		0.33	0.43	1.96

3. Effect of the Conjugation Path and Molecular Volume on the Calculated Hyperpolarisability

A series of calculations have been carried out to explore the effect of increasing the conjugation path length on the hyperpolarisabilities of substituted polyphenyls (II) and polyenes (III) containing an electron donating dimethylamino-group at one end of the chain and an electron attracting nitro-group at the other[5]. The values obtained are critically dependent on the molecular structures adopted for calculation purposes. An examination of the Cambridge Crystallographic Database shows that although some substituted biphenyls are nonplanar, most unsubstituted polyphenyls such as terphenyl and quaterphenyl are generally planar. Similarly, in the polyene series most examples in the database show a planar polyene backbone as in dimethyl trans-1-(N,N-dimethylamino)hexa-1,3,5-triene-6,6-dicarboxylate (see [5] for sources of experimental data). Substituent geometries in these series were derived from N,N-dimethyl-4-nitroaniline for the polyphenyls and N,N-dimethyl-2-nitroethenamine for the polyenes. The molecules were assumed to be fully planar to obtain the maximum possible effect and the results obtained show that an extension of the conjugation path length between the dimethylamino-group and the nitro-group generally results in a sharp increase in the intrinsic hyperpolarisability. The effect, however, is almost two orders of magnitude greater for the polyenes (Fig. 1).

However, if the molecular volume is considered the hyperpolarisability density ,or the effective value per unit volume, for the polyphenyls is greatest for the terphenyl (Fig. 2) though the polyene series does not peak until approximately 20 double bonds are present (Fig. 3). The values obtained for the polyenes are calculated to be an order of magnitude larger than any other class of compound yet studied, though no experimental data is available for confirmation. The number of states included in the calculation is of

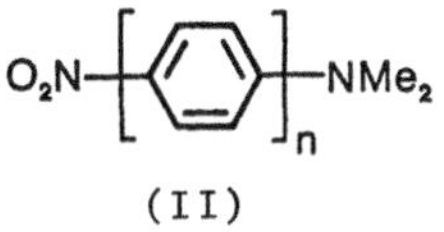

(II)

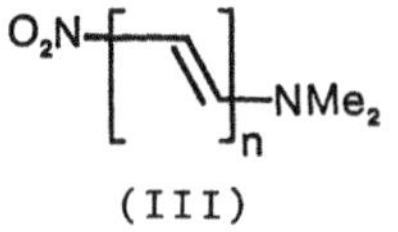

(III)

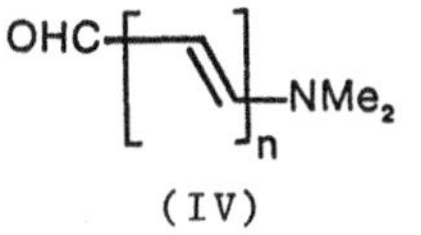

(IV)

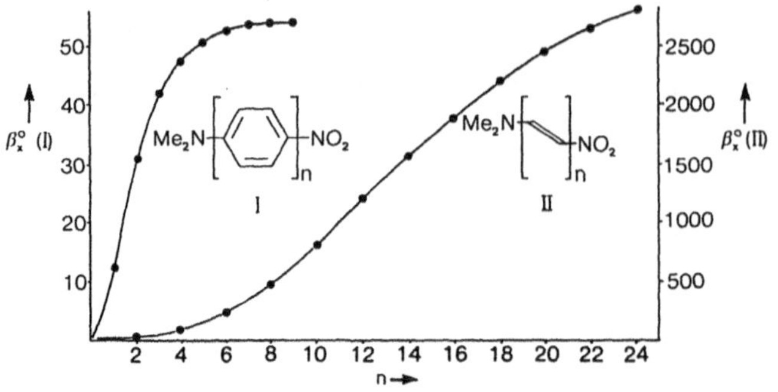

Figure 1. Effect of increasing chain length (n) on the calculated hyperpolarisabilities (β_x^o) of polyphenyls (II) and polyenes (III)

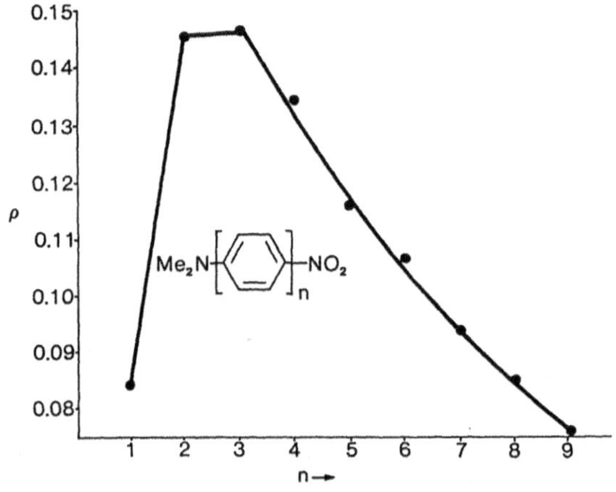

Figure 2. Effect of increasing chain length (n) on the calculated hyperpolarisability density (ρ) of polyphenyls (II)

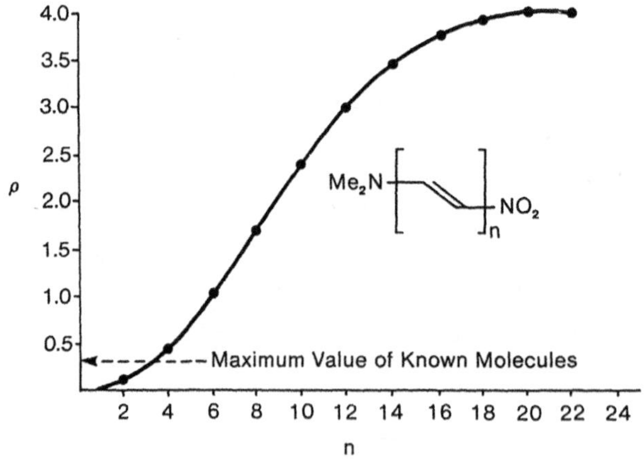

Figure 3. Effect of increasing chain length (n) on the calculated hyperpolarisability density (ρ) of polyenes (III)

considerable importance as the number of pi-electrons present increase with increasing values of n. For the lower members of the series 50 states are adequate but for the higher members, particularly in the polyphenyl series, 200 states are required before the hyperpolarisability perturbation expansion converges to a steady value. In the corresponding polyenals (IV), however, a similar increase in the hyperpolarisability is observed with increasing chain length, but here, limited experimental data is available for comparative purposes from dipole moment measurements[13] and absorption spectra[14]. An analysis between this data and the calculated properties shows a reasonable correlation for the lower members of the series, for n =1 to 3, but thereafter, the values diverge rapidly as the experimental data levels out. This effect may be attributable either to theoretical inaccuracies derived from the approximations adopted by the semi-empirical method or because of experimental error arising from decreasing solubility and cis isomerism as the series is ascended[5]. Further calculations at the ab initio STO-3G and 3-21G levels on the same structures, however, show a steady increase in the value of the dipole moment with increasing chain length though the former surprisingly underestimates the overall value. The 3-21G results, however, show an excellent correlation with the experimental data for the early members of the series ,but thereafter, they follow the same trends as the semi-empirical results[5].

The effect of other conjugation paths has been explored for substituted butatrienes (V) and butadienes (VI) which contain an electron donating amino-group at one end of the molecule and an electron attracting nitro-group at the other, and where the conjugation extends in two planes. Little structural data is available from the Cambridge Database on these systems and accordingly the molecular structures were generated by STO-3G optimisations from empirical ones. In addition, the structure of the corresponding substituted butadiene (VII) was generated by the same method for comparative purposes and the three molecules compared with 4-nitroaniline (VIII) using the crystal structure in this case.

The calculated structures of the butadiene (VII) and the butatriene (V) show similar bond lengths (given in angstroms) for the C1-C2 and C3-C4 distances but the central C3-C4 bond in the latter is considerably shorter than that of the former (Table 2). In contrast, the first distances are considerably shorter in the butadiyne (VI) though the central bond length is approximately midway between the two extremes. The calculated hyperpolarisabilities of these systems are shown in Table 2, where μ_g and μ_e are the ground and major excited state dipole moments (in D), and β^o_x is the static field hyperpolarisability which lies along the direction of the molecular dipole moment (the values are given in units of 10^{-30} cm^5 e.s.u.). The results show a number of interesting features in that the butatriene has a value which is comparable to 4-nitroaniline but only half that of either the butadiene or butadiyne value.

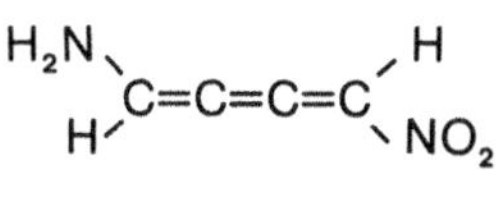

(V)

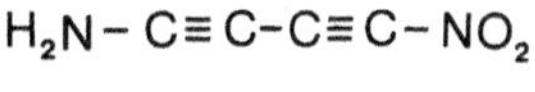

(VI)

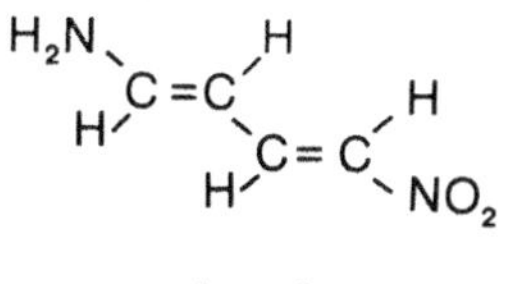

(VII)

Table 2. Calculated bond lengths, dipole moments and hyperpolarisabilities for substituted butadienes, butatrienes, butadiynes, and 4-nitroaniline

	Bond Distances					
Structure	C1-C2	C2-C3	C3-C4	μ_g	μ_e	β_x^o
VII	1.33	1.46	1.32	9.07	17.5	12.1
V	1.32	1.24	1.31	10.1	14.3	6.28
VI	1.18	1.39	1.17	8.37	18.9	11.3
VIII				8.15	15.0	7.61

This behaviour is clearly a reflection of the degree of delocalisation of the pi-electrons which are more restricted in the former cases as indicated by the smaller change in the dipole moment on excitation from the ground state to the major excited state (Table 2).

4. Calculations on Polar Hydrocarbons

The majority of organic materials with large hyperpolarisabilities consist of a conjugated system containing a donor and acceptor group positioned at either end of the conjugation (see above). Unfortunately, the presence of polarising groups of this type usually causes the molecules to crystallise in a centrosymmetric fashion and the prized molecular effect is much reduced or even lost. Furthermore, in one important nonlinear optical application which involves frequency doubling from a fundamental wavelength of 830 nm to a second harmonic of 415 nm, polar molecules of this type show a strong absorption either near to or in the visible region of the spectrum producing destructive absorption of the required wavelength. In contrast, aromatics such as the azulenes (IX) or sesquifulvalene (X) are more likely to crystallise in the desired way because of their low ground state dipole moments, and yet polarise on excitation to form charge separated systems with large hyperpolarisabilities. In addition the absorption of these molecules is generally weak in the visible range with typical extinction coefficients of 100-300 in the 580-700 nm range and little absorption in the second harmonic region[15]. Stable derivatives of both systems are known but there is little crystallographic data available for comparative studies with the exception of azulene itself which is disordered[16] and 2-aminoazulene which is indeed non-centrosymmetric[17].

The influence of the number of included states (see equation 1) on the calculated hyperpolarisability of sesquifulvalene and azulene is illustrated (Fig. 4) and at least 25 excited states are required before the value converges to an optimum[18]. The third excited state is the major contributor to the overall value for azulene though its magnitude is reduced with the inclusion of further states, while the

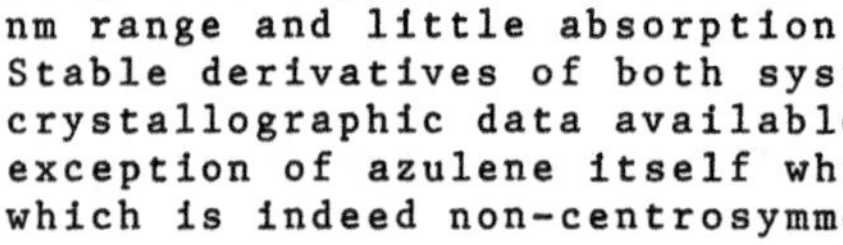

(IX)

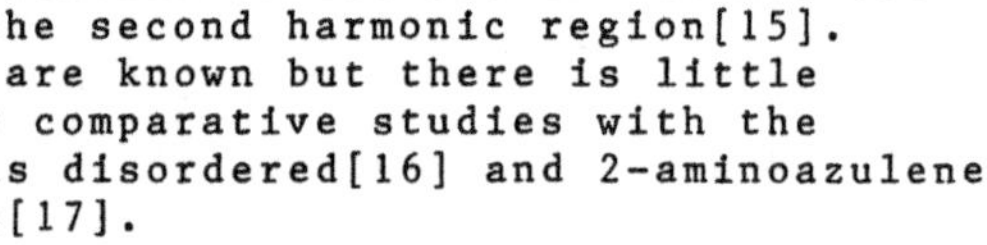

(X)

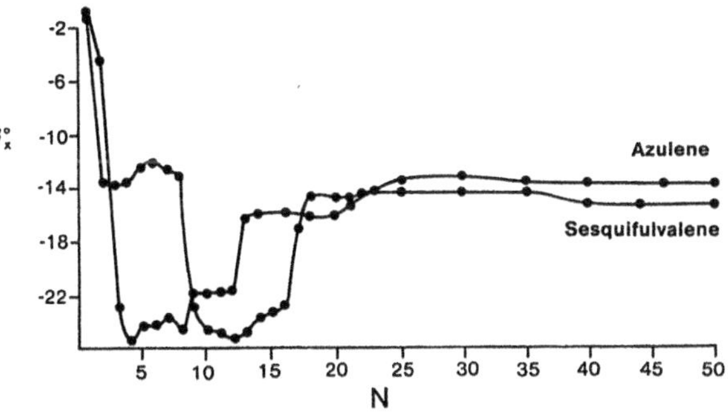

Figure 4. Effect of the number of included states on the calculated hyperpolarisability of azulene (IX) and sesquifulvalene (X)

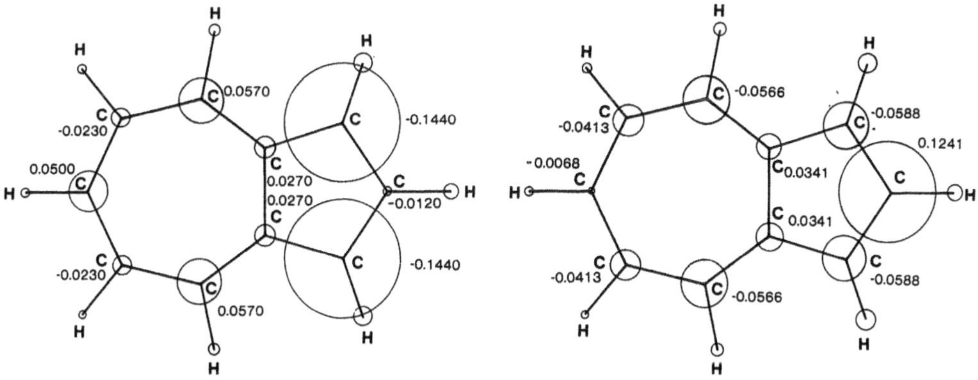

Figure 5. Calculated ground state (left) and excited state charges (right) for azulene (IX) and sesquifulvalene (X)

second and ninth excited states are dominant in sesquifulvalene. An analysis of the charge distribution in azulene shows dramatic changes between the ground and third excited state (Fig. 5) which leads to a reversal in the sign of the x-component of the dipole moment from -3.06 to 1.93D (Table 3) in line with results reported by the Huckel method[19] and more recently by ab initio techniques[20]. If the charges at the 9- and 10- position of azulene are apportioned equally between the seven (A) and five (B) membered rings, the overall ring charges for each respectively changes from +0.2069 (A) and -0.2069 (B) in the ground state to -0.1769 (A) and +0.1769 (B) in the excited state.

A similar transfer of electrons occurs from the small ring to the large ring on excitation from the ground to the second and ninth excited states in sesquifulvalene but the magnitude in each case is much less and insufficient to cause a reversal in the sign of the dipole moment[18]. The change in direction of electron transfer on excitation results in a negative sign for the hyperpolarisabilities of each molecule at the CNDO level of approximation (Table 3). A similar result has been reported recently for sesquifulvalene using the finite field procedure at the MNDO level[21]. The overall magnitude of the calculated hyperpolarisability is comparable to the positive values calculated for highly polar molecules such as 4-nitroaniline although the origin of the effect is different with the latter showing the same

Table 3. Calculated dipole moments and hyperpolarisabilities of azulenes and sesquifulvalene

Structure	Substituents	x-component μ_g	μ_e	β^{o}_{x}
IX	None	-3.06	1.93	-13.94
	6-NMe_2	-6.64	-5.74	-4.43
	2-NO_2	-9.10	-5.13	-9.46
	2-NMe_2	1.20	8.50	39.77
	6-NO_2	3.08	9.13	16.88
	2-NO_2,6-NMe_2	-13.37	-12.19	2.55
	2-NMe_2,6-NO_2	7.92	12.18	59.89
X	None	-6.31	-4.60	-15.29

direction of electron transfer from the donating amino-group through the aromatic ring to the nitro-group in both ground and excited states.

Because the hyperpolarisability is derived in part from the transition moments between ground and excited states (see equation 1) any substituent which is able to stabilise the latter would be expected to enhance its value. It follows that the introduction of an electron donor into the large ring or an electron attractor into the small ring of azulene would be expected to preferentially stabilise the ground state but destabilise the major contributing excited state resulting in a increase and decrease respectively in the magnitude of the dipole moment.

This effect is clearly seen in 6-dimethylaminoazulene where the magnitude of the ground state dipole moment increases sharply from that of azulene and only decreases slightly for the excited state with a resulting fall in the value of the hyperpolarisability. A similar effect is calculated for 2-nitroazulene though less pronounced. In contrast, the introduction of an electron attractor into the large ring or an electron donor into the small ring would be expected to stabilise the excited state and destabilise the ground state resulting in an increase and decrease respectively in the magnitude of the dipole moment. Thus 2-dimethylaminoazulene and 6-nitroazulene show a sharp increase in the magnitude of the excited state dipole moment from the ground state value which arises from a reversal in the direction of charge transfer across the rings. The introduction of both an electron attractor and donor into each end of the molecule produces the expected enhancement of the hyperpolarisability for 2-dimethylamino-6- nitroazulene and reduction for 6-dimethylamino-2-nitroazulene in line with the arguments developed above.

5. Calculations on 2-Pyrazolines

Interesting structural effects are observed in the 1,3-diphenyl-2-pyrazolines (XI) where the calculated hyperpolarisability are highly dependent on the position of ring substituents[22]. 1,3-Diphenyl-2-pyrazoline itself has a fairly large calculated hyperpolarisability which is greatly enhanced by the presence of an electron donating 4-methoxy-group in the left hand ring and a corresponding electron

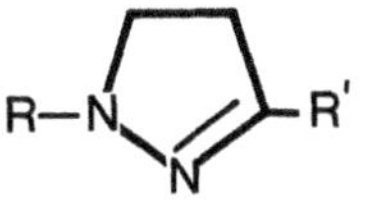

(XI)

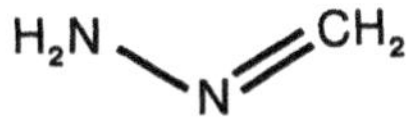

(XII)

attracting 4-nitro-group in the right hand ring. In contrast the corresponding isomer obtained by reversing the position of the two substituents has a greatly reduced value which is smaller than the unsubstituted molecule (Table 4).

The explanation for these dramatic changes almost certainly lies with the change of electron distribution within each molecule on excitation which follow the same pattern in the dipole moments as those reported experimentally for the related substituted 1,3-diphenyl-2-pyrazolines containing cyano- and methoxyl- substituents[23]. In general, because one excited state dominates the hyperpolarisability value in this series of molecules, calculations have been carried out on the parent 2-pyrazoline itself (XI; R=R =H) and the related formaldehyde hydrazone (XII) using structures generated from STO-3G optimisations, to determine the general trends in the electron transfer on excitation[22].

The calculated structures of each molecule are similar with respect to the conjugated bond lengths, but the bond angle N-N-C which is somewhat constrained in the pyrazoline is some 7 degrees larger in the hydrazone. The calculated hyperpolarisability of the latter is smaller than the pyrazoline suggesting that the electron donating inductive effect of the attached sp^3 carbon atoms in the pyrazoline is greater than the effect produced by relieving the strain at the N1-N2-C3 angle (Table 4).

The total charge distribution in 2-pyrazoline in the ground state changes dramatically on excitation to the dominant excited state (Fig. 6) with considerable movement of charge from the sp^2 nitrogen atom at the 1-position which becomes positively charged through to the second nitrogen and also to the carbon at the 3-position where this centre now becomes negatively charged. The increase in the internal angle and loss of two alkyl groups in the change from the pyrazoline to the hydrazone does little to the calculated charges and the same trend

Table 4. Calculated hyperpolarisabilities of 1,3-substituted pyrazolines and formaldehyde hydrazone

Structure	R	R'	μ_g	μ_e	β^o_x
XI	Ph	Ph	3.60	7.86	18.48
	4-MeOPh	4-PhNO_2	7.24	18.45	60.26
	$4\text{-O}_2\text{NPh}$	4-PhOMe	9.11	9.09	14.45
	H	H	2.88	6.90	1.52
	H_2N	Ph	3.67	10.38	28.64
	H_2N	4-PhNO_2	9.68	18.76	50.59
	Ph	2-ENO_2	7.81	16.70	39.50
	Ph	4-BNO_2	8.10	19.31	75.10
	Ph	4-PhNO_2	7.77	16.48	38.33
XII			2.37	5.88	1.24

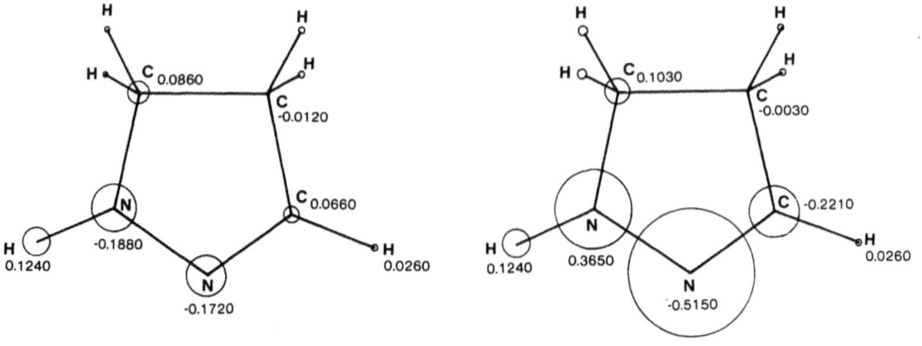

Figure 6. Calculated ground state (left) and excited state charges (right) for 2-pyrazoline (XI)

emerges in both cases with the exception of the magnitude ot the charge at the 3-position which is somewhat reduced for the latter. The results strongly suggest that those substituents which are able to stabilise the excited state by donating electrons at the 1-position and/or attracting electrons at the 3-position should possess large hyperpolarisabilities. Conversely, those substituents which have opposite effects at these positions would be expected to have small values in line with the results obtained for the substituted 1,3-diphenyl-2-pyrazoline discussed earlier. These conclusions are supported by calculations on other substituted pyrazolines such as 1-amino-3-phenyl-2-pyrazoline where the electron donating amino-group substantially increases the calculated hyperpolarisability. The value increases further if an electron attractor is placed in the phenyl ring in line with the arguments developed above.

Although the conjugation provided by the phenyl group to delocalise the charge at the 3-position in the excited state for the unsubstituted molecule is very important as far as the calculated hyperpolarisability is concerned, other conjugated pathways such as the ethenyl group or the butadienyl group (represented as E and B in Table 4) produce a similar if not enhanced effect. Indeed in the last cases the molecular hyperpolarisability value is conserved in the solid state as the molecule crystallises in a noncentrosymmetric space group.

References

1. See for example: Nonlinear Optical Properties of Organic Molecules and Crystals, ed. by D.S. Chemla and J. Zyss, (Academic Press, New York, 1987).
2. V.J. Docherty, D. Pugh, J.O. Morley, J.Chem.Soc., Faraday Trans.2, 81, 1179 (1985).
3. J. Ward, Rev.Mod.Phys., 37, 1 (1965).
4. J.O.Morley and D.Pugh, reference [1], p193.
5. J.O.Morley, V.J. Docherty, and D. Pugh, J.Chem.Soc., Perkin Trans. 2, 1351 (1987).
6. J.O.Morley, V.J. Docherty, and D. Pugh, J.Chem.Soc., Perkin Trans. 2, 1357 (1987).
7. J.O.Morley, V.J. Docherty, and D. Pugh, J.Chem.Soc., Perkin Trans. 2, 1361 (1987).
8. B.J.Orr and J.F. Ward, Mol.Phys., 20, 513 (1971).
9. See for example, R.D.Topsam, Acc.Chem.Res., 16, 292 (1983); Prog.Phys.Org. Chem., 12, 1 (1976); W.J.Hehre, R.W.Taft, and R.D.Topsam, ibid., 2, 159 (1976), and references therein.

10. W.J.Hehre, L.Radom, and J.A.Pople, J.Am.Chem.Soc., 94, 1496 (1972).
11. J.Hine, J.Am.Chem.Soc., 82, 4877 (1960).
12. R.W.Taft, E.Price, I.R.Fox, I.C.Lewis,K.K.Andersen, and G.T. Davies, J.Am.Chem.Soc., 85, 3146 (1963); R.T.C.Brownlee, R.E.J.Hutchinson, A.R.Katritzky, T.T.Tidwell, and R.T.Topsam, ibid., 90, 1757 (1968).
13. H.H.Hutchinson and L.E.Sutton, J.Chem.Soc., 4382 (1958).
14. S.S.Malhotra and M.C.Whiting, J.Chem.Soc., 3812 (1960).
15. See for example:W.W.Robertson and A.D.King,Jr., J.Chem.Phys., 34, 190 (1961).
16. G.S.Pawley, Acta Crystallogr., 18, 560 (1965).
17. Y.Takaki, Y.Sasada, and I.Nitta, J.Phys.Soc.Jpn., 14, 771 (1959).
18. J.O.Morley, J.Am.Chem.Soc., in the press.
19. R. Pariser, J.Chem.Phys., 25(6), 1112 (1956).
20. R.J. Buenker and S.D. Peyerimhoff, Chem.Phys.Lett., 3(1), 37 (1969).
21. G.R.J.Williams, J.Mol.Struct., 151, 215 (1987).
22. J.O.Morley, V.J. Docherty, and D. Pugh, J.Mol.Electronics, in press.
23. H.Gusten, G.Heinrich, and H.Fruhbeis, Ber.Bunsen-Ges., 51, 810 (1977).

CNDO/S-CI Calculations of Hyperpolarizabilities. I: Substituted Benzenes, Polyaza Compounds and Related Species

M. Nakano, K. Yamaguchi, and T. Fueno

Faculty of Engineering Science, Osaka University, Toyonaka, Osaka 560, Japan

The second-order hyperpolarizabilities (β) for monosubstituted benzenes, donor (D)-acceptor(A) disubstituted π-conjugated systems and polyaza aromatic compounds were calculated by the use of the CNDO/S-CI method. The β values calculated for aniline, nitrobenzene and p-nitroaniline are in good agreement with the experimental data. The β values for D-A disubstituted systems exhibit no additive property and are proportional to the strengths of the charge-transfer (CT) interaction between the D and A groups. For diphenylpolyethylenes the calculated β is proportional to the conjugated bond length. Introduction of the N atoms into the aromatic ring raises the transition energy. Designings of conjugated compounds which may have large β values have been attempted. For one such compound the β value is calculated to be as high as 325.5 x 10^{-30} esu. The third-order hyperpolarizabilities (γ) have also been calculated for benzene, nitrobenzene, aniline and p-nitroaniline.

1. INTRODUCTION

Materials which may exhibit nonlinear optical properties are receiving intense interest in recent years. We wish here to report the results of CNDO/S-CI calculations of the hyperpolarizabilities for some donor(D)-acceptor(A) disubstituted π-conjugated systems and related species. Our primary objective is to search for certain π-conjugated compounds having a large second-order hyperpolarizability (β). The third-order hyperpolarizabilities (γ) have also been calculated for benzene, nitrobenzene, aniline and p-nitroaniline.

2. CALCULATION METHOD

The microscopic polarizability p induced by an external field is generally expressed as

$$p = \mu_{tot} - \mu_o = \alpha E + \beta E^2 + \gamma E^3 + \cdots\cdots \quad . \quad (1)$$

Here, μ_{tot} is the total electric dipole moment, μ_0 is the permanent dipole moment and the terms that are proportional to E^n ($n \geq 2$) give rise to nonlinear properties. β and γ are the second- and third-order hyperpolarizability tensors, respectively. If molecules are centrosymmetric, their even-order hyperpolarizabilities should vanish. Therefore, we will consider only noncentrosymmetric molecules here.

The hyperpolarizabilities can be calculated by evaluating the expectation values of the electric dipole moments if the wavefunctions of polarized states are obtained. The second- and third-order hyperpolarizabilities (β , γ) can be derived from the time-dependent perturbation theory [1]. Details will be described elsewhere. In order to evaluate the hyperpolarizabilities, it is necessary to obtain the transition moments (μ_{n0}), the differences between excited state and ground state dipole moments ($\Delta\mu$) and the transition energies (E_{n0}). In this work, all these quantities were calculated by the CNDO/S-CI method using the original Jaffe parameters.

Springer Proceedings in Physics, Vol. 36
Nonlinear Optics of Organics and Semiconductors
Editor: T. Kobayashi

3. CALCULATIONS OF β FOR SOME ORGANIC MOLECULES

In order to design organic molecules which may have a large β, we assume the following conditions:

(1) The molecules are non-centrosymmetric.
(2) The molecules are π-conjugated systems.
(3) The molecules have charge transfer (CT) structures.

Figure 1 shows some typical organic compounds that satisfy these conditions. The β values for these compounds were calculated by the CNDO/S single-excitation CI procedure involving 60 excited states. Table 1 summarizes the results obtained for aniline, nitrobenzene and p-nitroaniline, together with their experimental data. The calculated β values are in agreement with the observed values.

Table 2 gives the calculated results for the disubstituted compounds (I-2). From Table 2, it is seen that the sums of the β values for monosubstituted benzenes are uniformly smaller in magnitude than the values calculated for the corresponding donor(D)-acceptor(A) paradisubstituted benzenes. The results for compounds (I-3) and (I-5) appear to indicate that the magnitude of β increases with increasing strength of the D-A interaction.

Figure 2 shows the effect of the number of π-conjugated bonds (N) on β for compounds (I-4). It can be seen that |β| is nearly proportional to the square of N . The trend is in line with the experimental results obtained by DULCIC et al.[2].

We examined several D-A conjugated compounds as shown in Fig.3. The calculated results are given in Table 3. Because these compounds have not only a relatively small E_{n0} but a relatively large μ_{n0} and $|\Delta\mu|$, their β values are large in magnitude. However, E_{n0} for (III-2) and (III-4) are smaller than 2.34 eV (the energy of the SHG with YAG laser), so that lasers having wavelengths longer than 1.06 μm would be needed to effect the SHG.

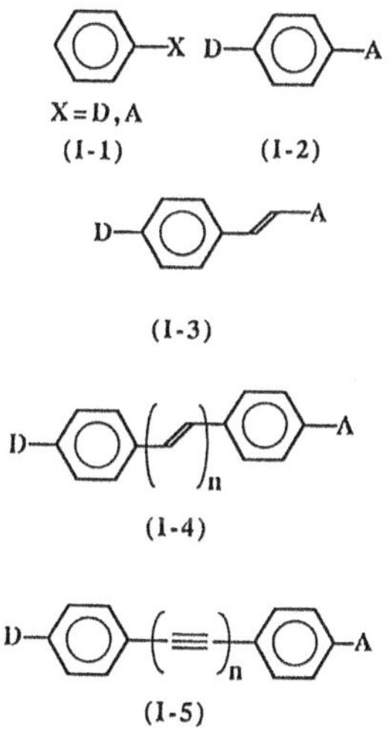

D-	-A
H_3C-	-CN
H_3CO-	$-COCH_3$
H_2N-	-CHO
$(H_3C)_2N$-	$-NO_2$

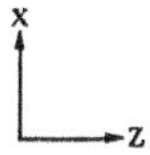

Fig. 1. π-Conjugated systems studied

Table 1. Calculated and experimental values[a] of β

Compounds	$\beta_{exp} \times 10^{30}$[esu]	$\beta_{calc} \times 10^{30}$[esu]
aniline	1.1	1.62
nitrobenzene	-4.6	-6.14
p-nitroaniline	-16.2 - -17.2	-12.69

[a] Reference 3

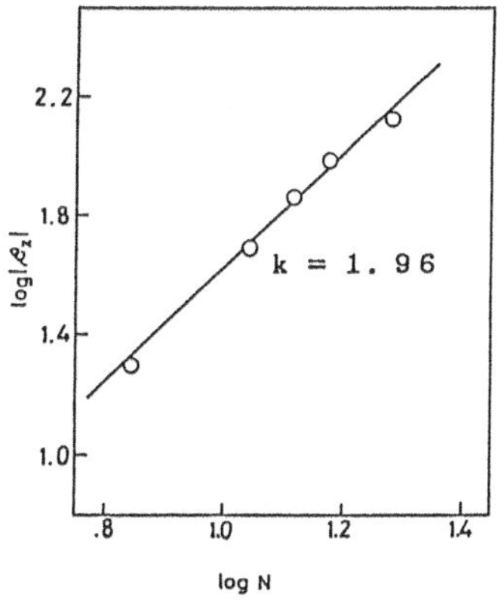

Fig2. Dependence of β for (I-4) on the number of conjugated bonds (N)
$D=NH_2, A=CHO$
k ; slope

Table 2. Calculated $\beta_{zzz} \times 10^{30}$[esu] for various disubstituted systems

(I-2) p-disubstituted benzenes [a]

A \ D	CH_3	OCH_3	NH_2	$N(CH_3)_2$
CN	-3.282 (-2.695)	-5.520 (-3.474)	-5.616 (-3.857)	-10.761 (-5.484)
$COCH_3$	-3.765 (-2.598)	-5.689 (-3.377)	-7.330 (-3.760)	-9.367 (-5.387)
CHO	-3.981 (-2.685)	-6.007 (-3.464)	-8.033 (-3.847)	-13.299 (-5.474)
NO_2	-8.361 (-6.597)	-9.954 (-7.376)	-12.691 (-7.759)	-23.041 (-9.386)

[a] The values in the parentheses are the vector additions of β_{zzz} for the corresponding monosubstituted benzene

(I-3) styrene derivatives

A \ D	CH_3	OCH_3	NH_2	$N(CH_3)_2$
CN	-14.124	-9.980	-11.603	-7.500
$COCH_3$	-12.149	-14.746	-17.217	-20.070
CHO	-12.989	-16.695	-19.735	-23.898
NO_2	-26.421	-28.595	-33.062	-47.590

(I-5) diphenylpolyacetylenes

i) n=1

A \ D	CH_3	OCH_3	NH_2	$N(CH_3)_2$
$COCH_3$	-25.937	-33.406	-43.185	-60.875
CHO	-28.268	-53.540	-56.874	-65.591
NO_2	-56.428	-65.860	-72.720	-114.524

ii) n=2

A \ D	CH_3	OCH_3	NH_2	$N(CH_3)_2$
$COCH_3$	-63.580	-60.055	-65.664	-89.741
CHO	-58.368	-66.982	-71.899	-97.877
NO_2	-194.261	-211.368	-221.350	-258.673

Figure 4 shows p-nitroaniline and some of its aza derivatives. The β values calculated for these compounds are given in Table 4. The β value for (IV-1) is larger than the values for any other compounds, and (IV-4) has the largest E_{n0} and a large $|\beta|$. Since the introduction of the N atoms into aromatic ring does not necessarily lower the transition energy, effects of this substitution on β cannot be expressed in simple terms.

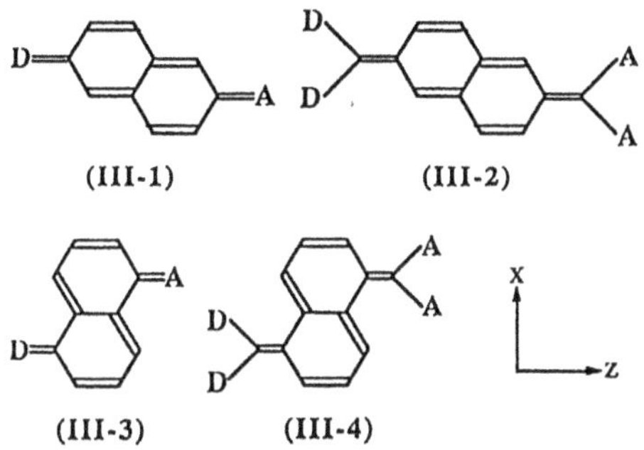

Fig.3. π-Conjugated systems having relatively large |β| values

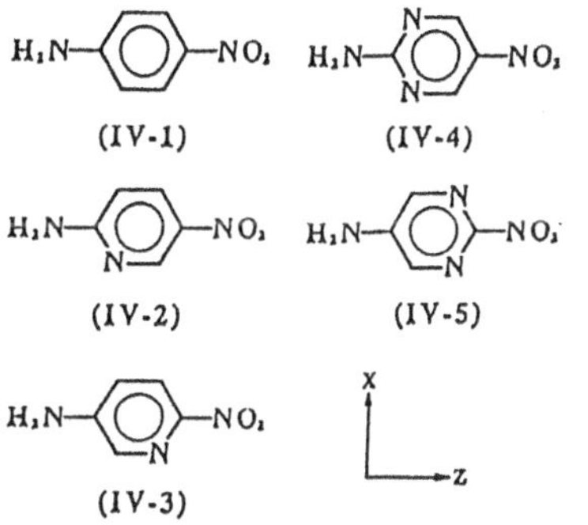

Fig.4. p-Nitroaniline and heterocyclic compounds studied

Table 3. Calculated values of β_{zzz} for (III-1)-(III-4) [a]

Molecules	ΔE [eV]	$\|\mu_{gn_z}\|$ [D]	$\Delta\mu_z$[D]	$\beta_{zzz}\times 10^{30}$ [esu]
(III-1)	3.6961	2.2520	-6.5589	-18.14
(III-2)	2.3148	10.4263	6.9345	325.5
(III-3)	3.2432	4.5054	-7.4132	-25.70
(III-4)	1.9288	6.8690	7.3961	65.02

[a] (III-1)(III-3) ... D=NH_2, A=NO_2
(III-2)(III-4) ... D=NH_2, A=CN
ΔE ; transition energy, μ_{gn_z}; transition moment
$\Delta\mu_z$; the difference between excited state and ground state dipole moments

Table 4. Calculated values of β_{zzz} for (IV-1)-(IV-5) [a]

Molecules	ΔE [eV]	$\|\mu_{gn_z}\|$ [D]	$\Delta\mu_z$[D]	$\beta_{zzz}\times 10^{30}$ [esu]
(IV-1)	4.5855	5.0653	-10.7717	-12.58
(IV-2)	4.5038	3.7844	-3.524	-3.035
(IV-3)	4.3105	3.9189	-4.8164	-6.481
(IV-4)	4.9229	5.0983	-9.2787	-9.875
(IV-5)	4.5239	4.8739	-11.7006	-9.849

[a] ΔE ; transition energy, μ_{gn_z} ; transition moment
$\Delta\mu_z$; the difference between excited state and ground state dipole moments

4. CALCULATIONS OF γ FOR SOME ORGANIC MOLECULES

The third-order hyperpolarizabilities (γ) for benzene, nitrobenzene, aniline and p-nitroaniline were calculated by the CNDO/S-CI method based on the perturbation theory. The results are shown in Table 5, together with their experimental results. For benzene and aniline, the experimental γ values are positive, whereas the calculated values are negative in sign. For p-nitroaniline, the sign of the calculated value is in agreement with that of the experimental data, although the difference between the calculated and the experimental values is rather large. The γ value calculated for nitrobenzene agrees well with the experimental value. The disagreements seem to be attributable to the inherently approximate character of the perturbation method used; the equation of γ based on the perturbation theory apparently gives a negative value for benzene which is known experimentally to have a positive value [4].

Table 5. Calculated and experimental values of γ

Compounds	$\gamma_{exp} \times 10^{36}$ [esu]	$\gamma_{calc} \times 10^{36}$ [esu]
benzene	2.34 [a]	-4.70
nitrobenzene	7.6 [b]	7.42
aniline	4.9 [b]	-5.65
p-nitroaniline	48 [b]	13.03

[a] Reference 4 [b] Reference 5

5. REMARKS

The present CNDO/S-CI calculations still involve the following imperfections:

(1) The number of excited states necessary for the convergences of β and γ values are not known.

(2) Knowledge of the excited states is limited in many cases.

Practically, however, the β and γ are found to be nearly proportional to E^{-2} and E^{-3}, respectively, so that their convergences can be regarded to be sufficiently good. In this study, the number of excited states has been taken to be 60, an expedience which is reliable enough to assure the convergence of β. In order to gain more accurate information of the excited states, recourse to ab initio CI procedure would be required.

REFERENCES

1. J.F.Ward, B.J.Orr: Mol. Phys. 20, 513 (1971),
   M.Nakano, K.Yamaguchi, T.Fueno: Kinouzairyo (Japanese) 8(6), 54 (CMC, Tokyo, Japan 1988)
2. A.Dulcic, et al.: J. Chem. Phys. 74, 1559 (1981)
3. S.J.Lalama, A.F.Garito: Phys. Rev. A20, 1179 (1979),
   C.C.Teng, A.F.Garito: Phys. Rev. Lett. 50, 350 (1983),
   J.A.Morrell, A.C.Albrecht: Chem. Phys. Lett. 64, 46 (1979),
   V.J.Docherty, D.Pugh, J.O.Morley: J. Chem. Soc. Faraday Trans. II. 81,1179 (1985),
   C.C.Teng, A.F.Garito: Phys. Rev. B28, 6766 (1983)
4. B.F.Levine, C.G.Bethea et al.: J. Appl. Phys. 50, 2523 (1979)
5. J.L.Oudar, D.S.Chemla et al.: J. Chem. Phys. 67, 1626 (1977)

CNDO/S-CI Calculations of Hyperpolarizabilities. II: Clusters of Formaldehyde, Urea and Nitrobenzene

M. Nakano, K. Yamaguchi, and T. Fueno

Faculty of Engineering Science, Osaka University, Toyonaka, Osaka 560, Japan

The hyperpolarizabilities for the face-to-face (FF) and the head-to-tail (HT) stacking clusters of formaldehyde, urea and nitrobenzene were calculated by the use of the CNDO/S-CI method. The β values for the FF clusters exhibit little change with the intermolecular distance, but those for the HT clusters change suddenly in the vicinity of 2.0 A. The hydrogen-bonding interaction is no doubt responsible for this trend. The effect of the number of the excited states on β has been investigated. Calculations by small-scale CI procedures are incapable of giving reasonably good results in the region of short intermolecular distances.

1. INTRODUCTION

The nonlinear optical susceptibilities of organic materials are known to be sensitive to crytal structures. In this paper, molecular clusters with assumed geometries are investigated in order to elucidate the contribution of the intermolecular interactions to the second-order hyperpolarizabilities (β). The computational method used is the same as that described in the preceding paper. However, the so-called size-consistency problem inherent in the calculation of clusters has to be coped with. For this purpose, the orbital space used in the CNDO/S-CI is truncated judiciously. The results of calculation by this limited CI formalism are discussed in relation to the experimental data.

2. CNDO/S-CI CALCULATIONS OF β FOR CLUSTERS OF ORGANIC MOLECULES

2.1. Clusters of Formaldehyde

The clusters considered in this study are shown in Fig.1. It is well known that the reliability of the CI results depends on the number of the excited states included in CI. In the first place, the clusters of formaldehyde were examined with full-single CI (FSCI). Two different geometries were assumed; one was the face-to-face (FF) stacking (I-1), while the other was the planar head-to-tail (HT) stacking (I-2). The dependences of β on the intermolecular distance (R) are shown in Figs. 2 and 3. From these results β for the FF cluster is seen to be insensitive to a small change of the intermolecular distance (R), whereas that of the HT cluster changes suddenly in the vicinity of R=2.0 A. These results indicate that the intermolecular interaction is a minor factor for the FF cluster, while it is essentially influential on the HT cluster probably because of the hydrogen-bonding interactions. The change of the total energy for the FF dimer is shown in Fig.4. It increases gradually as the distance R decreases from 15 to 4 (A), below which it decreases sharply by virtue of the complete neglect of the exchange-repulsion terms in the CNDO approximation.

In the limited CI treatment, the truncation of orbital space was used so as to satisfy the size consistency of β . The pattern of truncation (n) is shown in Fig.5. In order to investigate the effects of the number of excited states to β , FCI and some CI's limited by orbital spaces (4x4CI, 6x6CI, 8x8CI) were carried out for both the FF and HT clusters of formaldehyde. The results are

Springer Proceedings in Physics, Vol. 36
Nonlinear Optics of Organics and Semiconductors
Editor: T. Kobayashi © Springer-Verlag Berlin, Heidelberg 1989

shown in Figs. 2 and 3. Both FSCI and nearly FSCI provide reasonable results, while small CI calculations give inadequate results.

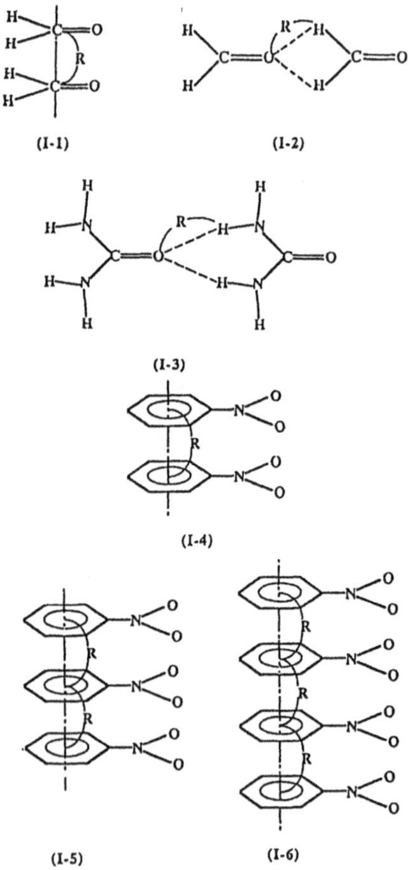

Fig.1. Clusters

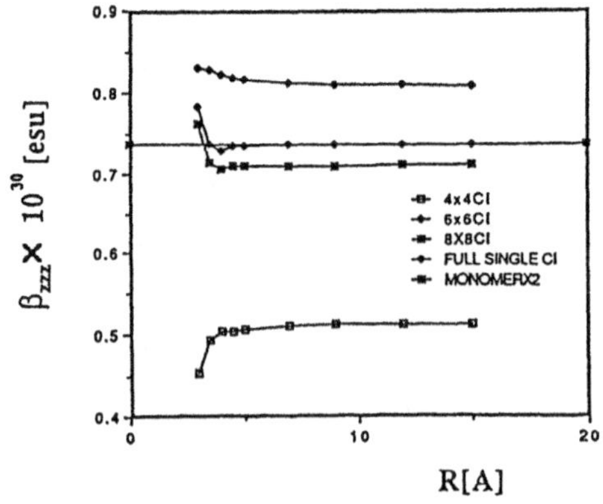

Fig.2. Variations of β_{zzz} for (I-1) with the change in intermolecular distance R

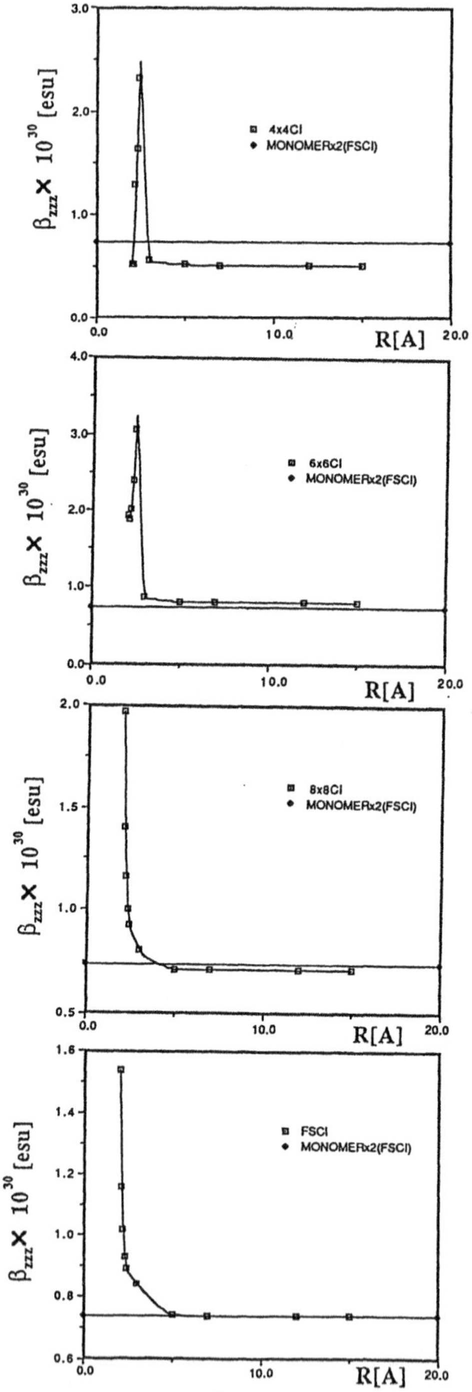

Fig.3. Variations of β_{zzz} for (I-2)

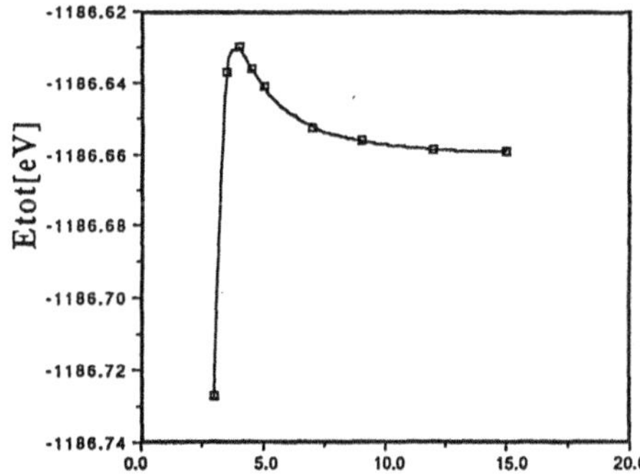

Fig.4. Total energy Etot (FSCI) for (I-1) as the function of the intermolecular distance R

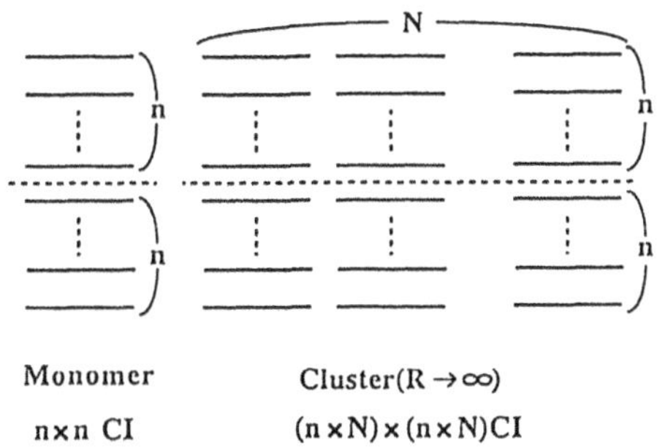

Fig.5. Configuration interaction

2.2. Cluster of Urea

The urea crystal has the structure (I-3). By the hydrogen bonding interactions, the second-order susceptibility of the crystal is expanded and the structure is stabilized[2]. The calculated values of β for (I-3) at various distances (R) are shown in Fig.6. It shows the same tendency as has been noted for the clusters of formaldehyde (Fig.3.).

2.3. Clusters of Nitrobenzenes

In the cases of such large molecules as nitrobenzene, it is extremely cumbersome to carry out FSCI. Thus, the limited CI is desirable. As has been mentioned above, the orbital space used in the limited CNDO/S-CI should be truncated appropriately in order to satisfy the size consistency condition. We have carried out the (3x3CI) for the monomer, (6X6CI) for the dimer, (9x9CI) for the trimer and (12x12CI) for tetramer. The results are shown in Fig.7. The patterns of the variation of β, particularly for the case of the trimer (9x9CI), in the vicinity of R=4.0 A are unreasonable. Probably, this is due primarily to a deficiency of the CNDO/S-CI method used with too limited a number of configurations.

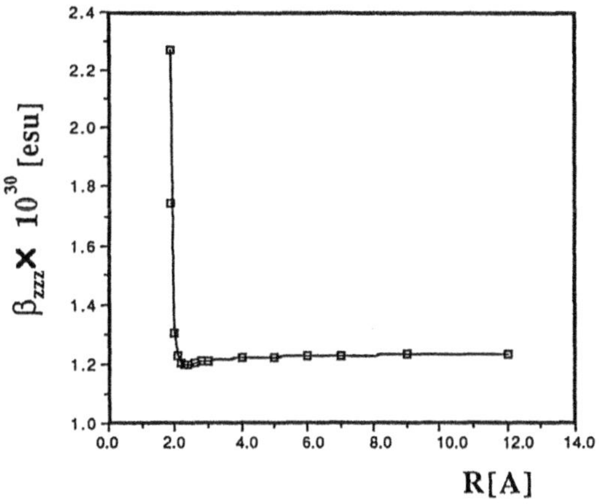

Fig.6. Variation of β_{zzz} for (I-3) with the distance R

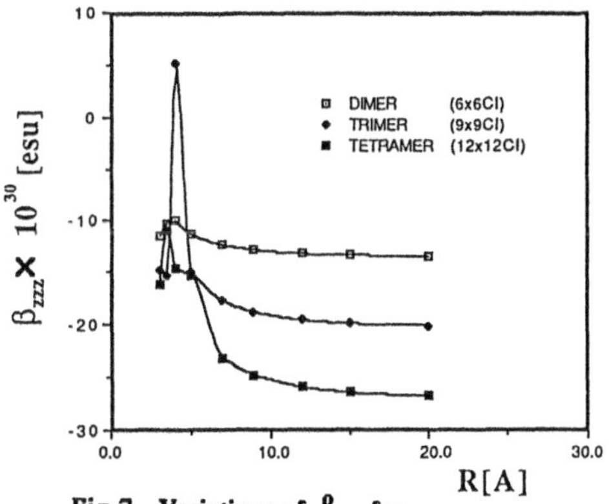

Fig.7. Variations of β_{zzz} for (I-4),(I-5) and (I-6) with the distance R

3. CONCLUDING REMARKS

The preceding[3] and present results are summarized as follows.

(1) In D-A systems, β is proportional to the strength of the intramolecular CT interaction between the D and A groups.
(2) In D-A systems, β exhibits no additive property.
(3) β varies in proportion to the square of the double bond length.
(4) The introduction of N atoms into an aromatic ring tends to increase the excitation energy.
(5) In the calculation of γ , the perturbation theory gives unreasonable results in several cases.
(6) In order to cope with the "size consistency" problem, the limited CI using appropriately truncated orbital space seems to be useful.
(7) Either full-single CI or nearly full-single CI is expected to provide reasonable results for clusters.
(8) The clusters with large intermolecular interactions such as hydrogen-bonding interactions exhibit drastic changes of β in the region of the intermolecular distances where these interactions are dominant.
(9) In the region of small intermolecular distances, the CNDO/S-CI method is fatally unreasonable by virtue of the innate entire neglect of the exchange-repulsion terms.

REFERENCES

1. M.Nakano, K.Yamaguchi, T.Fueno: Kinouzairyo (Japanese) 8(7), 48 (CMC,Tokyo, Japan 1988)
2. J.Zyss, G.Bertheir: J. Chem. Phys. 77(7), 3635 (1982)
3. M.Nakano, K.Yamaguchi, T.Fueno: In the preceding paper

Part IV

Molecules

Complex Hyperpolarizability and Solvatochromic Effects in Merocyanine Solutions

F. Kajzar

Centre d'Etudes Nucléaires de Saclay, IRDI/D.LETI/DEIN/LPEM, F-91191 Gif-sur-Yvette Cedex, France

1. INTRODUCTION

The rapidly developing optical image processing, tuned light sources technology, optical telecommunications, logic systems and optical or electrooptical sensors create a large demand for highly efficient nonlinear optical materials. Among the various different candidates, organic materials present several advantages. They are characterized by large second- and third- order hyperpolarizabilities and a high damage threshold. They offer also a large diversity of easily processable materials and an ease of modification of their physico-chemical properties through so-called molecular engineering. It is well known that for the second-order effects the best are charge transfer molecules crystallizing in a noncentrosymmetric system [1]. The presence of hydrogen bonding enhances the hyperpolarizability of such molecules [2]. For third-order nonlinear optical processes the most promising are one- dimensional conjugated π electron systems [3].

One of the principal tasks which a materials scientist faces is the relatively precise and quick determination of nonlinear susceptibilities. The most direct technique is harmonic generation from powders [4]. This technique however lacks precision in the hyperpolarizability determination. The generated harmonic intensity depends on grain size, absorption and the wavelength of operation. The growth of good optical quality single crystals is costly and slow. In the case of centrosymmetric materials a relatively easy determination of cubic susceptibility $\chi^{(3)}$ can be obtained from thin film measurements, provided that such films with sufficiently low light scattering are available.

A precise determination of molecular hyperpolarizabilities can be realized for soluble molecules by coupling the electric field induced second harmonic generation (EFISHG) with third harmonic generation (THG) techniques. The first technique yields (for a noncentrosymmetric molecule)

$$\gamma = \gamma(-2\omega;\omega,\omega,0) + \frac{\mu\beta\ (-2\omega;\omega,\omega)}{5kT} \tag{1}$$

where ω is the incident laser frequency, μ is the permanent dipole moment and β and γ are second- and third-order molecular hyperpolarizabilities. In the THG technique one determines directly $\gamma(-3\omega;\omega,\omega,\omega)$ which far off resonance is close to $\gamma(-2\omega;\omega,\omega,0)$. Coupling these two techniques one can determine β and γ independently (off resonance).

For the fundamental (or its multiples) and for harmonic frequencies close to those corresponding to transition energies between the ground state and one of the excited levels the molecular hyperpolarizabilities β and γ are complex and depend

Springer Proceedings in Physics, Vol. 36
Nonlinear Optics of Organics and Semiconductors
Editor: T. Kobayashi

significantly on ω. In this case $\gamma(-3\omega;\omega,\omega,\omega)$ is no longer equal to $\gamma(-2\omega;\omega,\omega,0)$ and the separation of β and γ implies a use of wave dispersed harmonic generation techniques (both EFISHG and THG) and also a use of quantum mechanical formulas for $\gamma(-3\omega;\omega,\omega,\omega)$ and $\gamma(-2\omega;\omega,\omega,0)$. The procedure leading to the determination of the phase of $\gamma(-2\omega;\omega,\omega,0)$ in the EFISHG technique was described in a recent paper [5]. In the present work we deal mostly with the determination of the phase of $\gamma(-3\omega;\omega,\omega,\omega)$ in THG measurements. The formalism developed here we apply to the γ phase determination for the merocyanine (4-[(1-methyl-4(14)-pyrydinylidene)ethylidene]-2-5-cyclohexadien-1-one hydrate) molecule solubilized in ethanol and in propanol.

An exact knowledge of the phase of γ is important from both a fundamental and a practical point of view. In fact, the quantum mechanical formulas for the cubic susceptibility $\chi^{(3)}(-3\omega;\omega,\omega,\omega)$ derived by the time-dependent perturbation method read [6-8] (in dipolar approximation)

$$\chi^{(3)}_{xxxx}(-3\omega;\omega,\omega,\omega) \propto \frac{e^4}{8\hbar^3} \sum_{n,f,n'} Q_{gn} Q_{nf} Q_{fn'} Q_{n'g}$$

$$\times\left[\frac{1}{(E_{ng}-\omega-i\Gamma)(E_{fg}-2\omega-i\Gamma_f)(E_{n'g}-3\omega-i\Gamma_{n'})}+\frac{1}{(E_{ng}+\omega+i\Gamma_g)(E_{fg}-2\omega-i\Gamma_f)(E_{n'g}-\omega-i\Gamma_{n'})}+\frac{1}{(E_{ng}-\omega-i\Gamma_g)(E_{nf}+2\omega+i\Gamma_f)(E_{n'g}+\omega+i\Gamma_{n'})}+\frac{1}{(E_{ng}+\omega+i\Gamma_g)(E_{nf}+2\omega+i\Gamma_f)(E_{n'g}+3\omega+i\Gamma_{n'})}\right] \quad (2)$$

where n,n' are allowed and f forbidden for one-photon transition levels, respectively, Q_{ij} are the dipolar transition matrix elements between $|i\rangle$ and $|j\rangle$ states, and E_{ij} are corresponding energy differences in $\hbar$ units.

In equation (2) we have introduced a priori the damping factors $\Gamma_i(i=n,n',f)$ showing that when the fundamental frequency, its double or triple approaches the energy difference between the ground state and one (corresponding) excited state, $\chi^{(3)}(-3\omega;\omega,\omega,\omega)$ will be complex and resonantly enhanced. Of special interest is a two-photon resonance, falling at the laser frequency equal to one half of the energy difference E_f-E_g ; thus in general in the transparency range for the fundamental wave. The observation of such resonances and knowledge of the phase of $\chi(-3\omega;\omega,\omega,\omega)$ are useful in the determination of the position of these levels that are forbidden for one-photon transitions, leading to so-called nonlinear spectroscopy.

On the other hand, at large incident laser light intensity I a refractive index variation is induced whose variation is given by

$$n = n_0 + n_2 I \quad (3)$$

where

$$n_2 = 12\pi^2 \chi^{(3)}(-\omega;\omega,\omega,-\omega)/n_0^2 c$$

Although the Kerr susceptibility $\chi^{(3)}(-\omega;\omega,\omega,-\omega)$ is not the same as that responsible for the THG process ($\chi^{(3)}(-3\omega;\omega,\omega,\omega)$) the one- and two-photon resonances will occur simultaneously (cf. refs [9] and [10]).

In the special case of the $\chi^{(3)}(-\omega;\omega,\omega,-\omega)$ phase being equal to π the index of refraction n will decrease with increasing light intensity (outside the fundamental wave absorption range). For a complex Kerr $\chi^{(3)}$ an intensity- dependent fundamental beam absorption will occur through a four photon process (two-photon absorption - TPA).

The procedure for γ phase determination is described in Section 2. Section 3 is devoted to the experimental determination of the γ phase for the merocyanine solutions in ethanol and in propanol showing a solvatochromic effect, at 1.0642 μm and 1.907 μm.

2. HARMONIC GENERATION AND NONLINEAR PROPAGATION IN A WEDGED LIQUID CELL

For THG measurements we use a wedged liquid cell with thick input and output windows. The principle of operation of such a liquid cell together with the procedure for γ determination for pure liquids is described elsewhere [11 - 12]. Here we recall briefly fundamental formulas and generalize them for solutions absorbing at harmonic frequency ; we also take account of complex $\chi^{(3)}$. We present also the procedure leading to the determination of real and imaginary parts of γ for a molecule in solution.

The resultant harmonic field generated in the wedged liquid cell is a sum of three contributions :

$$E_R^{3\omega} = T_{G1}E_{G1}^{3\omega} + T_L E_L^{3\omega}(x) + T_{G2}E_{G2}^{3\omega} \tag{4}$$

where G1 and G2 refer to the input and output windows, respectively and L to the liquid compartment (cf. Fig. 1). T's are corresponding total transmission factors. E_{G1} and E_{G2} are bound waves given by [11]

$$E_{G1}^{3\omega} = \pi \left[\frac{\chi^{(3)}}{\Delta \epsilon}\right]_G (E_\omega)^3 \, t_{3\omega}^{GL} \, t_{3\omega}^{LG} \, t_{3\omega}^{GO} \, e^{i\psi_H} \tag{5}$$

$$E_{G2}^{3\omega} = -\pi \left[\frac{\chi^{(3)}}{\Delta \epsilon}\right]_G (E_\omega)^3 \left[t_\omega^{GL} \, t_\omega^{LG}\right]^3 t_{3\omega}^{GO} \, e^{i\Delta\psi} \, e^{i\psi_H} \tag{6}$$

where the subscripts (or superscripts) G refer to the window material and L to the liquid (solution) respectively. $t_{\omega(3\omega)}^{ij}$ are transmission factors between media i and j at frequency $\omega(3\omega)$ and for linearly polarized incident light (a general requirement in these measurements)

$$t_{\omega(3\omega)}^{ij} = \frac{2n_{\omega(3\omega)}^{j}}{n_{\omega(3\omega)}^{i} + n_{\omega(3\omega)}^{j}} \quad . \tag{7}$$

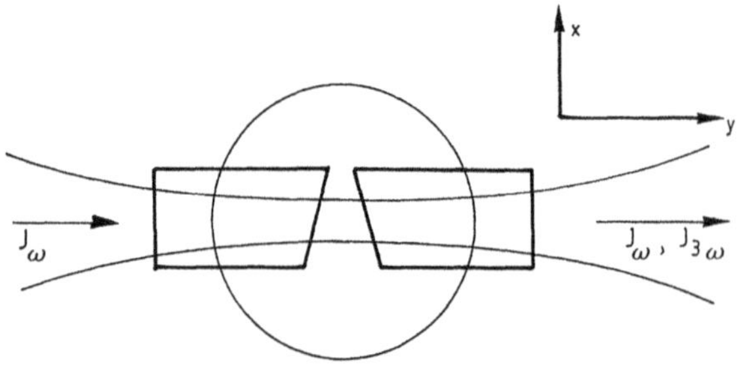

Fig. 1 Schematic representation of liquid cell for THG measurements

$\Delta\epsilon$ in Eqs. (5)-(6) is the dielectric constant dispersion and $\Delta\psi$ is the phase mismatch between the fundamental and harmonic waves

$$\Delta\psi = \psi_F - \psi_H = 3\omega(n^L_\omega - n^L_{3\omega})\ell(x)/c \qquad (8)$$

where $\ell(x)$ is the liquid compartment thickness depending on translation x (cf. Fig. 1).

$$\ell(x) = 2x \tan\frac{\alpha}{2} + \ell_0 \qquad (9)$$

where α is the wedge angle and ℓ_0 its initial thickness.

In Eqs. (7)-(8) we have assumed the cosines of the incidence and refraction angles toward the liquid cell to be equal to 1. In fact these angles are close to zero for our liquid cell with a wedge angle of about 0.7°.

The harmonic field generated in the liquid cell is given by (outside the input window)

$$E^{3\omega}_L(x) = \pi \left[\frac{\chi^{(3)}}{\Delta\epsilon}\right]_L \ell^{i\psi_H}(\ell^{i\Delta\psi} - 1)(E_\omega t^{GL}_\omega)^3\, t^{GO}_{3\omega} \quad . \qquad (10)$$

We notice here that only the $E^{3\omega}_L$ field depends on x, leading to Maker fringes of harmonic intensities when the liquid cell is translated (cf. Fig. 2).

For a liquid absorbing at the fundamental and/or harmonic frequency the corresponding refractive indices in Eqs. (5)-(8), are complex :

$$n_{\omega(3\omega)} = n^r_{\omega(3\omega)} + i\,K_{\omega(3\omega)} \qquad (11)$$

with imaginary parts related to the linear absorption coefficients $\alpha_{\omega(3\omega)}$ through the following relation :

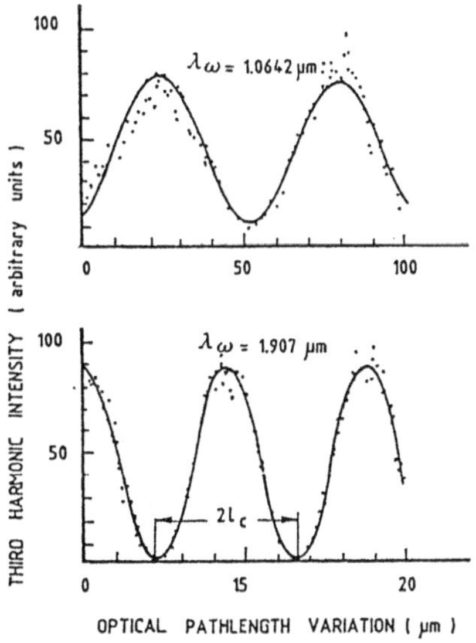

Fig. 2 : Harmonic intensity as a function of the cell translation for merocyanine solutions in ethanol for the cases of absorbed (λ_ω = 1.064 μm) and nonabsorbed (λ_ω = 1.907 μm) harmonic light

$$K_{\omega(3\omega)} = \frac{\lambda_{\omega(3\omega)}\,\alpha_{\omega(3\omega)}}{4\,\pi} \qquad (12)$$

where $\lambda_{\omega(3\omega)}$ is the corresponding wavelength.

For a solution the cubic susceptibility $\chi^{(3)}$ is related to the molecular hyperpolarizability γ through the following formula :

$$\chi_L^{(3)} = N_s\gamma_s f_s + N_p\gamma_p f_p \qquad (13)$$

where s refers to solvent and p to solute molecules, N's are corresponding numbers of molecules and f's are local field factors which in the case of THG and within the Lorentz-Lorentz approximation read

$$f_{s(p)} = \left[\frac{n_\omega^2 + 2}{3}\right]^3_{s(p)} \left[\frac{n_{3\omega}^2 + 2}{3}\right]_{s(p)} . \qquad (14)$$

Denoting by C the mass concentration ($C = m_p/m_s$), Eq. (13) can be rewritten in the following form :

$$\chi_L^{(3)} = \frac{f(C)\,d(C)}{1 + C} \left[\frac{C\,\gamma_p\,N_A}{M_p} + \frac{\gamma_s\,N_A}{M_s}\right] \qquad (15)$$

where d(C) is the solution density, N_A Avogadro's number and M_p, M_s corresponding molecular masses.

In Eq. (15) we have used the same local field factor f(C) for solvent and solute molecules in agreement with Oudar's [13] assumption for EFISHG in solution. Oudar [13] has analyzed local field factors for several liquid-liquid mixtures taking exact refractive indexes of components and has found a minor difference (less than 5 %) between local field factors calculated separately and those using the same for both molecules. We believe that the same arguments hold also in the case of THG.

As we mentioned before, close to one-, two- or three-photon resonances (cf. Eq. (2)) the γ_p hyperpolarizability is complex :

$$\gamma_p = \gamma'_p + i\gamma''_p \quad . \qquad (16)$$

Assuming γ_s real and positive as it is customarily done (otherwise it can be assured by an appropriate choice of solvent) we get for harmonic intensity as a function of concentration and incident light the intensity following expression :

$$J_{3\omega}(C) = \frac{64\,\pi^4}{c^2} \left[\frac{\chi^{(3)}}{\Delta\epsilon}\right]^2_G \frac{\left[t_{3\omega}^{GO}\right]^2}{\beta^2} \left| t_{3\omega}^{GL}\, t_{3\omega}^{LG} - \alpha\beta e^{i\Delta\psi}\left[t_\omega^{GL} t_\omega^{LG}\right]^3 \right.$$

$$\left. + \rho(C)\,\frac{(\Delta\epsilon)_G}{(\Delta\epsilon)_L}\,(\alpha\beta e^{i\Delta\psi} - 1)\,(t_\omega^{GL})^3\, t_{3\omega}^{LG} \right|^2 J_\omega^3 \qquad (17)$$

where

$$\alpha = e^{3\omega\,K_\omega \ell/c} \qquad (18)$$

and

$$\beta = e^{3\omega} K_{3\omega} \ell/c \qquad (19)$$

take account of absorption at fundamental and harmonic frequencies, respectively, and

$$\rho(C) = \frac{\chi_L^{(3)}}{\chi_G^{(3)}} = \frac{f(C)\, d(C)\, N_A}{(1+C)\, \chi_G^{(3)}} \left[\frac{C(\gamma_p' + i\gamma_p'')}{M_p} + \frac{\gamma_s}{M_s} \right] . \qquad (20)$$

We note here that for our experimental situation (liquid compartment thickness of about 150 μm) and for the maximum absorbance at harmonic frequency less than 2 the imaginary parts of $K_{3\omega}$ are of the order of 10^{-3} and consequently $K_{3\omega}$ can be neglected with respect to $n_{3\omega}^{r}$. The same is also valid for K_{ω} which is even smaller ($K_{\omega} \lesssim 10^{-5}$ at 1.907 μm and negligible at 1.0642 μm). But one cannot neglect the absorption of the harmonic wave (cf. Eq. (19)) which is quite large at some concentrations (cf. Fig. 3).

The normalized harmonic intensities with respect to that of pure solvent measured at the same geometry and input light intensity are given by

$$J_N(C) = J_{3\omega}(C)/\, J_{3\omega}(0) = \frac{1}{\beta^2} \left\{ \left[(A+B(C)) \left[\frac{C\gamma_p'}{M_p} + \frac{\gamma_s}{M_s} \right] \right]^2 \right. \qquad (21)$$

$$\left. + C^2 \left[\frac{\gamma''}{M_p} B(C) \right]^2 \right\} \Big/ \left[A + B(0) \frac{\gamma_s}{M_s} \right]^2$$

where

$$A = t_{3\omega}^{GL}\, t_{3\omega}^{LG} - \alpha\beta\, e^{i\Delta\psi} \left[t_{\omega}^{GL}\, t_{\omega}^{LG} \right]^3 \qquad (22)$$

and

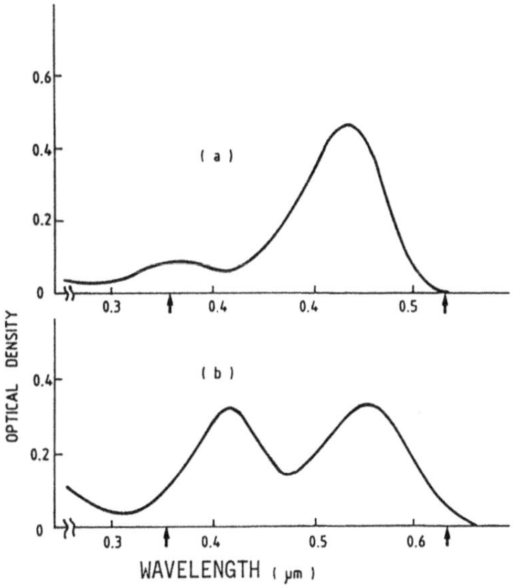

Fig. 3 : Optical absorption spectra of studied merocyanine solutions in propanol (a) and ethanol (b)

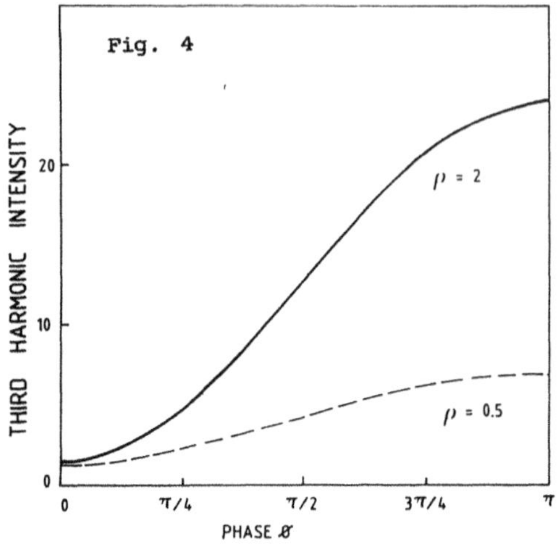

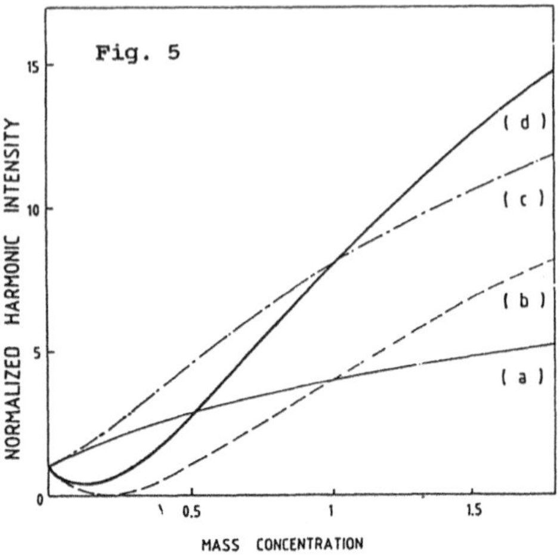

Fig. 4 : Calculated harmonic intensity as a function of liquid $\chi^{(3)}$ phase (constant modulus ρ)

Fig. 5 : Normalized harmonic intensity as a function of mass concentration for different values of real and imaginary parts of $r = \chi_P^{(3)} / \chi_G^{(3)}$. a–r' = 4, r" = 0 ; b–r' = −4, r" = 0 ; c–r' = 4, r" = 4 ; d–r' = −4, r" = 4 (ρ = 2)

$$B(C) = \frac{f(C)\, d(C)\, N_A}{(1+C)\, \chi_G^{(3)}} \left(\alpha\beta e^{i\Delta\psi} - 1\right) \left(t_\omega^{GL}\right)^3 t_{3\omega}^{LG} \frac{(\Delta\epsilon)_G}{(\Delta\epsilon)_L} \,. \quad (23)$$

Equation (21) shows (in the absence of absorption) a quadratic variation of normalized intensity versus concentration for a complex γ_p. Influence of γ'_p on $J_N(C)$ is more complex. In Fig. 4 we have displayed simulated concentration dependence $J_N(C)$ for different phases Φ = a tan (γ''_p/γ'_p) assuming for simplicity the same refractive indices for liquid and window material. A strong dependence, especially at small concentration is seen.

Impact of γ_p phase on harmonic intensities is also seen in Fig. 5. In this figure we have shown the calculated harmonic intensities as a function of phase Φ, assuming for simplicity an infinite concentration C (pure solute). This figure shows the importance of the γ_p phase in data reduction and the error perpetrated if this is not taken into account. Especially one can see from this figure that there exist two solutions for γ_p in the case of $\phi = 0$ and 2π. Another point is that the sign of γ'' is undetermined (for real and positive γ'_s).

3. EXPERIMENTAL PROCEDURE

The THG measurements have been done with the apparatus shown schematically in Fig. 6. The light source is a Q-switched Nd:YAG laser (Quantel model Y581C) operating at 1.0642 µm fundamental wavelength. The 1.907 µm wavelength was obtained by a stimulated Raman scattering of the fundamental beam in a high pressure hydrogen cell.

This wavelength was separated from the fundamental and anti-Stokes beams by a Pellin-Broca prism. After filtering of visible light and correcting for polarization losses the incident light was focused on the liquid cell. The generated

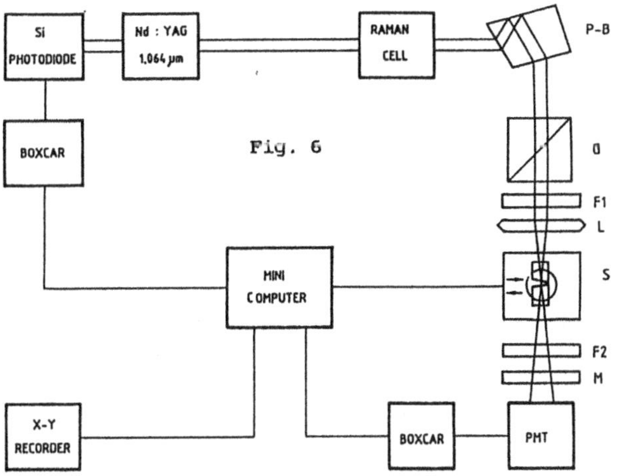

Fig. 6

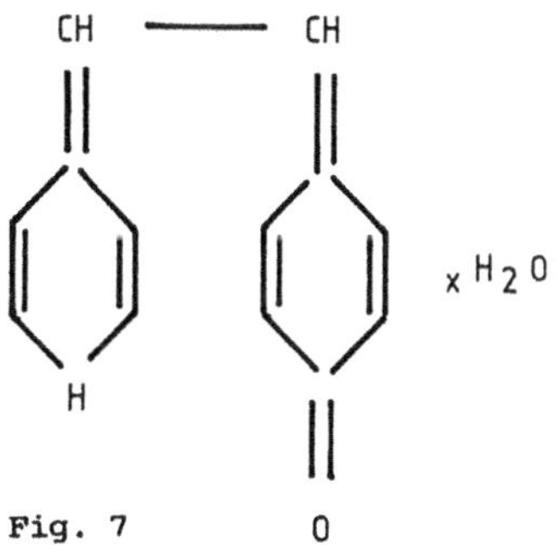

Fig. 7

Fig. 6 : Schematic representation of experimental arrangement: P-B-Pellin-Broca prism, G-Glan polarizer, F1, F2-filters, L-lens, S-sample, M-monochromator

Fig. 7 : Molecular structure of studied merocyanine

harmonic light was filtered with the help of a KG3 filter (absorbing entirely the fundamental light) and a Jobin-Yvon monochromator. The measurements were performed alternatively for the solution and host solvent and the fluctuations of the fundamental beam were monitored by a rapid Si photodiode. This procedure allows one to calibrate the measured harmonic intensities and to test the quality of the liquid cell clearing after each solution measurement. The last procedure was done in situ in order to keep the experimental geometry unchanged during THG measurements.

The merocyanine we studied is a commercial one (Aldrich) and its chemical structure is shown in Fig. 7. Its absorption spectra in propanol and in ethanol are shown in Fig. 3. An important absorption shift toward higher wavelengths is observed in ethanol.

In Figs. 8 and 9 we have shown coherence length variation with merocyanine concentration at 1.0642 μm and 1.907 μm fundamental wavelength in ethanol and methanol, respectively.

Both solutions show a similar dependence of the coherence length on concentration. At 1.0642 μm the coherence length increases and at 1.907 μm decreases with increasing solute concentration. Such behavior is typical for anomalous and normal dispersion : in the first case harmonic intensity lies above the absorption band and in the second case below (cf. Fig. 3).

In Figs. 10 and 11 we have displayed the concentration variation of normalized harmonic intensity at 1.0642 and 1.907 μm fundamental wavelength, respectively, for merocyanine solutions in ethanol. The 1.0642 μm data show a strong, nearly quadratic dependence of $J_N(C)$ on merocyanine, concentration, indicating a complex γ. This dependence is much weaker at 1.907 μm. By a least squares fit of Eq. (21) to experimental data we have determined the real and imaginary parts of γ, which are listed in Table I. In the fitting procedure we have used for ethanol the published γ

value (12) and for silica $\chi^{(3)} = 2.8 \times 10^{-14}$ e.s.u. at 1.907 μm as determined by Meredith et al. [14] and $\chi^{(3)} = 3.11 \times 10^{-14}$ e.s.u. at 1.0642 μm ; the last value obtained from Miller's rule.

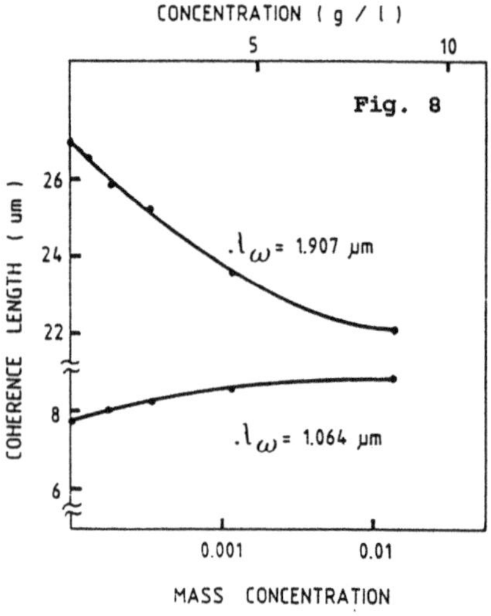

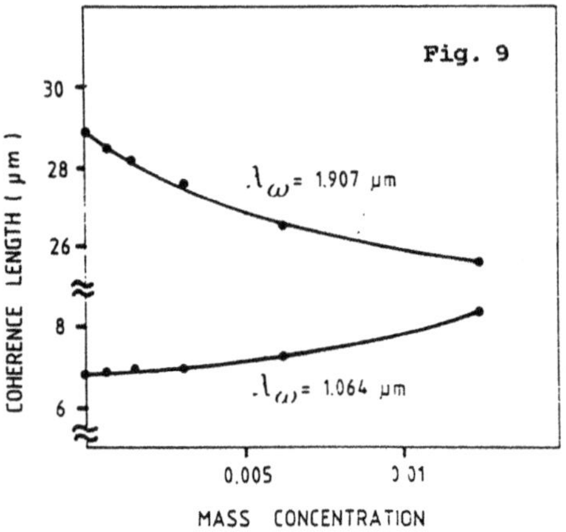

Fig. 8 Coherence length variation as a function of merocyanine contration in ethanol. Points represent experimental values

Fig. 9 Coherence length variation as a function of merocyanine concentration in propanol

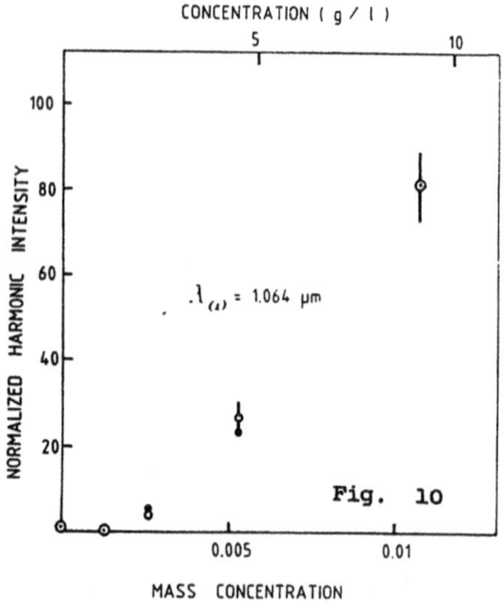

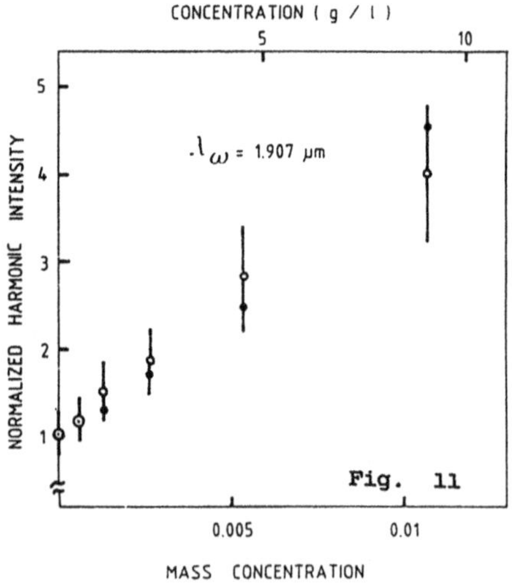

Fig. 10 : Normalized harmonic intensity as a function of merocyanine concentration in ethanol at 1.064 μm. Open circles are experimental values and dots the calculated ones

Fig. 11 : Normalized harmonic intensity as a function of merocyanine concentration in ethanol at 1.907 μm

Table I

Real (γ') and imaginary (γ'') parts of molecular hyperpolarizabilities in 10^{-36} e.s.u. units for studied molecules at 1.0642 μm and 1.907 μm

Compound	λ_ω = 1.0642 μm		λ_ω = 1.907 μm	
	γ'	γ''	γ'	γ''
Acetone[a]	1.80	0	1.88	0
Ethanol[a]	1.27	0	1.49	0
Propanol	1.72	0	1.82	0
Silica[b]	4.02	0	3.73	0
Merocyanine in ethanol	-2.3×10^3	$\pm1.6\times10^3$	0.8×10^3	0
Merocyanine in propanol	-0.6×10^3	$\pm2.1\times10^3$	0.4×10^3	0
Polydi-acetylene TS-12[c]	-1.7×10^3	-1.1×10^3	1.2×10^3	0

a- Ref. 12

b- Ref. 14 (the 1.0642 μm value is obtained from the 1.907 μm value [14] using Miller's rule)

c- Ref. 15

At 1.0642 μm γ is complex with negative real part whereas at 1.907 μm it is real and positive. The absolute value of γ at first wavelength is about four times larger than at 1.907 μm, indicating a resonance enhancement.

Very similar behavior is shown by the merocyanine solutions in propanol. The corresponding concentration variation of normalized intensity is shown in Figs. 12 and 13 at 1.0642 μm and 1.907 μm, respectively and the fitted γ values are also listed in Table I. The γ value for propanol was determined using acetone as a standard. The molecular hyperpolarizability of merocyanine is complex with negative real part at 1.0642 μm and is real and positive at 1.907 μm. Again the absolute value of γ at 1.0642 μm is larger than at 1.907 μm.

The $|\gamma|$ values are larger for merocyanine solutions in ethanol than in propanal. This result shows the existence of a solvatochromic effect on γ ; in ethanol the absorption bands are closer to the harmonic frequency and γ is more significantly enhanced through presumably a three-photon resonance.

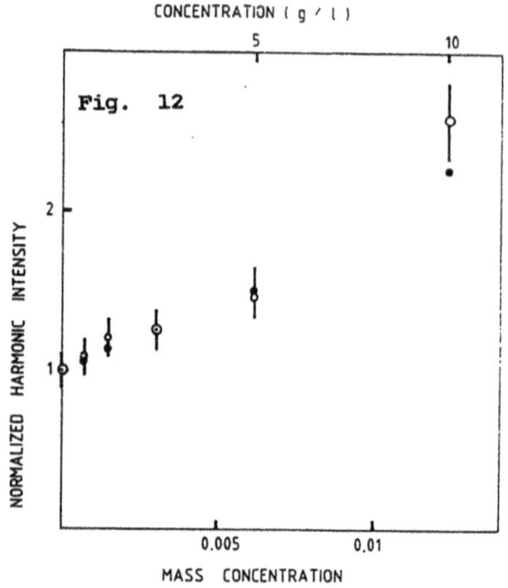

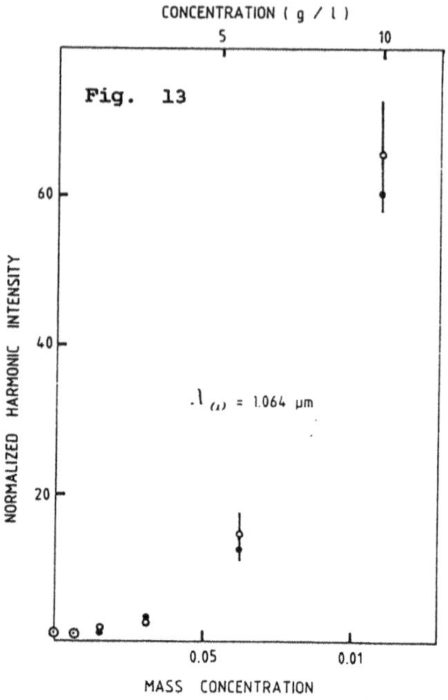

Fig. 12 : Normalized harmonic intensity as a function of merocyanine concentration in propanol at 1.064 μm

Fig. 13 : Normalized harmonic intensity as a function of merocyanine concentration in propanol at 1.907 μm

4. DISCUSSION

We have determined the real and imaginary parts of the molecular hyperpolarizability γ of merocyanine molecule in two solvents : ethanol and propanol, and at two fundamental wavelengths of operation : 1.0642 μm and 1.907 μm. The results are given in Table I. At 1.907 μm γ is real and positive whereas at 1.0642 μm it is complex with negative real part. This behavior can be well accounted for in terms of a two-level model and a three-photon resonance using Eq. (2), as is shown in Fig. 14.

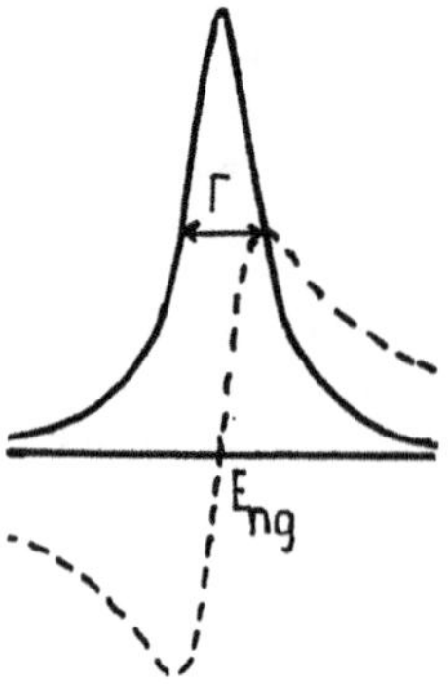

Fig. 14 : Real (broken line) and imaginery (solid line) hyperpolarizability γ variation around a three photon resonance.

From Table I it is also seen that the absolute value of γ is about four times larger in both cases at 1.0642 μm than at 1.907 μm, showing a stronger resonant enhancement at the former wavelength. Also the absolute values of γ are larger for solutions in ethanol than in propanol ; in former cases the harmonic wavelength is closer to the absorption band, thus again, there is a stronger resonant enhancement. In Table I we have reported also the γ values for polydiacetylene TS-12 per monomer unit [15]. These are very close to those of merocyanine, showing a similar π electron delocalisation in both molecules.

REFERENCES

1. For a general review see e.g. Nonlinear Optical Properties of Organic Molecules and Crystals, D.S. Chemla and J. Zyss, eds., Academic Press, Orlando 1987, Vol. 1
2. J. Zyss and J.L. Oudar, Phys. Rev. A26, 2028 (1982)
3. K.C. Rustagi and J. Ducuning, Opt. Commun. 10, 258 (1974)
4. S.K. Kurtz and J.J. Perry, J. Appl. Phys. 37, 3798 (1968)
5. F. Kajzar, I. Ledoux and J. Zyss, Phys. Rev. A36, 2210 (1987)
6. J.F. Ward, Rev. Mod. Phys. 37, 1 (1965)
7. P.W. Langhoff, S.T.E. Epstein and M. Karplus, Rev. Mod. Phys. 44, 602 (1972)
8. D.C. Hanna, M.A. Yuratich and D. Cotter, Nonlinear Optics of Free Atoms and Molecules, Springer-Verlag, Berlin 1979
9. N. Bloembergen, H. Lotem and R.T. Lynch, Ind. J. Pure Appl. Phys. 16, 151 (1978)
10. F. Kajzar, L. Rothberg, S. Etemad, P.A. Chollet, D. Grec, A. Boudet and T. Jedju, Thin Solid Films, 160, 373 (1988)
11. F. Kajzar and J. Messier, Rev. Sci. Instrum. 58, 2061 (1987)
12. F. Kajzar and J. Messier, Phys. Rev. A32, 2352 (1985)
13. J.L. Oudar, J. Chem. Phys. 67, 446 (1977)
14. G.R. Meredith, B. Buchalter and C. Hanzlik, J. Chem. Phys. 78, 1533 (1983)
15. F. Kajzar and J. Messier, in Polydiacetylenes. Synthesis, Structure and Electronic Properties NATO ASI Series, Series E. Applied Sciences n° 102, D. Bloor and R.R. Chance eds Martinus Nijhoff Publ., Dordrecht 1985, p. 325

Nonlinear Optical Properties of Retinal Derivatives

Y. Kawabe, H. Ikeda, T. Sakai, and K. Kawasaki

Central Research Laboratories, Idemitsu Kosan Co. Ltd.,
Sodegaura, Kimitsu, Chiba 299-02, Japan

1. Introduction

Conjugated organic materials have been attracting attention because of their large and fast optical nonlinearities. New compounds with long conjugated chains and electron acceptors at one end of the chains, were derived from retinal, and their optical nonlinear properties were measured by electric-field-induced second harmonic generation (EFISHG) and phase conjugation.

2. Synthesis

Two kinds of retinal derivatives were synthesized by replacing aldehyde substituents with acceptors, dicyano substituent 1 or pyridinium substituent 2, on Knoevenagel reaction. Their chemical structures are shown in Fig.1.

1

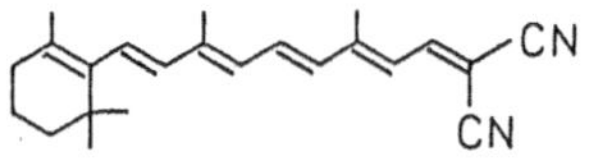

2

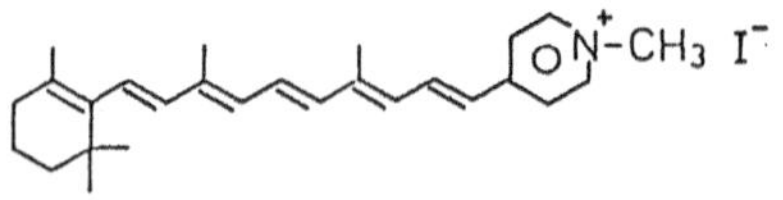

Fig. 1. Chemical structures of retinal derivatives

3. Electric-Field-Induced Second Harmonic Generation

Second order optical nonlinear coefficients of molecules can be determined by the wedge Maker fringe experiment under electric field in solution [1-3]. Figure 2 shows the experimental apparatus. The fundamental wave of a Q-switched Nd^{3+}:YAG laser (1.06 μm) (Spectra Physics DCR-3D) was focused into the cell polarized parallel to the electric field of 3MV/m which was synchronized with the laser. The energy of the laser was about 0.5 mJ per pulse and pulse duration was 10ns.

The generated harmonics were filtered through glass filters and a monochromator. Signals detected by a photomultiplier tube (Hamamatsu R928) were amplified,

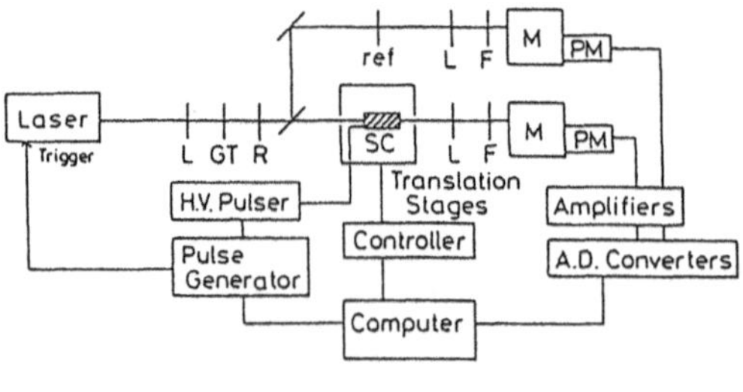

Fig. 2. Experimental set-up for the EFISHG experiment. GT is the Glan prism; R, the $\lambda/2$ retarder; L, the lens; F, the glass filters; SC, the sample cell; M, the monochromator; PM, the photomultiplier tube; ref, the reference crystal powder of 2-methyl-4-nitroaniline.

Springer Proceedings in Physics, Vol. 36
Nonlinear Optics of Organics and Semiconductors
Editor: T. Kobayashi

digitized, and transferred to a microcomputer. Samples were dissolved in dimethylsulfoxide (DMSO) in concentrations of $0 \sim 1 \times 10^{-2}$ mol/l and were contained in a sample cell made of quartz glass.

Maker fringes were observed by the change of optical path length. But because of the optical absorption of this sample at the wavelength of second harmonics (Fig.3), the Maker fringes became obscure with increasing the concentration of the sample. Solving the propagation equation for the second harmonic wave under the boundary conditions of the electric and magnetic fields, SHG intensity and its fringing pattern can be formulated. Macroscopic nonlinear optical coefficient can be determined from the fringe patterns with the nonlinear least-squares-fit.

The macroscopic nonlinear coefficient of solution can be expressed as [1-3]

$$\Gamma = f \, \Sigma_i \, N_i(\gamma_i + \mu_{0i} \, \beta_i/5kT) \tag{1}$$

where γ_i and β_i refer to the purely electronic third and second order nonlinear optical coefficients, respectively, and μ_{0i}, the permanent dipole moment. N_i is the number of molecules per unit volume, and f is the total local field correction. In most cases, the third order term can be neglected.

Concentration dependence of the square root of SHG intensity for the sample 1 is shown in Fig.4. That the curve cannot be expressed by a linear function results from the contribution of the imaginary part of Γ as KAJZAR et al. [4] claimed for their EFISHG experiment for poly-diacetylene solution. Because the sample has large optical absorption at the SHG wavelength, it is reasonable that the imaginary part of β which originates from the two photon resonance, contributes to SHG. Real and imaginary parts of the nonlinear optical coefficients were determined from the dependence of Γ on the concentration, those are $Re(\mu_0\beta)=1.3 \times 10^{-45}$ (esu) and $Im(\mu_0\beta)=1.5 \times 10^{-45}$ (esu). These values were much larger than that of 2-methyl-4-nitroaniline, a typical organic nonlinear optical material, whose value was determined to be $\mu_0\beta = 1.8 \times 10^{-46}$ (esu) with the same method. For the sample 2, it was impossible to measure by this method, because it was ionized in solution.

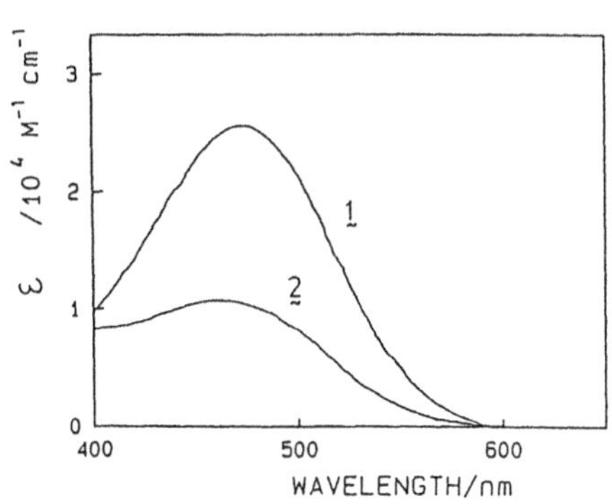

Fig. 3. Absorption spectra of retinal derivatives.

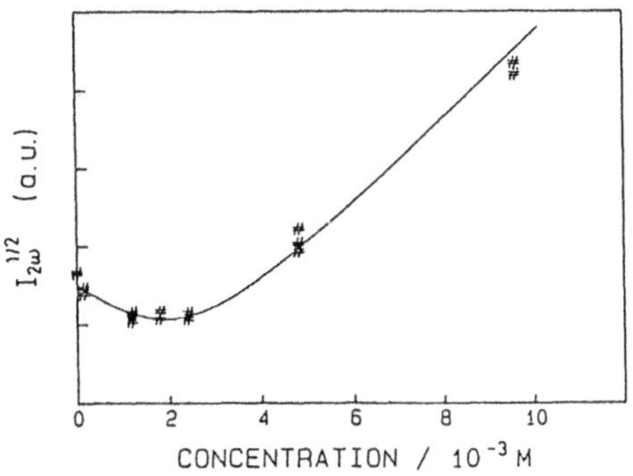

Fig. 4. Square root of the EFISHG intensity in the DMSO solution of the sample 1. Solid line is a fitting curve.

4. Phase Conjugation

A phase conjugate wave was observed in DMSO solution of the sample **2** (~ 5×10^{-3} mol/l) with an experimental set-up for the degenerate four wave mixing (Fig.5) [5]. The second harmonic wave of Nd^{3+}:YAG laser (532nm) was separated into two strong pump beams (I_{pump}~ 100MW/cm^2) and a weak probe beam (I_{probe}~ 3MW/cm^2).The pulse width was about 10ns. The spectral line width of the laser was claimed to be 30 GHz, so the correlation time of the light was about 30ps. The sample was held in a quartz cell with 1mm optical path length. Both of the pump beams were vertically polarized and the polarization plane of the probe beam could be varied with a $\lambda/2$ retarder. To polarize each beam, Glan prisms were used whose extinction ratio were 10^{-5}. A delay line was inserted in the optical path of the forward pump beam. The backward beam proceeded the other two beams by 1ns, so that it had no correlation with the others.

The signal intensities as a function of delay time of the forward pump beams are depicted in Fig.6. When all beams were vertically polarized, a relatively strong signal was observed and its reflectivity was 6×10^{-4}. Because of the presence of absorption at the wavelength, the dominant origin of the signal was considered to be the thermal grating. The width of the peak was 25ps, that corresponds to the correlation time of the laser.

When the probe beam was horizontally polarized and the pump beams vertically, effects from the thermal grating were eliminated, and the signal was originated from electronic processes. In this case, the ratio between the peak and background intensities was about 3, which corresponded to the theoretical prediction [6]. The reflectivity was 5×10^{-6}, that was very small compared with the former case. But even this value was five times as large as that of CS_2 which was measured

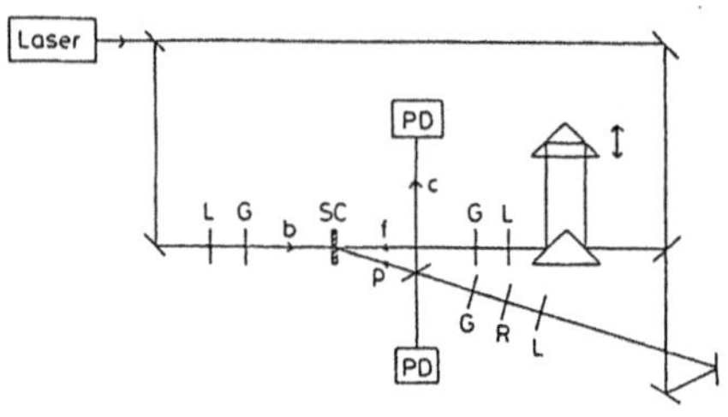

Fig. 5. Optical set-up for the observation of the phase conjugate wave. PD is the photodiode; f, b, p and c are forward, backward, probe and conjugate waves, respectively.

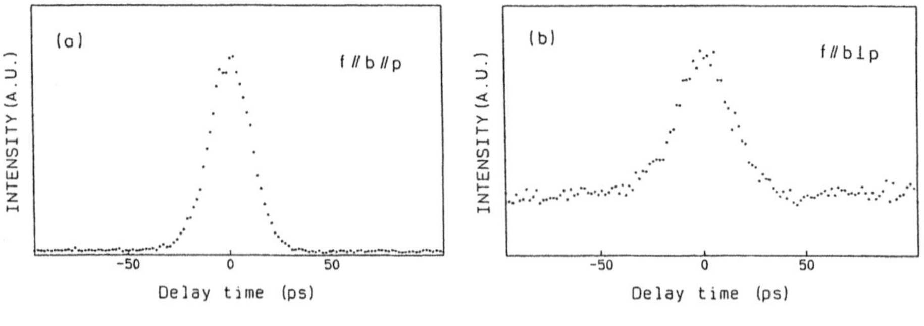

Fig. 6. Signal intensity as a function of delay time, in the case of (a) f // b // p and (b) f // b ⊥ p.

by the same experimental set-up. When the optical absorption is not saturated and the reflectivity is small, the reflectivity can be represented as [7],

$$R \propto |\chi^{(3)}|^2 I_{pump}^2 \exp(-\alpha L) [1 - \exp(-\alpha L/\cos\vartheta)]^2/\alpha^2 . \quad (2)$$

In this formula, α is the absorption constant, L, the cell thickness, and ϑ, the angle between the probe and the forward pump beam. In this case, $\chi^{(3)}$ can be determined as 6.2×10^{-11} (esu) compared to 1.6×10^{-12} of CS_2 as the standard [8]. Using the values of the concentration and the refractive index, the third order nonlinear optical coefficient of this molecule is known to be 5.2×10^{-30} (esu). From this value, a crystal of this sample is predicted to have the value $10^{-9} \sim 10^{-8}$ (esu) as $\chi^{(3)}$ which is comparable to that of poly-diacetylene.

Acknowledgment

The authors wish to thank Professor T.Kobayashi, Univ. of Tokyo for useful discussion.

References

1. B.F.Levine and C.G.Bethea: J. Chem. Phys. 63, 2666 (1975).
2. J.L.Oudar: J. Chem. Phys. 67, 446 (1977).
3. K.D.Singer and A.F.Garito: J. Chem. Phys. 75, 3572 (1981).
4. K.Kajzar, I.Ledoux and J.Zyss: Phys. Rev. A36, 2210 (1987).
5. D.M.Pepper: Opt. Eng. 21,156 (1982).
6. T.Kobayashi: private communication.
7. R.G.Caro and M.C.Gower: IEEE J. Quantum Electron. QE18, 1376 (1982).
8. D.M.Pepper and A.Yariv: In Optical Phase Conjugation, ed. by R.A.Fisher (Academic Press, Orlando, 1983) p.23

Part V

Polymers

Nonlinear Optics in Doped Amorphous Polymers

C. Broussoux, S. Esselin, P. Le Barny, J.P. Pocholle, and P. Robin

Thomson-CSF/LCR, BP 10, F-91401 Orsay, France

INTRODUCTION

For many years, the new optical signal processing systems have needed new concepts, new optical components, new structures and new materials permitting high signal speed or a blue laser beam to be reached. These new components can be made with inorganic crystals such as $LiNbO_3$, KDP, BBO or $LiIO_3$ or semiconductors such as GaAs, AlGaAs, ZnSe or Ag_3AsS_3. The former are now commercialized and $LiNbO_3$ electrooptic modulators can be bought. The latter are still under development in laboratories. The second harmonic properties can be described by a figure of merit and a comparison between inorganics and organic materials made (see Fig. 1). It is obvious that the organic materials are the best in the frequency range VIS-NIR and specially the doped polymers.

It would be an oversimplification to describe the second harmonic properties by the figure of merit. We have also to take into account the processability, the

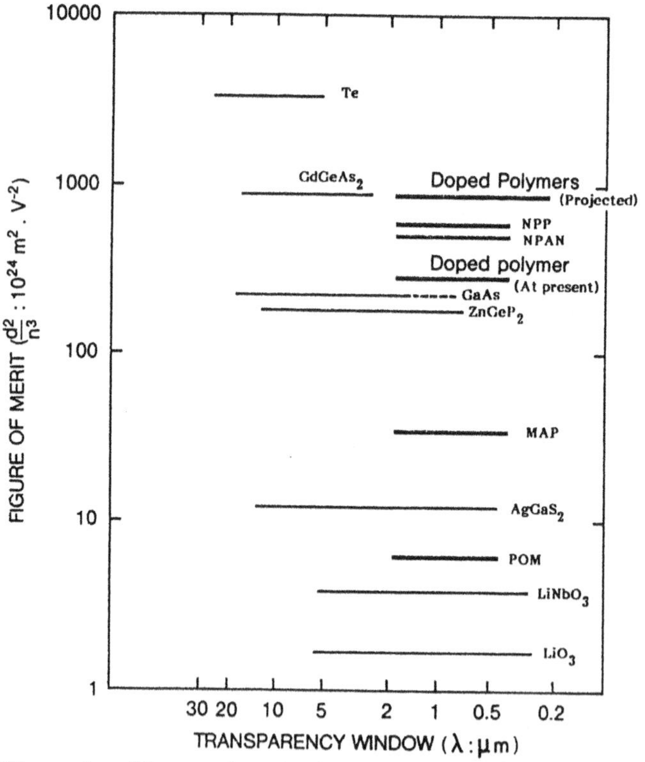

Figure 1: Figure of merit for nonlinear materials versus wavelength λ.

Springer Proceedings in Physics, Vol. 36
Nonlinear Optics of Organics and Semiconductors
Editor: T. Kobayashi

optical damage threshold, the transparency and the ageing. It is well known that organics can resist a high power beam (≥ $GW.cm^{-2}$). This property is important for the second harmonic generation efficiency, which is proportional to the square of the incident power and for the high power beam modulation, specially in integrated optical applications.

We know now that the transparency depends on the choice of molecule and it can be adjusted according to the application. Today the problem is the ageing in storage conditions and in a continuous laser beam. It seems that it is still necessary to study more and more the organic materials'ageing.

Concerning the processability, we have to separate organics into two classes. The first one concerns the crystals: in this case the main difficulty is the crystal's growth, as for the inorganic materials. The second one concerns the doped polymers deposited in thin layers by spin-coating. At present many laboratories are involved in optimizing the processability and the second harmonic properties of doped polymers.

This paper concerns the doped polymers with active molecules in a guest-host system and ferroelectric polymers that we can consider as an amorphous matrix doped with micro-active-crystals. After a general presentation of the second harmonic effect illustrated with examples, we present the electrooptic effect and we describe the principal properties required for an electrooptic modulator.

1. SECOND HARMONIC GENERATION

In the general case of an electric field applied to a molecule, the dipole moment can be expanded in a power series /1/:

$$\vec{p}(t) = \vec{\mu}_0(t) + \overline{\overline{\alpha}}\,\vec{E}(t) + \overline{\overline{\overline{\beta}}}\,\vec{E}(t)\,\vec{E}(t) + \overline{\overline{\overline{\overline{\gamma}}}}\,\vec{E}(t)\,\vec{E}(t)\,\vec{E}(t)\ldots \qquad (1)$$

where $\overline{\overline{\alpha}},\overline{\overline{\overline{\beta}}},\overline{\overline{\overline{\overline{\gamma}}}}$ are the static molecular polarizability tensors.

It is more convenient to write this equation in frequency-dependent molecular polarizabilities. In the case of a monochromatic field, we obtain

$$\vec{p}(\omega) = \overline{\overline{\alpha}}(\omega)\vec{E}(\omega) + \overline{\overline{\overline{\beta}}}(-\omega;\omega_1,\omega_2)\vec{E}(\omega_1)\vec{E}(\omega_2) + \overline{\overline{\overline{\overline{\gamma}}}}(-\omega;\omega_1',\omega_2',\omega_3')\vec{E}(\omega_1')\vec{E}(\omega_2')\vec{E}(\omega_3') + \ldots (2)$$

where $\omega = \omega_1 + \omega_2$; $\omega = \omega_1' + \omega_2' + \omega_3'$.

Immediately, we can see that in a symmetric molecular system [β] vanishes as [β] is an odd tensor.

In this paper, we focus our interest on non-centrosymmetric macroscopic systems containing N polarizable molecules. We can describe such a system taking a thin film of amorphous polymer doped with N molecules. Just after the spin-coating deposition, the film is centrosymmetric and all odd-order tensors are equal to zero. In order to obtain a non-centrosymmetric system, an electric field must be applied to orient the molecules (Fig. 2).

In such a system, the macroscopic polarization $\vec{P}(\omega)$ can be written

$$\vec{P}(\omega) = N\,\vec{p}(\omega) = \vec{P}_L(\omega) + \vec{P}_{NL}(\omega) = N\,\overline{\overline{\alpha}}\,\vec{E}(\omega) + N\,\overline{\overline{\overline{\beta}}}\,\vec{E}_L(\omega_1)\vec{E}_L(\omega_2) \qquad (3)$$

where the index L means "local" defined by the Onsager relations.

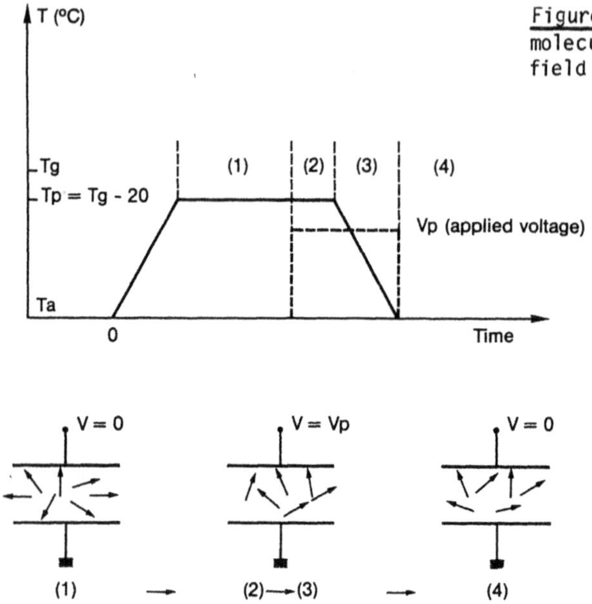

Figure 2: Schematic representation of molecular orientation under an electric field

Assuming that the tensors [α], [β] and [γ] commute, one obtains a macroscopic second-order nonlinear susceptibility function of [β]:

$$\overline{\overline{\overline{\chi}}}^{(2)}(-\omega;\omega_1,\omega_2) = f(-\omega)\ f(\omega_1)\ f(\omega_2)\ N\ \overline{\overline{\overline{\beta}}}(-\omega;\omega_1,\omega_2) \tag{4}$$

where $f(\omega_n)$'s are the local Onsager relations. Then each tensor's elements can be expressed in terms of molecular second-order susceptibility β_{npq} /2,3/:

$$\chi^{(2)}_{ijk}(-\omega;\omega_1,\omega_2) = N\ f_i(-\omega) f_j(\omega_1) f_k(\omega_2)\ \langle\beta_{npq}(-\omega;\omega_1,\omega_2)\rangle_{ijk} \tag{5}$$

where $\langle\ \ \rangle$ means the average value calculated in our case in the presence of a poling field E_p and

$$\langle\beta_{npq}(-\omega;\omega_1,\omega_2)\rangle_{ijk} = \int d\Omega\ F(\Omega.E_p)\beta_{npq}(-\omega;\omega_1,\omega_2) a_{ni} a_{pj} a_{qk} \quad . \tag{6}$$

In the relation (6), $\overleftrightarrow{a}$ is the transformation tensor matrix and $F(\Omega,Ep)$ a distribution function with solid angle Ω and poling field E_p.

If, now, we consider a non-centrosymmetric point group ∞ mm, the three vector components of the macroscopic nonlinear polarization P_{NL} can be written using the tensor relation $2d_{ijk} = \chi^{(2)}_{ijk}$

$$P^{NL}_i(-\omega;\omega_1,\omega_2) = d_{ijk}(-\omega;\omega_1,\omega_2) E_j(\omega_1) E_k(\omega_2) \quad . \tag{7}$$

Contracting the tensor's indices /4/, the tensor [d] is

$$[d] = \begin{pmatrix} 0 & 0 & 0 & 0 & d_{15} & 0 \\ 0 & 0 & 0 & d_{24} & 0 & 0 \\ d_{31} & d_{32} & d_{33} & 0 & 0 & 0 \end{pmatrix}$$

According to Kleinman symmetry /5/, we obtain

d_{33} ; $d_{31} = d_{32}$; $d_{24} = d_{15}$, and taking $\omega = 2\omega$ and $\omega_1 = \omega_2 = \omega$ we get

$$\begin{cases} P_1^{NL}(2\omega) = 2\, d_{15}(-2\omega;\omega,\omega)\; E_1(\omega)\; E_2(\omega) \\ P_2^{NL}(2\omega) = 2\, d_{24}(-2\omega;\omega,\omega)\; E_3(\omega)\; E_2(\omega) \\ P_3^{NL}(2\omega) = d_{31}(-2\omega;\omega,\omega)\left(E_1^2(\omega) + E_2^2(\omega)\right) + d_{33}(-2\omega;\omega,\omega)\; E_3^2(\omega)\ . \end{cases} \tag{8}$$

If, now, we consider a material composed of N active molecules per unit volume dispersed in an isotropic matrix and oriented parallel to axis 3 with an electric field, we can assume that $\chi_{33}^{(2)}$ can be written /6/

$$\chi_{33}^{(2)}(-2\omega;\omega,\omega) = N\ f(2\omega)\ f^2(\omega)\ \beta_{zz}(-2\omega;\omega,\omega)\ \langle\cos^3\Theta\rangle_{3z} \tag{9}$$

where $\langle\cos^3\Theta\rangle = \dfrac{\int_0^\pi 2\pi\,\sin\Theta\,\cos^3\Theta\,F(\Theta,Ep)d\Theta}{\int_0^\pi 2\pi\,\sin\,\Theta\,F(\Theta,Ep)d\Theta}$.

For N dipolar molecules with a static dipolar moment μ_z carrying a β_{zz} component (z = molecular axis), we can describe the statistical distribution of the molecular axis in interaction with an applied electric field E_p by a Boltzmann law, so we obtain

$$\chi_{33}^{(2)}(-2\omega;\omega,\omega) = N\ f(2\omega)f^2(\omega)\ \beta_{zz}(-2\omega;\omega,\omega)\ \frac{\mu_z Ep}{5kT}\ .\ \varepsilon\,\frac{(n^2+2)}{n^2+2\varepsilon} \tag{10}$$

where $\langle\ \cos^3\Theta\ \rangle_{3z} = \dfrac{\mu_z Ep}{5kT}\ .\ \varepsilon\,\dfrac{(n^2+2)}{n^2+2\varepsilon}$

and the Onsager relations are

$$f(\omega) \sim f(2\omega) = \frac{n^2+2}{3}\ ;\ f(0) = \varepsilon\,\frac{(n^2+2)}{n^2+2\varepsilon}\ .$$

The second-order coefficient $d_{33}(-2\omega;\omega,\omega)$ can be measured on poled systems under a laser Nd^{3+}YAG beam at 1.06 µm or 1.32 µm according to the absorption resonant band. In transmission mode, the second harmonic generation is a function of the angle between the incident fundamental beam and the film normal. In general the second harmonic intensity is given by /7,8/

$$I_{2\omega} = (512\ \pi^3)t_\omega^4\ T_{2\omega}\ d^2(\Theta)I_\omega^2\ \frac{1}{(n_\omega^2 - n_{2\omega}^2)^2}\ \sin^2\psi(\Theta) \tag{11}$$

where I_ω is the incident density intensity, t_ω and $T_{2\omega}$ Fresnel-like transmission factors, n_ω and $n_{2\omega}$ the refractive indices at ω and 2ω and $\psi(\Theta)$ the angular dependence of the second harmonic intensity coming from interference between the free and bound waves. In the case where the thickness ℓ is small compared with the coherence length $\ell_c = \lambda/4(n_\omega - n_{2\omega})$, we can expand $\sin^2\psi(\Theta)$ for small $\psi(\Theta)$ and we get

$$I_{2\omega} = (512\ \pi^3)\ t_\omega^4\ T_{2\omega}\ d^2(\Theta)\ I_\omega^2 \frac{1}{(n_\omega^2 - n_{2\omega}^2)^2} \times \left(\frac{\pi}{2}\frac{\ell}{\ell c}\frac{\bar{n}}{(N^2 - \sin^2\Theta)^{\frac{1}{2}}}\right)^2 \tag{12}$$

where $\bar{n} = (n_\omega + n_{2\omega})/2$ and $N^2 = \bar{n}^2 + [(n_\omega - n_{2\omega})/2]^2$.

The expression of $d^2(\Theta)$ depends on the incident laser light polarization p or s. In our case, according to Kleinman symmetry /5/ on films deposited by spin coating and electrically oriented, we find $d_{31} = d_{32}$; d_{33} ; $d_{24} = d_{15}$, and we get

$$d_p(\Theta) = (d_{31} \cos^2 \Theta_\omega + d_{33} \sin^2 \Theta_\omega)\sin \Theta_{2\omega} + 2d_{31} \cos \Theta_\omega \sin \Theta_\omega \cos \Theta_{2\omega}\ , \tag{13}$$

$$d_s(\Theta) = d_{31} \sin \Theta_{2\omega}\ . \tag{14}$$

Taking $d_{33} = 3d_{31}$, the value of d_{33} was determined at 1.32 µm by comparison with d_{11} of quartz and at 1.06 µm with comparison to d_{31} of $LiNbO_3$ taking into account the absorption of $I_{2\omega}$ in the layers. The samples are described in Table 1.

Table 1: Description of experimental samples (* and **: see Figure 3).

Matrix	Dopant	$N(m^{-3})$	E_p(MV/cm)	T_p (°C)	λ_{max}(nm)
PMMA (Polymethylmethacrylate)	Sc 113172*	0.14 to 1.29 10^{26}	up to 1	80 - 100	470
75/25 PVDF/TrFE (polyvinylidene fluoride/ trifluorethylene)	None	3.7×10^{28}	up to 1	25	< 380
Copolymer** PLBP47	Sc 113172	3.3×10^{26}	up to 0.88	85	470

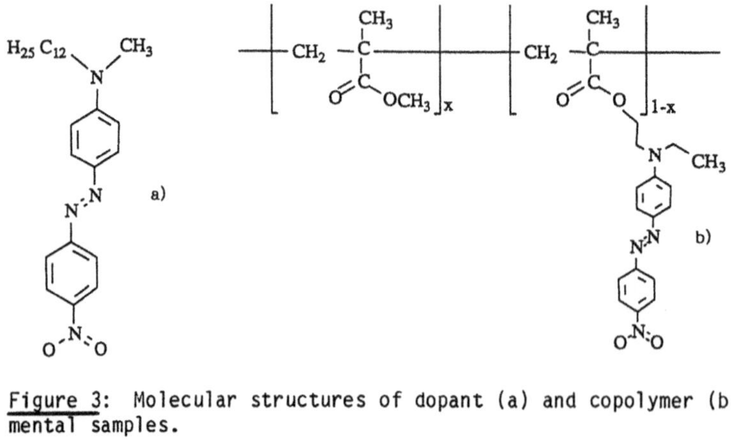

Figure 3: Molecular structures of dopant (a) and copolymer (b) used in the experimental samples.

Curve 1 in Fig. 4 corresponds to a guest-host amorphous system poled at 0.88 $MV.cm^{-1}$ at 80 °C and curve 2 corresponds to a ferroelectric polymer poled at about the same field. The peak width is due to the molecular distribution: large for guest-host system according to the $\langle\cos\Theta\rangle$ value equal to 0.24 and narrow for ferroelectric system in which $\langle\cos\Theta\rangle$ is equal to 0.86.

In Figs. 5 and 6, d_{33} is reported versus the applied voltage and the dopant density. In Fig. 6, the projected d_{33} value has been computed in the case of a large β_{zz} value (10^{-27} esu) and a large $\langle\cos^3\Theta\rangle$ value two times larger than the actual value (0.14 x 2). For about 5×10^{26} m^{-3} active molecules, we get a d_{33} = 120 to 130 $pm.V^{-1}$ comparing to d_{31} = 6.2 $pm.V^{-1}$ for $LiNbO_3$ or d_{14} = 9.7 $pm.V^{-1}$ for POM.

In the case of ferroelectric polymers, an example of typical ferroelectric behaviour is the hysteresis curve obtained between the electrical polarization P and the applied field E_p. In accordance with the theory giving the susceptibility X as a function of P, the $I^{2\omega}$ signal shows in Fig. 7 a butterfly shape with a maximum value at the maximal polarization P_s and a zero value at the zero polarization. On this kind of polymer, we have obtained a value of d_{33} equal to 1 ± 0.5 $pm.V^{-1}$, larger than the value measured by Hill et al. /9/. Boyd /10/ has determined the β value of the CF_2-CH_2-dipolar group equal to 0.9×10^{-31} esu.

The values reported in Table 2 have permitted us to calculate the d_{33} values according to the relation (10) and to compare them with the amorphous doped polymer's values.

Finally, we report in Fig. 8 the temperature behaviour of the $I^{2\omega}$ signal for the two types of polymers. The $I^{2\omega}$ signal decreases up to the Curie temperature for the ferroelectric polymer and up to the glass temperature for the amorphous polymer. One can notice that the $I^{2\omega}$ signal is reversible for the ferroelectric polymer (up to 80 °C) which is not the case for the amorphous one.

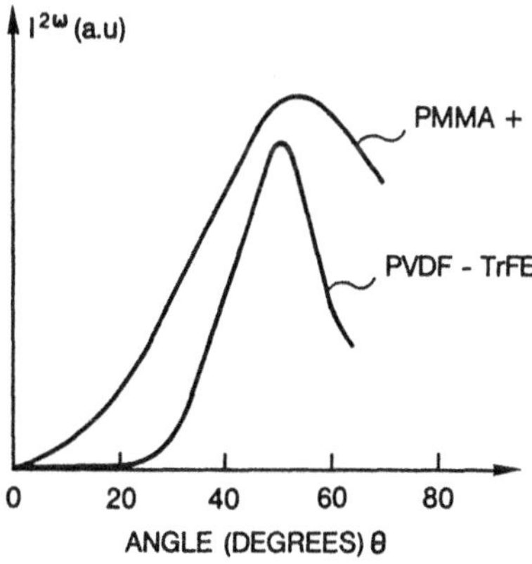

Figure 4: Second harmonic signal versus the Θ incident angle for a doped amorphous polymer PMMA+dye and a ferroelectric polymer PVDF-TrFE.

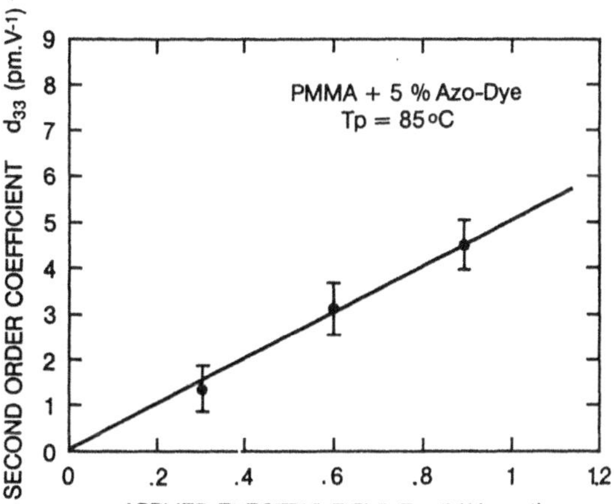

Figure 5: Second-order coefficient d_{33} versus electric field applied at 85 °C to a PMMA doped with 5% azo-dye.

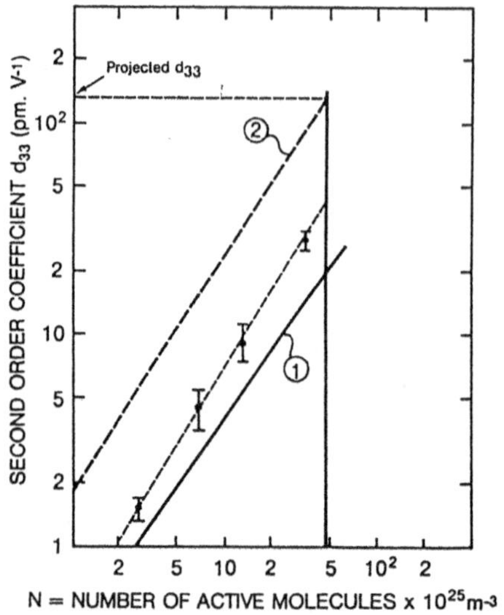

Figure 6: Second-order coefficient d_{33} versus number of active molecules.
Curve (1) has been calculated with present β_{zz} and $\langle\cos\Theta\rangle$ values. Curve (2) has been calculated taking the projected β_{zz} and $\langle\cos\Theta\rangle$ values.

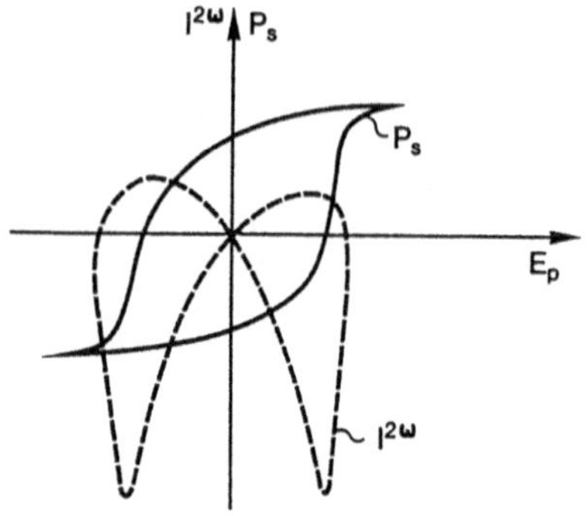

Figure 7: Hysteresis curve P_s versus E_p obtained on a ferroelectric polymer PVDF-TrFE with its second harmonic signal simultaneously recorded.

Table 2: Nonlinear optical properties of an amorphous doped polymer compared to a ferroelectric undoped polymer.

	Amorphous doped polymer	Ferroelectric undoped polymer
N (m^{-3})	0.68×10^{26}	3.7×10^{28}
$f(2\omega) \sim f(\omega);f(0)$	1.4;1.5	1.3;1.8
β_{zz} (esu)	308×10^{-30}	0.9×10^{-31}
μ_z (D)	8.4	2.1
$\langle\cos\Theta\rangle$ meas.	0.24	0.86
d_{33}^{cal} (pm/V)	2.6	1.5
d_{33}^{mea} (pm/V)	4.5 ± 2	1 ± 0.5
Age (months)	< 1	> 10^2
Stability (Tmax in °C)	60	100

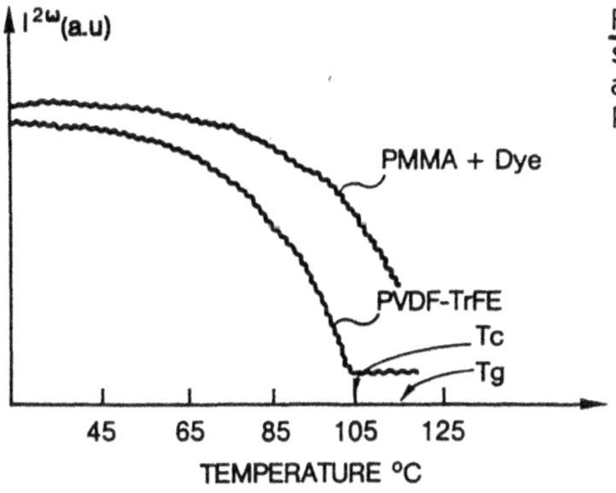

Figure 8: Second harmonic signal versus temperature for a doped PMMA and an undoped PVDF-TrFE.

2. ELECTROOPTIC MODULATION

The electrooptic effect was the earliest nonlinear phenomenon discovered. It is defined through the change in the electric impermeability induced by an applied electric field /11/:

$$B_{ij}(\vec{E}) - B_{ij}(\vec{0}) = r_{ijk}\, E_k + s_{ijk\ell}\, E_k\, E_\ell + \dots \tag{15}$$

where B_{ij} is the impermeability tensor, E_k the applied electric field and r_{ijk} and $s_{ijk\ell}$ respectively the linear electrooptic coefficient and the quadratic coefficient.
Non-centrosymmetric materials with high $\chi^{(2)}_{ijk}$ values show also large electrooptic properties. In the general case, under an external stress (electrical, mechanical...) the variation of $(1/n^2)_{ij}$ describing the optical ellipsoid is described by the following two equations:

$$\Delta B_{ij} = \Delta \left(\frac{1}{n^2} \right)_{ij} = r^x_{ijk}\, E_k + p^E_{ijk\ell}\, x_{k\ell} \tag{16}$$

$$= r^X_{ijk}\, E_k + \pi^E_{ijk\ell}\, X_{k\ell}$$

with r^x_{ijk} the electrooptic coefficient at constant deformation x_{ij}

r^X_{ijk} the electrooptic coefficient at constant stress X_{ij}

$p^E_{ijk\ell}$ the photoelastic coefficient at constant electric field E_k.

$\pi^E_{ijk\ell}$ the piezooptic coefficient at constant electric field E_k.

Combining these two expressions, we obtain

$$\Delta \left(\frac{1}{n^2} \right)_{ij} = r^X_{ijk}\, E_k + p^E_{ijk\ell}\, s^E_{k\ell pq}\, X_{pq} \tag{17}$$

where $r^X_{ijk} = r^x_{ijk} + p_{ijk\ell}\, d_{pq\ell}$

with $d_{pk\ell}$ the piezoelectric coefficient.

The first term r^x_{ijk} corresponds to the primary electrooptic effect due to electronic, acoustic and optical phonons' contribution and the second term $p_{ijk\ell}\, d_{pk\ell}$ corresponds to secondary effect due to the piezoelectricity.

In the case of an amorphous material in the glassy state, we can assume that [d] is small, giving $r^x_{ijk} = r^X_{ijk}$. This is not the case in a ferroelectric polymer, where some of the piezoelectric tensor components are large, i.e. more than 30 pC/N for d_{33}. So, the secondary effect will only depend on the $p_{ijk\ell}$ value. Broussoux and Micheron /12/ have shown that in certain configurations, the secondary effect measured on a ferroelectric polymer could be greater than the primary effect.

As the tensor [d] is a third-order tensor, we can write in a ∞ mm point group symmetry [r] as

$$[r] = \begin{pmatrix} 0 & 0 & r_{13} \\ 0 & 0 & r_{23} \\ 0 & 0 & r_{33} \\ 0 & r_{42} & 0 \\ r_{51} & 0 & 0 \\ 0 & 0 & 0 \end{pmatrix} .$$

If, now, we consider a planar slab waveguide as represented in Fig. 9 with a confining medium of thickness e and length ℓ, we can calculate the electrooptic coefficient with respect to the propagating mode of the light.

In this structure, the confining medium is deposited between two buffer layers of thickness e_B themselves covered by two conductive electrodes. The two directions 1 and 2 are equivalent and the indices n_1 and n_2 are equal to n_o, the ordinary index. $n_3 = n_e$ is the extraordinary index.

For a TM propagating mode, we can write

$$2\, \xi_1^2 \left(\frac{1}{n_o^2} + r_{13} E_3 \right) + \xi_3^2 \left(\frac{1}{n_e^2} + r_{33} E_3 \right) = 1 \quad . \tag{18}$$

The dephasing between E_3 and E_1, induced by the electric field component of the beam, is

$$d\phi = \frac{\pi}{\lambda} n_e^3 \left(r_{33} - \frac{n_o^3}{n_e^3} r_{13} \right) E_3 \quad . \tag{19}$$

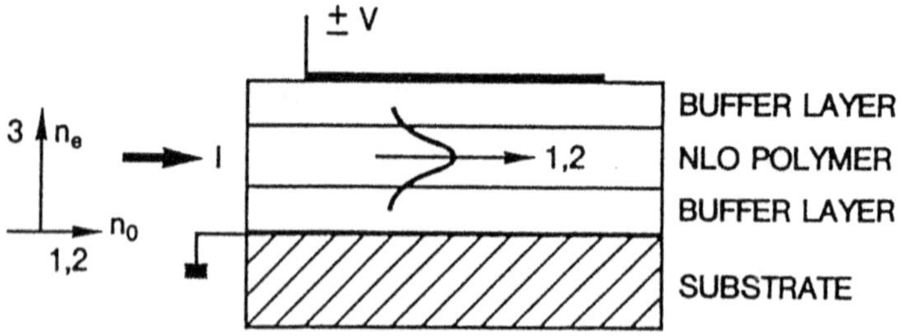

Figure 9: Planar strip waveguide structure composed of two buffer layers sandwiching a nonlinear optical layer.

From this expression, we can deduce the voltage necessary for total extinction

$$V_\pi = \frac{\lambda}{n_e^3 \left(r_{33} - \frac{n_o^3}{n_e^3} r_{13}\right)} \cdot \frac{e}{\ell} \quad . \tag{20}$$

If we assume that $(n_o^3/n_e^3)\, r_{13} \ll r_{33}$, we obtain

$$V_\pi \sim \frac{\lambda}{n_e^3 r_{33}} \cdot \frac{e}{\ell} \quad . \tag{21}$$

With this relation, we can see that it is necessary to minimize the ratio e/ℓ, i.e. to have $\ell \gg e$. This is the case in a wave-guided structure where e/ℓ can be equal to 10^{-4}.

If we now consider the relation existing between $r_{ijk}(-\omega;\omega,0)$ and $d_{ijk}(-2\omega;\omega,\omega)$ in the case of an electronic contribution and $n^{2\omega} \sim n^{\omega}$, we obtain far from the resonance ω_0 /1,6/:

$$r^{X}_{33}(-\omega;\omega,0) = -\frac{12 d_{33}(-2\omega;\omega,\omega)}{n_e^4} \cdot \frac{\varepsilon_{33}}{n_e^2 + 2\varepsilon_{33}} \quad . \tag{22}$$

For experimental values $d_{33} = 28 \times 10^{-12}$ m/V, $\varepsilon_{33} = 3$ and $n_e = 1.53$ obtained on a guest-host system with 30% dopant, we get $r^{X}_{33} = 22 \times 10^{-12}$ m/V in agreement with the first electrooptic measurements at 0.63 µm performed on this layer by the plasmon method /13/ giving a <r> between 16 and 20×10^{-12} m/V. For the corresponding V_π at this value, for λ = 0.96 µm on a waveguide structure with $e/\ell = 5 \times 10^{-4}$, we obtain V_π = 8.35 volts. This value is comparable with those obtained with a $LiNbO_3$ modulator. In the case of a ferroelectric polymer such as polyvinylidene fluoride (PVDF) or its copolymers (PVDF-TrFE), the electrooptic effect is the sum of two terms whose frequency behaviour is different. The first one, corresponding to the electronic contribution, exists at all frequencies far from the resonance. The second one, corresponding to the piezoelectric contribution, is limited at low frequencies except for the resonant structures. The electrooptic measurements on PVDF-TrFE have been made with the surface-plasmon method on a guided mode /13/. This new method gives us a value of the average electrooptic coefficient <r> equal to 16×10^{-12} m/V. The determination of the electronic part of this coefficient gives a value between 1 and 3×10^{-12} m/V in accordance with the calculated d_{33}. We are still working in this field using other plasmon configurations.

3. ELECTROOPTIC MODULATOR

Many structures /14,15/ can be used to modulate the beam light, such as the ordinary strip structure represented in Fig. 9, or a structure using a Mach-Zehnder interferometer structure proposed by De Martino et al. /15/ and represented in Fig. 10.

In general an electrooptic modulator defined as a waveguide structure can be characterized by four principal criteria: modulating voltage, bandwidth, power requirements and propagating losses.

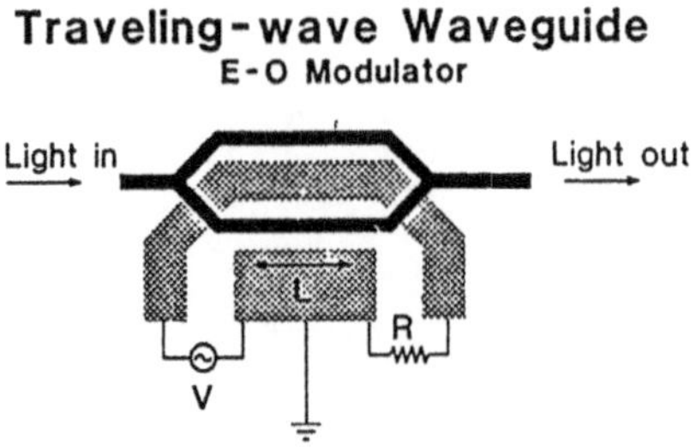

Figure 10: Device proposed by R. De Martino et al. /15/.

3.1 Modulating Voltage

To calculate the modulating voltage, we have to define the waveguide parameters such as the thickness of the guide in order to get a monomode propagation and the length and the width in order to get a great modulation efficiency.

Taking the V_π expression, we introduce a confining factor η such that

$$V_\pi = \frac{\lambda}{n_e^3\, r_{33}} \cdot \frac{e}{\ell} \cdot \frac{1}{\eta} \quad .$$

At λ = 1.32 µm with r_{33} = 100 pm/V corresponding to a d_{33} = 130 pm/V, e = 5 µm, ℓ = 1 cm and η = 0.8, we obtain V_π = 2.4 volts. Another parameter can define the modulation efficiency:

$$\phi = \frac{180°}{V_\pi \times \ell} = 7.5°/(V/mm).$$

3.2 Bandwidth

One advantage of an organic guide is that we can optimize the mismatch between the microwave pulse velocity and the light velocity respectively defined by $c/\sqrt{\varepsilon_e}$ and c/n_e where ε_e and n_e are the effective dielectric constant and the effective optical index due to the surroundings.

Without a buffer layer, we get

$$n_e = \sqrt{\frac{1+\varepsilon_r}{2}} = 1.414$$

$$n_o = 1.51$$

and the bandwidth at 3 dB (for ℓ = 1 cm) is

$$B_{-3dB} = \frac{0.443\ c}{\ell . |n_e - n_o|} = 138\ \text{GHz} \quad .$$

3.3 Power Requirements

One problem is the power required to get a π phase shift: with a modulator using an organic layer, we can expect to obtain a low power. For a modulation length of 2 cm and V_π = 1.2 volt with electrodes calculated to obtain 50 Ω we obtain a power

P = 14 mW (11 dBm). This value must be compared with those of $LiNbO_3$ equal to 300 mW (+ 25 dBm). It is obvious that the organic modulation is more interesting.

3.4 Optical Losses

The optical loss is an important parameter for low signals. In this field $LiNbO_3$ has very low absorption losses ($\leq$ 0.1 dB/cm). For organic materials, we can easily expect to obtain losses $\leq$ 0.5 dB/cm /14,15/.

In Table 3, we report different parameters for an organic modulator and a $LiNbO_3$ modulator.

Table 3: Electrooptic properties of a conventional $LiNbO_3$ modulator compared to the projected polymer modulator.

	$LiNbO_3$ modulator	Organic modulator
r_{33} (pm V^{-1})	38	100
Optical losses (dB/cm)	$\leq$ 0.1	$\leq$ 0.5
Length (mm)	10	20
Maximum frequency (GHz)	6 to 10	70
Power (mW) at 10^6 Hz	300 (25 dBm)	11 (11 dBm)
Factor of merit (pm/V) $\frac{n^3 r}{\varepsilon_r}$	13	110
ε_{33}	30	3
n_{33}	2.2	1.5
Modulation efficiency °/(V.mm)	2-3	7.5

4. STABILITY AND AGEING

In Fig. 11, we report two curves corresponding to the d value reported versus the time in days.

The decay curves suggest two relaxation processes as already observed by Ye et al. /16/. It is suggested that the short-term process may involve rapid dopant reorientation into a box or local free volume created by lateral groups of PMMA. The long-term process may involve a low box-walls rearrangment due to the existence of a secondary β transition, near room temperature. The lower decrease observed on grafting dye on PMMA suggests that a decrease of the correlation length between two consecutive lateral PMMA groups is diminishing the β transition effect.

As the ferroelectric polymers are very stable (more > 10^2 months), we have to fix the active molecules mechanically, electrically or chemically as previously mentioned by Ye et al. /16/.

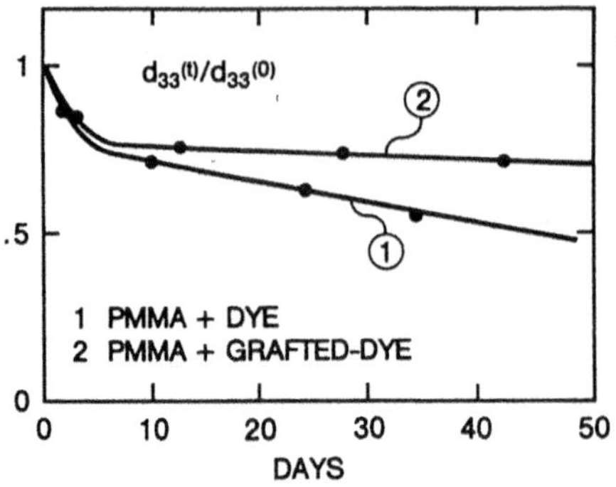

Figure 11: Temporal behaviour of the second harmonic coefficient at room temperature for a PMMA doped with dye (1) and a dye grafted on PMMA (2).

5 CONCLUSION

We have shown that amorphous doped polymers and ferroelectric undoped polymers are good materials for integrated optical devices. Amorphous polymers show high second harmonic efficiency and good electrooptical properties compared to ferroelectrics which have good temporal stability. Further work is needed to understand relaxation processes in order to study the means of freezing the molecular movments. We have shown that the grafting of dyes on the main polymers chain stabilizes the second harmonic signal.

For second harmonic generation, new structures will have to be studied permitting a good phase matching in order to reach a great efficiency in the blue.

For electrooptic modulation, the expected values seem to be very interesting, especially the low required power, large bandwidth and simple process, in spite of the optical absorption.

Although the organic materials could be used in future optical components, we have to consider new inorganic crystals and new II-VI compounds which will be serious rivals.

ACKNOWLEDGEMENTS

This work was supported by the EEC under ESPRIT 443.

REFERENCES

1. D.S. Chemla and J. Zyss (Eds.): Nonlinear Optical Properties of Organic Molecules and Crystals (Academic Press, New York, 1986)
2. J. Zyss and J.L. Oudar: Phys. Rev. A26, 2028 (1982).
3. S. Kielich: IEEE J. Quant. Electron. QE-5, 562 (1969)
4. J.F. Nye: Physical Properties of Crystals (Clarendon Press, London, 1957)
5. D.A. Kleinman: Phys. Rev., 126, 1977 (1962).
6. D. Singer, M.G. Kuzyk and J.E. Sohn: J. Opt. Soc. Am. B4, 968 (1987)
7. J. Jerphagnon and S.K. Kurtz: J. Appl. Phys. 41, 1667 (1970)
8. P.D. Maker, R.W. Terhune, N. Nisenoff and C.M. Savage, Phys. Rev. Lett. 8, 21 (1962)
9. J.R. Hill, P.L. Dunn, G.J. Davies, S.N. Olivier, P. Pantelis and J.D. Rush: Elect. Lett. 13, 700 (1987).
10. G.T. Boyd: Thin Solid Films, 152, 295 (1987)

11. I.P. Kaminow: An Introduction to Electrooptic Devices (Academic Press, New York, 1974)
12. D. Broussoux and F. Micheron: J. Appl. Phys. 51, 2020 (1980)
13. Y. Levy, M. Dumont and P. Robin: to be published.
14. J.I. Thackara, G.F. Lipscomb, R.S. Lytel and A.J. Ticknor: Nonlinear Optical Properties of Polymers (MRS Symposium Proceedings, Boston, 1987)
15. R. De Martino, D. Haas, G. Khanarian, T. Leslie, H.T. Man, J. Riggs, M. Sansone, J. Stamatoff, C. Teng and H. Yoon: Nonlinear Optical Properties of Polymers (MRS Symposium Proceedings, Boston, 1987)
16. C. Ye, N. Minami, T.J. Marks, J. Yang and G.K. Wong: to be published in NATO Symposium Proceedings Nice (1988).

Electro-Optic Constant of Doped Polymers by a Novel Method

T. Kobayashi, H. Uchiki, and K. Minoshima*

Department of Physics, Faculty of Science, University of Tokyo, 7-3-1 Hongo, Bunkyo-ku, Tokyo 113, Japan

1. Introduction

Several methods have been reported for the determination of electro-optic (EO) constants of materials [1]. They can be classified into the following three types. The first one is an ellipsometry to detect field-induced birefringence. The configuration is schematically shown in Fig. 1(a). A linearly polarized light which propagates through the sample is either perpendicular or parallel to the electric field (transverse or longitudinal scheme). Then the transmitted light intensity after the polarizer crossed with the first one (I_{out}) is detected. If only the field-induced change in the real part of the refractive index (birefringence) is taken into account, I_{out} is given as

$$I_{out} = I_{in}\sin^2[\omega/c(\Delta n_x(E)-\Delta n_y(E))L]. \tag{1}$$

Here Δn_x and Δn_y are the refractive index changes of light polarized along x and y, respectively, and L is the propagation length of light in the sample along the z axis.

The second method is for the determination of the change in the imaginary part of the refractive index. Figure 1(b) describes schematically the experimental equipment. A light beam linearly polarized along x passes through the sample, and the field-induced absorbance change is measured. In this case, the light intensity passing through the sample is

$$I_{out} = I_{in}\exp(-\alpha_x(E)L), \tag{2}$$

where $\alpha_x(E)$ and I_{in} are the field-dependent absorption coefficient of light polarized along x and incident light intensity, respectively. By this method, transverse dichroism can be measured using thick metal electrodes. Longitudinal dichroism can also be measured using transparent electrodes.

The transmittance change in the wavelength region near the absorption band is dependent on the change in both real and imaginary parts of the refractive index. By either the first or the second method, however, we cannot separate them. Furthermore, in the first method, only the difference between the two different tensor components can be obtained.

In addition, a high voltage supply of a few thousands volts is needed, and a pair of polarizers with high extinction ratio, which limits the precision in detection, must be used in the typical configuration for the measurement of the electric field induced birefringence. Especially if the sample has spatial dependence of birefringence over the sample area, it is difficult to compensate the anisotropy even by using a Babinet compensator.

*Present address: Department of Electrical Engineering, The Technological University of Nagaoka, Kamitomioka 1603-1, Nagaoka 940-21, Japan.

Springer Proceedings in Physics, Vol. 36
Nonlinear Optics of Organics and Semiconductors
Editor: T. Kobayashi © Springer-Verlag Berlin, Heidelberg 1989

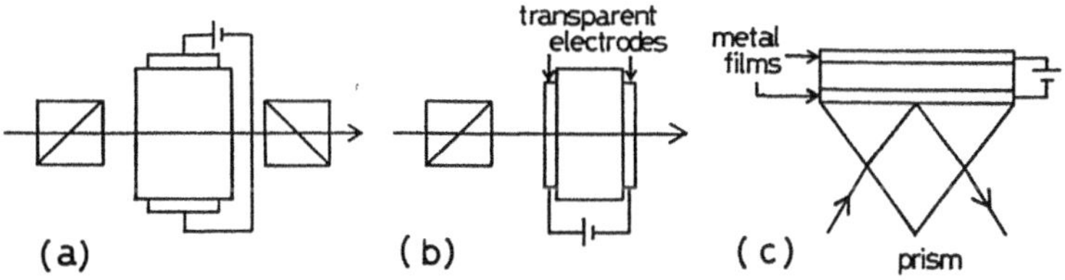

Fig. 1. Schematic of several ordinary methods for the measurement of EO constant: (a) transverse birefringence, (b) longitudinal dichroism, and (c) attenuated total reflection

The third method utilizes high sensitivity to the optical properties of surface plasmon waves or guided waves. To generate and study these waves, the attenuated total reflection (ATR) method has been used, as described by DUMONT and LEVY in this proceeding. A schematic of the sample configuration is shown in Fig. 1(c). Though this method is highly sensitive, it takes a long time to measure the wavelength dependence of the EO constant, since the angular dependence measurement of the reflected light intensity must be done for each wavelength to obtain the change in the refractive index.

Our proposed method [2] overcomes the difficulties associated with these ordinary methods. This method is characterized by the sample structure, which forms a Fabry-Perot (FP) cavity as shown in Fig. 2. The change in transmittance of monochromatic light was measured in the longitudinal configuration. The change in the imaginary part of the refractive index can be detected by the change in transmittance, as in the ordinary method of longitudinal dichroism. Here the pair of thin aluminum films were used not only as electrodes for voltage application, but also as mirrors of the FP cavity. Because of the high cavity resonance, the peaks in the transmittance spectrum become very sharp. Thus the change in the real part of the refractive index can be seen with high sensitivity as the shift of the cavity resonance. Hence both real and imaginary parts can be separately determined.

In this method a laser is not necessary, thus the wavelength dependence of the EO constant can be easily obtained. In addition, a low applied voltage is sufficient for thin film samples. If the sample has low loss due to either absorption or scattering, the measurement in the transverse configuration can be done using guide modes. Otherwise, if we use an inclined incidence in the longitudinal configuration, the other tensor components corresponding to the case where the applied field and the light polarization are parallel can also be measured. Furthermore, each tensor component of the corresponding EO constant can be obtained by using polarized probe light.

In this paper we will report the measurement of EO constants of thin polycarbonate (PC, Fig. 3(a)) films doped with 4-diethylamino-4'-nitrostilbene (DEANS, Fig. 3(b)). This molecule is expected to have a large optical nonlinearity due to a giant permanent dipole moment in the lowest excited singlet state [4]. To

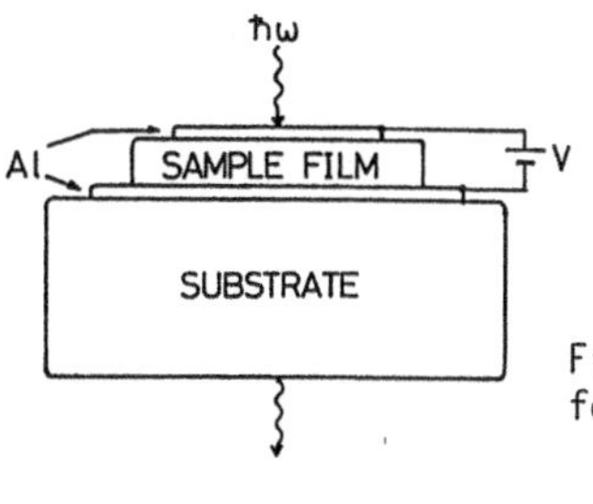

Fig. 2. Sample structure in FP configuration for the measurement of EO effect by our method

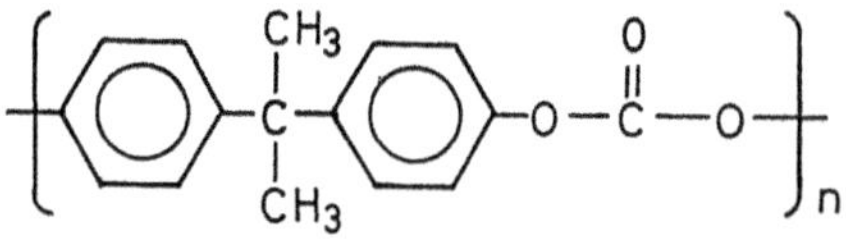

(a)

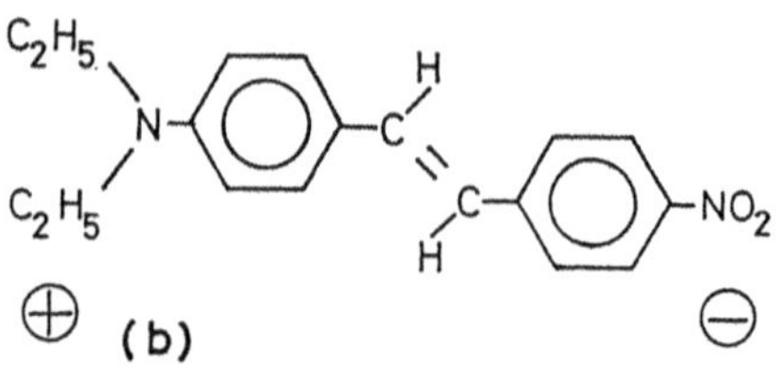

(b)

Fig. 3. Molecular structures of (a) polycarbonate (PC) and (b) 4-diethylamino-4'-nitrostilbene (DEANS). DEANS has both donor and acceptor groups in the same molecule, thus it has a giant permanent dipole moment

clarify the mechanism of the EO effect, we did experiments with several samples of different DEANS concentrations.

2. Experimental

The samples prepared for the measurement of the EO effect by our method were made by the following process. In the first step, an aluminum film was formed on a glass plate by evaporation in vacuum below 10^{-5} Torr. The thickness of this aluminum film was about 100 angstroms. In the second step, a mixture of DEANS and PC was dissolved in dioxane, and the solution was spin-coated on the aluminum film. A PC film containing DEANS molecules was formed by evaporation of dioxane. Then in the third step, another thin aluminum film was evaporated on the DEANS/PC film, so as not to make contact with the first aluminum film. These Al films were used as cavity mirrors and electrodes. For the measurement of the EO constants, an oscillatory electric field of rectangular or sinusoidal wave form was applied to the Al electrodes through copper leads. The transmittance change of the FP cavity modulated synchronously with the variation of the applied electric field was detected.

The variation of the transmittance was measured by the usual method. The light from a 100 W tungsten lamp was monochromated, and was used as a probe, whose spectral width (FWHM) was 2 nm. It was normally incident on the sample. The transmitted light was detected by a photodiode and its output was fed into a lock-in amplifier (PAR 5201). The output voltage of the amplifier was digitized and processed by a microcomputer.

3. Results and discussion

3.1 Transmittance change induced by applied field

We used four samples of PC films containing different concentration of DEANS. The experimental results of these samples are shown in Fig.4. The weight ratios of DEANS/PC in these samples were (a) 20/100, (b) 15/100, (c) 5/100, and (d) 0/100. Their corresponding molar concentrations were (a) 68, (b) 51, (c) 17, and (d) 0×10^{-3} mol%. The experimental conditions are summarized in Table 1. Sample (d) does not contain DEANS, and it was used to study the effect of an applied electric field on PC itself. Since the signal of the PC film is much weaker than that of DEANS, we have to apply a higher electric field to detect the signal. For this purpose, sample (d), a non-doped PC film, was made to have a thickness about a half that of the others.

The solid curves of Fig. 4 show the spectral dependence of the transmittance change (dT) divided by the transmittance (T), and the dashed curves show the spectra of output of PD without applied fields, which is proportional to the product of the transmitted light intensity (I) and the PD sensitivity (S). The latter ones show a typical Fabry-Perot resonance structure. From a comparison of

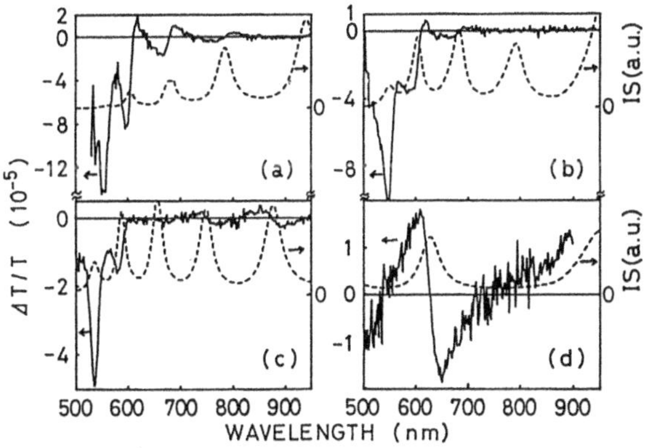

Fig. 4. Experimentally obtained spectra of the transmittance change due to applied field normalized by the transmittance (dT/T; solid curves) of the samples. Corresponding spectra of output of PD which is proportional to the product of the transmitted light intensity (I) and the PD sensitivity (S) without applied field are also shown (IS; dashed curves). The weight ratios of DEANS/PC are (a) 20/100, (b) 15/100, (c) 5/100, and (d) 0/100

Table 1. Experimental conditions for each sample

Sample	Weight ratio of DEANS/PC	Thickness of DEANS/PC film (μm)	Applied voltage
(a)	20/100	1.4	0 to +10 V : square
(b)	15/100	1.45	- 8 to + 8 V : sine
(c)	5/100	1.61	- 8 to + 8 V : sine
(d)	0/100	0.6	-10 to +10 V : sine

the transmission spectra with calculation, the variation of thickness over the area of the probe light beam on the sample was concluded to be less than a few percent.

In Fig. 4, we can see oscillatory structures in dT/T spectra. The amplitude and the off-set from dT=0 line of the oscillation increase with a decrease in wavelength. Here we can see in Fig. 5, DEANS molecules have strong absorption in the wavelength region shorter than 600 nm. Thus the increase in the amplitude and off-set can be explained in terms of the resonance effect on real and imaginary parts of refractive index change, respectively.

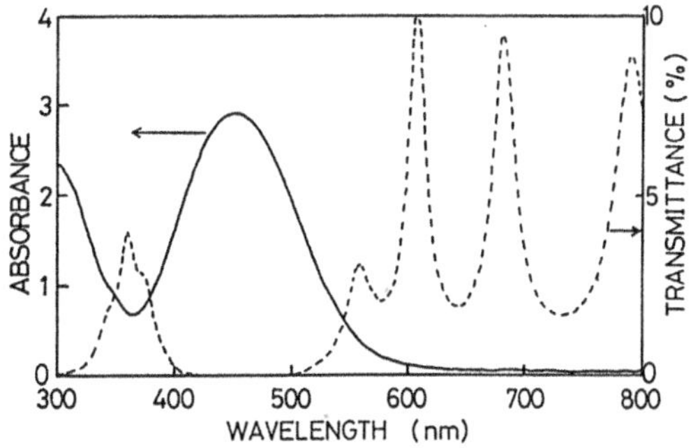

Fig. 5. Transmittance spectrum of sample (a) (solid line) with absorption spectrum of the 17 wt.% DEANS-doped PC film on the glass plate (dashed line)

All the samples showed quadratic dependence of dT on the applied field. This is reasonable since the orientation of doped molecules is randomly distributed, and hence the films have inversion symmetry.

3.2 Transmittance change of PC film without DEANS

As can clearly be seen in Fig. 4(d), PC itself causes a transmittance change on applying a field. The origin of this signal will be discussed in this section.

There are several possible origins of the transmittance change of the PC film induced by the applied electric field as follows: (I) The thickness change (dL) due to (Ia) electrostriction of PC, (Ib) transformation of the film by the electrostatic force between two electrodes, or (Ic) inverse piezoelectricity of PC. (II) The real part of the refractive index change (dn) due to the electro-optic effect. They are summarized in Table 2. In the case of the non-doped sample, the imaginary part of that can be neglected, since the absorption band of PC is far from the wavelength region of observation.

Table 2. Several possible mechanisms of electric field induced transmittance change of non-doped PC film

Mechanism		Dependence on E	Sign of the change
(1) Thickness change	(a) Electrostriction	quadratic	dL<0
	(b) Electrostatic force between two electrodes	quadratic	dL<0
	(c) Inverse piezoelectricity	linear	
(2) Kerr effect of PC		quadratic	dn>0
Experimental results		quadratic	dL<0 or dn<0

As seen in Fig. 4(d), the dT/T spectrum shows oscillatory structure, which turns from positive to negative with increasing wavelength at the transmittance peaks. This change in sign is explained by the reduction either in the thickness (mechanism (I)) or in the real part of the refractive index (mechanism (II)). On the other hand, in Fig. 4(a) dT/T in the wavelength region longer than 600 nm shows an oscillatory structure of opposite phase to that of PC. Namely it turns from negative to positive at the peak of T with wavelength. It is explained by positive dn, in contrast to PC. It is concluded that the contribution to dT of PC has the opposite sign to that of DEANS.

Polycarbonate has polar groups (Fig. 3(a)), which may contribute to the EO effect by orientation under the applied field. Thus the change in the refractive index is expected to be positive from calculation [1] contrary to the experimental result. Therefore the experimentally observed change in transmittance of PC can be attributed mainly to change in the film thickness.

Furthermore, mechanism (Ic) was ruled out by the dependence of dT on applied field. The transmittance change due to both (Ia) and (Ib) has quadratic dependence. In addition, the thickness change due to static force is negative, and dL due to electrostriction may also be negative. Thus the two mechanisms (Ia) and (Ib) are possible in the case of PC.

3.3 Calculation of EO constant

At each wavelength, experimentally observed dT/T was expanded by the change in real and imaginary parts of the refractive index, and in the thickness of the film (dn, dk, and dL, respectively) due to the applied field.

$$\frac{dT}{T} = \frac{1}{T}\frac{\partial T}{\partial n}dn + \frac{1}{T}\frac{\partial T}{\partial k}dk + \frac{1}{T}\frac{\partial T}{\partial L}dL. \tag{3}$$

Here we must consider that the refractive index changes given by dn and dk can also be induced by the thickness change dL.

The macroscopic susceptibility of dielectrics doped with molecules with permanent dipoles is described as

$$\chi^{(1)} = N\alpha f_{\omega} \tag{4}$$

where N is the number of molecules per unit volume, α is the mean polarizability of the molecule, and f_{ω} is a local field factor at the molecular position for the electric field at the light frequency. We adopted the Onsager local field factor for f_{ω} [5]:

$$f_{\omega} = \frac{n_C^2}{(1-A)n_C^2 + A - (1-A)(n_C^2-1)x}, \tag{5}$$

where $x=4\pi N\alpha/3$, n_C is the complex refractive index of the mixture, and A is the shape factor, which is 1/3 for the spherical molecule. When the size of the dielectric film changes, the concentration of molecules (N) also changes, but the molecular polarizability of the molecule (α) is considered not to change. Here we made an approximation that the change in size of the film caused by the electric field is unidirectional, since the size of the film perpendicular to the field is about 10^4 times larger than that along the field, and the film is held by the hard electrodes. Therefore the change in susceptibility of doped sample due to thickness change is given by

$$d\chi^{(1)} = -\frac{\chi^{(1)}}{L}\left[1 + \frac{4\pi\chi^{(1)}}{3}\frac{(1-A)(n_C^2-1)}{n_C^2}\right] dL. \tag{6}$$

Using the relation $\chi^{(1)}=(n_C^2-1)/4\pi=((n+ik)^2-1)/4\pi$, we obtained the change in refractive index of doped film due to thickness change as a function of dL. The total change in the refractive index on applying an electric field is described as the sum of the change due to thickness change and that due to the EO effect (n_{EO} and k_{EO}). Finally we obtained the following equation:

$$\left(\frac{dT}{T}\right)_{exp} = \frac{1}{T}\left(\frac{\partial T}{\partial n}\right)_{calc} dn_{EO} + \frac{1}{T}\left(\frac{\partial T}{\partial k}\right)_{calc} dk_{EO} + \frac{1}{T}\left(\frac{\partial T}{\partial L}\right)^{eff}_{calc} dL. \tag{7}$$

Where subscripts exp and calc denote the experimentally obtained and theoretically calculated values, and the last term on the right hand side includes the terms by the real and imaginary parts of the refractive index change due to the thickness change.

In the case of DEANS-doped PC, the second term on the right hand side of (6) is about a hundred times smaller than the first term and can be neglected, then

$$\left(\frac{\partial T}{\partial L}\right)^{eff}_{calc} = \frac{\partial T}{\partial L} - \frac{1}{2L(n^2+k^2)}\left[n(n^2+k^2-1)\frac{\partial T}{\partial n} + k(n^2+k^2+1)\frac{\partial T}{\partial k}\right]. \tag{8}$$

Using these equations, we obtained dn_{EO} and dk_{EO} as follows.

At first, n and k of the sample were determined from transmittance and reflectance at each wavelength, and the derivatives with the subscript calc were calculated using these values.

Next, dk_{EO} was obtained by substituting experimental dT/T and dL into the following equation at several wavelengths where $(\partial T/\partial n)_{calc}=0$ is satisfied.

$$dk_{EO}=\frac{\left(\frac{dT}{T}\right)_{exp}-\frac{1}{T}\left(\frac{\partial T}{\partial L}\right)_{calc}^{eff}dL}{\frac{1}{T}\left(\frac{\partial T}{\partial k}\right)_{calc}}. \tag{9}$$

The values of dk_{EO} at other wavelengths were estimated by interpolation with these values. In the third step, dn_{EO} was obtained as follows:

$$dn_{EO}=\frac{\left(\frac{dT}{T}\right)_{exp}-\frac{1}{T}\left(\frac{\partial T}{\partial L}\right)_{calc}^{eff}dL-\frac{1}{T}\left(\frac{\partial T}{\partial k}\right)_{calc}dk}{\frac{1}{T}\left(\frac{\partial T}{\partial n}\right)_{calc}}. \tag{10}$$

Then we obtained the EO constant R and the corresponding third-order susceptibility $\chi^{(3)}(-\omega;\omega,0,0)$,

$$\Delta\left(\frac{1}{(n+ik)^2}\right)=RE^2, \tag{11}$$

$$\chi_r^{(3)}+i\chi_i^{(3)}=\frac{(n+ik)(dn+idk)}{2\pi(E_{mz}^{0})^2}. \tag{12}$$

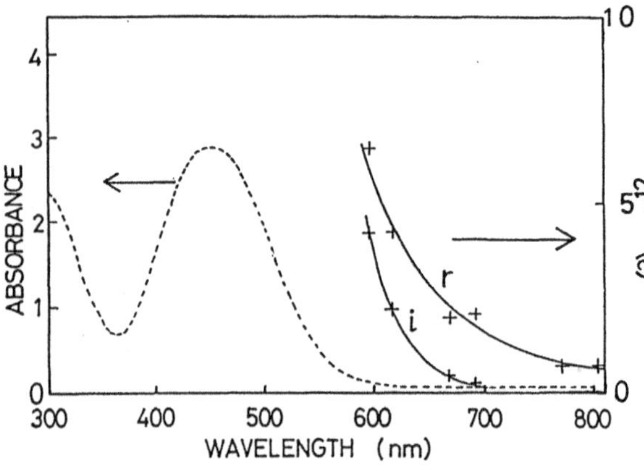

Fig. 6. Wavelength dependence of the real and imaginary parts (denoted by subscripts r and i, respectively) of the third-order nonlinear susceptibility of sample (a). Absorption spectrum of DEANS-doped PC film on glass plate (weight ratio of DEANS/PC is 20/100) is also shown

These values correspond to the contribution of DEANS to the EO effect. In Fig. 6, the values of $\chi_{r,i}^{(3)}$ of the sample (a) are plotted as functions of wavelength. In the case of the ordinary method to measure birefringence discussed in the previous section, only the real part of the refractive index without resonance effect is considered. However, Fig. 6 shows that the resonance effect of the real part and the value of the imaginary part cannot be neglected even at longer wavelengths than 600 nm where the absorbance is as small as one thirtieth of the peak observance.

3.4 Concentration dependence of $\chi^{(3)}$

The scanning electron microscope (SEM) image of sample (b) (13 wt.% doped sample) is shown in Fig. 7. In the image, there are particles of about 0.1 um diameter.

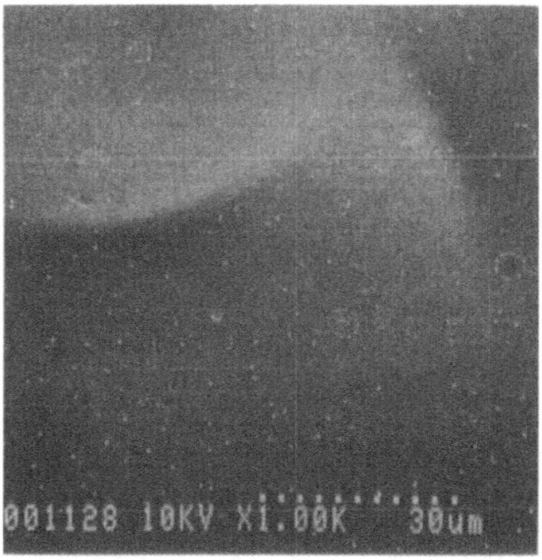

Fig. 7 The SEM images of sample (b). There can be seen particles of about 0.1 μm diameter

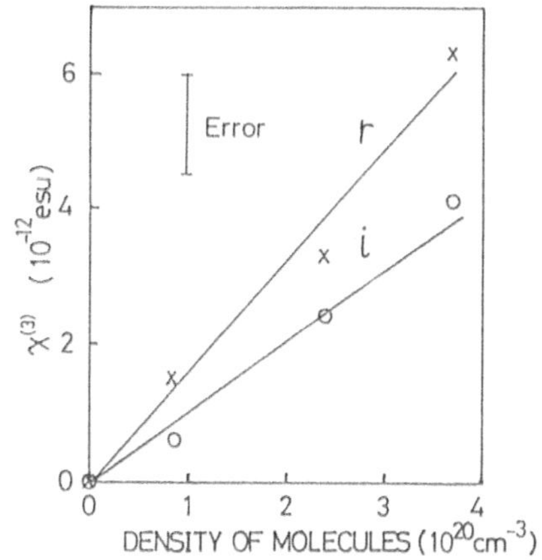

Fig. 8. The real (r) and imaginary (i) parts of $\chi^{(3)}$ at 600 nm of the DEANS-doped PC film after subtracting the contribution of PC as a function of number density of DEANS molecules dispersed uniformly in PC

They are most probably microclusters of DEANS which were formed at the surface associated with the evaporation of the solvent. However, the sum of the cross sectional area of the microclusters was estimated as from 1.5 to 6% of the total surface area. Therefore the observed absorbance of such films is mainly due to that of nonclustering DEANS molecules, under the condition that the absorbance of the uniform area is much larger than the ratio of the square size of clustering to nonclustering area.

Therefore the effective number of DEANS molecules per unit volume, N, of each sample can be estimated from the ratio of absorbance of the film to the molar absorption coefficient of benzene solution. Figure 8 shows that the real and imaginary parts of $\chi^{(3)}$ experimentally obtained are nearly proportional to the effective concentration of DEANS molecules, N. Therefore we concluded the observed EO effect is mainly attributed to the dissolved DEANS molecules in the PC film.

3.5 Microscopic mechanism of the EO effect

In this section, we will discuss the mechanism of the EO effect in terms of the motion of DEANS molecules dispersed uniformly in PC.

The Kerr constant of 4-dimethylamino-4'-nitrostilbene (DMANS) in solution, which has a similar electronic structure to DEANS, has been measured by KHANARIAN et al. [6] using the ordinary method for birefringence. They reported the molar Kerr constant of DMANS, $m^K=3.73\times10^{-8}$ esu at 633 nm at room temperature, which is given by

$$m^K = \frac{6n\lambda M}{(n^2+2)^2(\varepsilon+2)^2 Dw}\,(B(\text{solution})-B(\text{solvent})), \qquad (13)$$

where $B=Dn/\lambda E^2$, D is the density of the solvent, and M and w are the molecular weight and the weight fraction of solute in solution, respectively. In this case, the molar Kerr constant is the difference between the two components of parallel and perpendicular polarization to the applied field. Since molecules in the experiment of [6] can orient freely under the applied field, m^K for the polarized light perpendicular to the applied field was estimated as 1.87×10^{-8} esu from some calculations [1].

On the other hand, the molar Kerr constant of DEANS determined at 633 nm in the present study was $_mK = 3.3 \times 10^{-10}$ esu. This corresponds to the component where the light polarization is perpendicular to the applied field. Thus it is about 1/60 of that of DMANS in spite of the similarity of electronic structure of these two molecules. Therefore the orientational motion of DEANS molecules in the polymer matrix is expected not to be free under the applied field.

4. Conclusion

We have proposed a new convenient method for the determination of the EO constant and corresponding nonlinear susceptibilities. We applied this method to DEANS-doped PC thin films with several concentrations, and determined both real and imaginary parts of the EO constants and the corresponding nonlinear susceptibilities, which cannot be obtained separately by ordinary methods. We also obtained the wavelength dependence of $\chi^{(3)}(-\omega;\omega,0,0)$ in the spectral region between 600 nm and 800 nm. The contributions of DEANS and PC to the transmittance change induced by electric field have opposite signs, and we concluded that the latter is mainly caused by the thickness change of the PC film. The value of $\chi^{(3)}$ after subtracting the contribution of PC is nearly proportional to the concentration of DEANS molecules which were uniformly dispersed in PC. Therefore we concluded that the observed EO effect is mainly attributed to the dissolved DEANS molecules in the PC film.

5. Acknowledgments

The authors thank Dr. T. Murayama of the Research Center of Mitsubishi Chemical Industries Ltd., and Mr. A. Ueki for their help in sample preparation. We also thank Mr. A. Nagai and Mr. K. Sugano of Kureha Chemical Industry Co. Ltd. for their help in taking SEM photographs.

6. References

1. B.R. Jennings: In Molecular Electro-Optics, ed. by S. Krause, (Plenum, New York 1981) p.27
2. J.C. Loulergue, M. Dumont, Y. Levy, P. Robin, J.P. Pocholle, and M. Papuchon: Thin Solid Films, 160 399 (1988)
3. H. Uchiki and T. Kobayashi: J. Appl. Phys. 64(5) 2625 (1988)
4. T. Kobayashi, H. Ohtani, and K. Kurokawa: Chem. Phys. Lett. 121 356 (1985)
5. C.J.F. Bottcher: Theory of Electric Polarization, 2nd ed. (Elsevier, Amsterdam 1973)
6. G. Khanarian, A. Artigliere, R. Keosian, E.W. Choe, R. DeMartino, D. Stuetz, and C.C. Teng: In Molecular and Polymeric Optoelectronic Materials ed. by G. Khanarian, SPIE proc. Vol. 682 (SPIE, Bellingham 1986) p.153

Molecular and Crystal Engineering of Polydiacetylenes: Modification of Optical Properties

S.B. Clough[1], *S. Kumar*[1], *X.F. Sun*[1], *S. Tripathy*[1,*], *H. Matsuda*[2], *H. Nakanishi*[2], *S. Okada*[2], *and M. Kato*[2]

[1]Department of Chemistry, Polymer Program, University of Lowell, Lowell, MA 01854, USA
[2]Research Institute for Polymers and Textiles, 1-1-4 Yatabe-Higashi, Tsukuba, Ibaraki 305, Japan

1 Introduction

Polydiacetylenes have attracted a great deal of attention over the past decade due to their interesting electronic and optical properties [1,2]. It is also one of the few classes of polymeric materials that can be produced in the form of macroscopic single crystals [3]. These polymeric crystals are characterized by a tremendous richness of structure. A large number of side groups may be grafted onto the polymeric backbone stem which has a planar extended chain unsaturated structure (Fig. 1). The many different side groups (R and R') that can be part of the monomer structure have a strong influence on the crystalline organization and polymerization behavior in the solid state [4]. Depending upon the chemical nature of the side groups a number of crystalline phases may be present with different conformation of the side groups. Whether solid state polymerization in a particular crystalline phase will occur is dictated by the intermolecular spacings and precise arrangement of the adjacent diacetylene functionalities.

The one dimensional nature of the unsaturated polydiacetylene backbone implies novel electronic and optical properties [5]. Large and fast third order optical nonlinearities were predicted for these

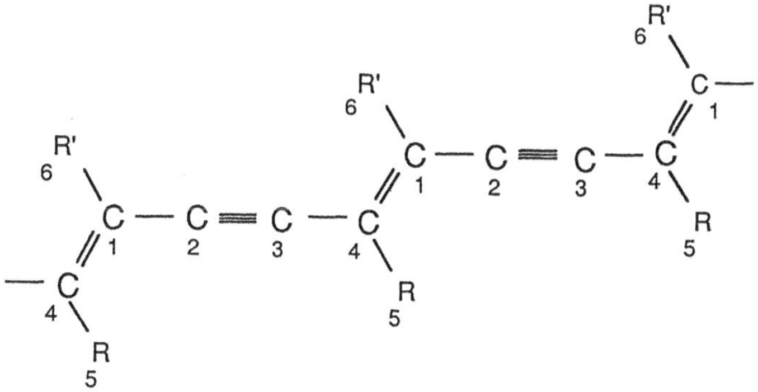

Fig. 1. Structure of the polydiacetylene backbone. The numbers label the carbon atoms on the backbone and the single masses representing the sidegroups R and R'

*Author to whom correspondence should be sent

Springer Proceedings in Physics, Vol. 36
Nonlinear Optics of Organics and Semiconductors
Editor: T. Kobayashi

materials and have subsequently been confirmed by a number of investigators [5]. These materials have been characterized by intense coloration, typically absorptive in the mid visible range with excellent transparency in the near infra-red. The absorption spectrum has essentially the same feature for most polydiacetylenes, viz., a sharp excitonic peak followed by the interband transition and they have been characterized as one dimensional semiconductors. However, the position of these features is very sensitive to the crystalline phase (organization of the side group). Significant thermochromic transition has been observed for a number of systems. It is believed that side group induced "stresses" in the backbone effecting its geometry, planarity and conformation in general, is responsible for the change in its electronic structure resulting in such thermochromic shifts. Solvatochromism also has been studied for soluble polydiacetylenes. Again, the extent and nature of solvation of the side groups indirectly influences the backbone geometry effecting its electronic structure [6].

The extended polydiacetylene backbone whether it is in a solution, a single crystal or a monolayer is confined in this geometry by the arrangement of the side groups. Its electronic structure is a sensitive function of this geometry. This is dramatically demonstrated in the case of substituted polyacetylenes. The extended chain geometry in polyacetylene is easily distorted as bulky side groups are substituted. This is reflected in the dramatic change of the electronic properties, and a ten orders of magnitude drop in the level conductivity is seen upon doping [7].

ORCHARD and TRIPATHY [6] in a recent publication investigated the influence of the coupling of the side group electronic structure to that of the backbone, in polydiacetylenes. The anticipated changes in the absorption spectrum when the saturated alkyl side groups are replaced with phenyl groups coplanar with the backbone are spectacular. The rhetorical question one asks, if the side groups as the mechanical tuners are replaced by electronic tuners, how will the backbone properties be modulated? The side groups may be a photochromic group or bridged donor-acceptor moietie, for example. Of course any such "molecular engineering" will necessarily have to be tempered with "crystal engineering". The backbone extended chain geometry should be maintained and side groups must find themselves in such a packing as to produce the maximum interaction. In the present paper, the crystal and molecular engineering aspects of a class of polydiacetylenes synthesized and processed with the above goal is discussed.

2 Methods

Symmetric diacetylene monomer with the substituents 4-n-Butyl, 2,3,5,6,-tetrafluoro phenyl (BTFP) was synthesized according to the method of Okuhara et al. [8]. The method of synthesis and purification procedures are listed elsewhere. This monomer is melt stable and polymerization in the monomer crystal may be carried out using UV radiation. Thin film single crystals were prepared between two quartz plates from both melt and solution. The crystal growth technique developed by THAKUR and coworkers was used in this process [9]. Polymerization in the thin film single crystal results in a deep blue film. Visible spectra from thin polymerized films were obtained.

The electronic structure of an analog of this polydiacetylene was calculated following the methods outlined by ORCHARD and TRIPATHY [6]. Full free valence geometry optimized structures of bis-phenyl diacetylenes were used in conjunction with the Valence Effective Hamiltonian technique [10] for the band structure calculation. These calculations were performed as a function of bond rotation about the bond linking the phenyl group to the polydiacetylene backbone. The torsion angles have been determined for BTFP from x-ray diffraction studies. Band gap for this geometry is compared with the spectral features seen for the thin film single crystals.

3 Results and Discussion

Short chain oligomers of poly bis-phenyl diacetylenes were used to obtain the optimized geometries. The total molecular energies were calculated for increasingly larger oligomers using the MNDO option in the chemical software package Chemlab-II (Molecular Design, Ltd.). As the size of the oligomer increases, the corresponding energies per repeat unit reach an asymptotic value. While the ends of the chain perturb the geometries of the penultimate units, once the energy/repeat unit has reached the asymptotic value geometries of the central unit may be used to construct the polymer for the calculation of the electronic band structure.

The MNDO optimized geometries for hydrogen and phenyl substituted polymers are listed in Table 1. For the symmetric diacetylene considered, two phenyl rings per repeat unit were used to determine the spatial dependence of the inductive effect along the polymer backbone. The electronic band structure was calculated using the Valence Effective Hamiltonian method developed by NICHOLAS and DURAND [10]. A portion of the electronic band diagram for phenyl substituted polydiacetylene is given in Fig. 2, when the phenyl group is coplanar with the polydiacetylene backbone. Figure 3 compares this with the band structure of hydrogen substituted polydiacetylene.

There is a significant narrowing of the gap as a result of the inductive effect from the coplanar phenyl side group. However, it is rarely possible that such idealized geometries will be realized in a real crystal. The substituent group may adopt one of a large number of conformations accessible to it, in order that a minimum in the packing energy is obtained. One may attempt to lock the side group in a specific conformation by appropriate substitution of the phenyl rings and this was the motive in selecting BTFP, yet the results are not straightforward. Crystal structure determination from bulk grown crystals of BTFP indicate the phenyl ring to be at 58° angle to the polydiacetylene plane.

The electronic structure calculations were repeated, with the phenyl ring incrementally rotated out of the plane of the polydiacetylene backbone. In Fig. 4 the band gap which is directly given by the position of the optical transition is plotted as a function of the torsional angle made by the phenyl ring with the plane of the polydiacetylene backbone. At 90° angle we see the band gap increasing toward the gap of a hydrogen substituted polydiacetylene. The π conjugation from the side group to the backbone has

Table 1. Optimized bond lengths and bond angles for polydiacetylenes with R=R'=H and R=R'=phenyl

		H	Phenyl
Bond Length [Å]	d2-3	1.2024	1.2041
	d3-4	1.4140	1.4159
	d4-1	1.3579	1.3989
Bond Angle [°]	∠3-4-1	123.9	123.9
	∠3-4-5	115.3	115.3

been lost. At 58° angle this conjugation has already significantly dropped off and the optical absorption is not expected to be very dissimilar from other known diacetylenes. At 30° the band gap appears to go through a minimum. The shift of the minimum in the band gap from 0° to 30° may be the result of the decrease in the strain in the backbone as the steric hindrance between the adjacent phenyl groups is decreased.

Figure 5 is a visible absorption spectrum taken from a thin single crystal of the polydiacetylene BTFP. The spectrum is indicative of a two phase system. MATSUDA and coworkers [11] have identified one of the crystal phases of this polydiacetylene and have carried out the detailed crystal structure. X-

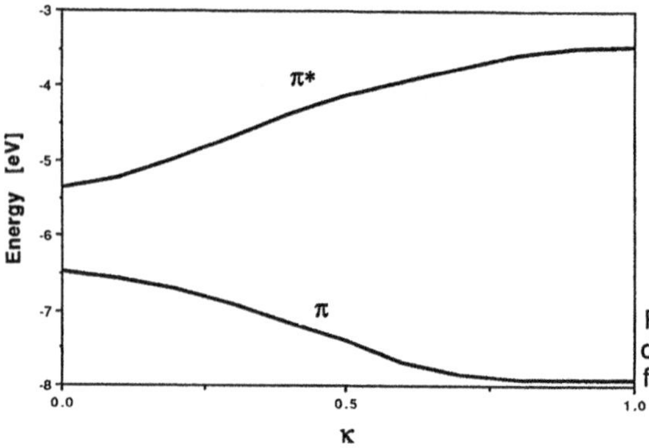

Fig. 2. VEH band structure for phenyl-disubstituted polydiacetylene showing band gap for π-π* transition(κ=π/a)

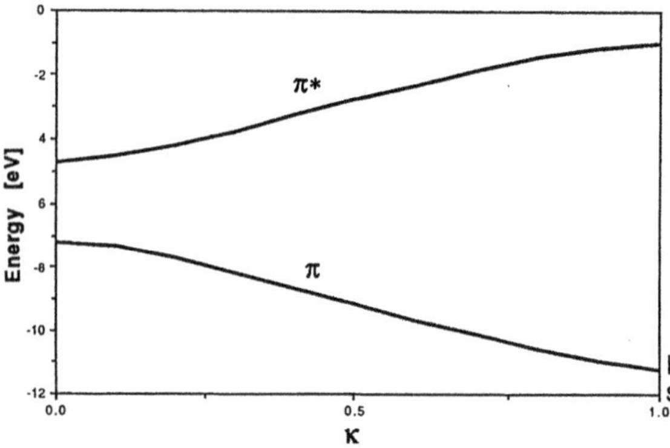

Fig. 3. VEH band structure for polydiacetylene showing band gap for π–π* transition(κ=π/a)

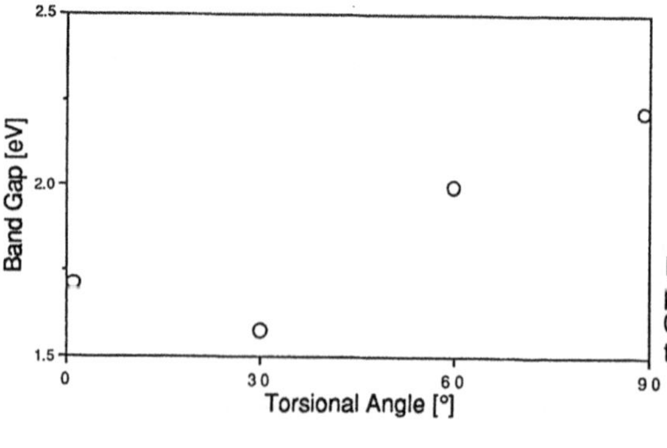

Fig. 4. Band gap of phenyl disubstituted polydiacetylene with rotation of phenyl groups. 0° indicates the phenyl rings are planar with the backbone

ray diffraction studies from the thin film single crystal will have to be carried out to identify the correct phase in these thin single crystals.

The theoretical calculations carried out on the model system are in excellent agreement with the experimental results. We believe our efforts at molecular engineering of the diacetylene system to possess electroactive side groups is appropriate. However, better ground rules for crystal engineering have to be established in order that optimal conjugation of the side group and backbone electronic structure may occur.

Figure 5 indicates that if the torsion angle of the phenyl ring could be reduced to about 25° a significant red shifting of the absorption edge will occur. This is the direction our present crystal and molecular engineering effort is taking. A number of approaches can be postulated where the electroactive side groups may be coaxed into a coplanar arrangement. This may be achieved, for example, through hydrogen bonding. Appropriately modified side groups to contain donors and acceptors to form a charge transfer bridge may be another approach. Side groups adjacent to each other may be modified to undergo dimerization in the solid state. This may be brought about simultaneously with the solid state polymerization through the diacetylene or in a two-step process [12].

The nonlinear optical properties are expected to be very sensitive to the position of the optical gap. In addition these properties may be modulated through side group induced processes. Assymetrically substituted surfactive diacetylenes suitable for monolayer growth may provide a more convenient vehicle for the molecular and crystal engineering concepts discussed here. This is expected to play a significant role in our continuing efforts in this area.

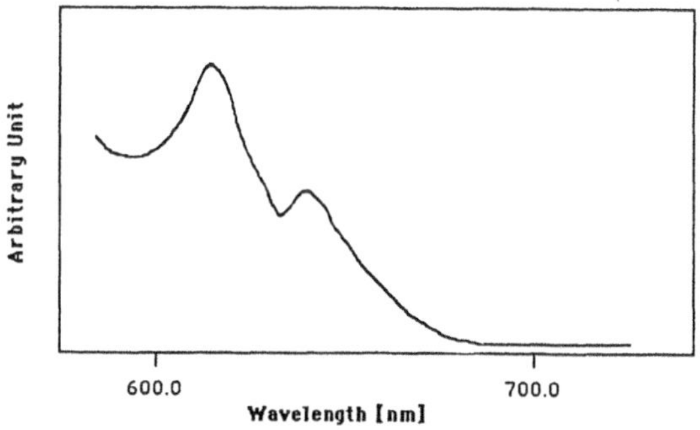

Fig. 5. Visible absorption spectrum for BTFP

Acknowledgement: Partial support for this work was provided by the office of Naval Research. S. T. acknowledges the support by the Research Institute for Polymers and Textiles.

4 References

1. D. Bloor, R. R. Chance eds.: Polydiacetylenes: Synthesis, Structure, and Electronic Properties (M. Nijhoff, Dordrecht, Boston 1985)
2. G. Lieser, B. Wegner: Thin Solid Films 68, 77 (1980)
3. M. Thakur, S. Meyler: Macromolecules 18, 2342 (1985)

4. V. Enkelman: Die Angew. Makromol. Chem. 109/110, 253 (1982)
5. G. M. Carter, Y. J. Chen, M. F. Rubner, D. J. Sadman, M. K. Thakur, S. K. Tripathy: In Nonlinear Optical Properties of Organic Molecules and Crystals, ed. by D. S. Chemla, J. Zyss, Vol. 2 (Academic Press 1987) p. 85
6. B. J. Orchard, S. K. Tripathy: Macromolecules 19, 1844 (1986)
7. M. A. Druy, D. Kitchen, S. K. Tripathy: Macromolecules 16, 190 (1983)
8. Presented at the 47th Annual Meeting of the Chemical Society of Japan at Kyoto, April 1-4, 1983
9. M. K. Thakur, D. J. Sandman and S. K. Tripathy, U.S. Patent #4,684,434 and 4,698,121
10. G. Nicholas, Ph. J. Durand: J. Chem. Phys. 70, 2020 (1979); 72, 453 (1980)
11. Presented elsewhere at this meeting
12. Private communication with D. J. Sandman, GTE Laboratories

Nonlinear Optical Properties of Polydiacetylenes with π-Conjugation Between the Main Chain and the Substituents

II. Nakanishi, II. Matsuda, S. Okada, and M. Kato

Research Institute for Polymers and Textiles,
1-1-4 Yatabe-Higashi, Tsukuba, Ibaraki 305, Japan

1. INTRODUCTION

There has been growing interest in the field of organic nonlinear optical materials and great progress has already been made both experimentally and theoretically [1]. Especially, polydiacetylenes, which can be obtained as peculiar single crystals of conjugated polymers by solid-state polymerization, have attracted much attention to their NLO properties because of the large third-order susceptibility $\chi^{(3)}$ [2] and the fast response [3], [4]. According to theoretical prediction [5], polydiacetylenes with increased numbers of π-electrons per repeating unit, such as those having aromatic rings directly bound to the main chain, or enlarged polarizability, as shown in Fig. 1, must be better candidates for extraordinarily large NLO properties. However, most of the well-known polydiacetylenes had methylene groups next to the conjugated main chain and few polymerizable diphenyldiacetylene derivatives were known [6]. For the materialization of polydiacetylene with aromatic rings directly bound to the main chain it is indispensable to have corresponding diacetylene monomers crystallized in a polymerizable stack, since polydiacetylenes are obtained by a topochemical solid-state polymerization. We have succeeded in the preparation of many such polydiacetylenes by proper molecular and crystal engineering [7], [8]. In this paper, in order to prove the enlarged $\chi^{(3)}$ of such polydiacetylenes, THG measurements of the fluorinated polydiphenyldiacetylenes, that is, bis-(4-n-butyltetrafluorophenyl)butadiyne (BTFP) and bis-(2,2',5,5'-tetraxistrifluoromethyl-phenyl)butadiyne (DFMP) [9], have been investigated.

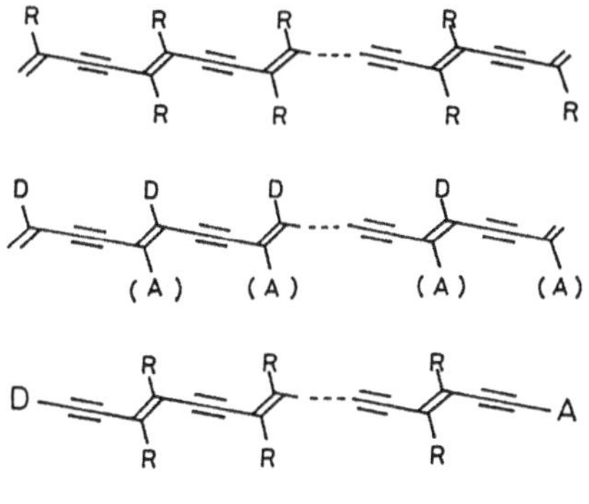

Fig. 1 Interesting polydiacetylene structures for NLO properties.

Springer Proceedings in Physics, Vol. 36
Nonlinear Optics of Organics and Semiconductors
Editor: T. Kobayashi

2. CRYSTAL STRUCTURES OF POLYDIACETYLENES

The solid-state polymerization scheme of diacetylene and the chemical structures of substituents of BTFP and DFMP are shown in Fig. 2, together with that of the well-known PTS. Polymer single crystals of BTFP and DFMP for x-ray crystallography were obtained by the following procedures: Monomer single crystals were grown by slow evaporation of their ethanol solutions, and after cutting the crystals for x-ray crystallography they were subjected to γ-ray polymerization to give polymer single crystals. Cell parameters and full intensity data were derived from measurements on a 4-circle diffractometer at room temperature. The structures were determined and refined by use of SHELX 76. R values of the final stage of anisotropic refinement including isotropic hydrogens were 0.12 for BTFP polymer and 0.13 for DFMP polymer. The relatively high values of R's were due to the small portions of disordering. Tables 1 and 3 show the crystallographic data of BTFP and DFMP polymers, respectively. Final atomic coordinates of BTFP and DFMP polymers are listed in Tables 2 and 4, respectively, and their molecular structures are shown in Fig. 3, in comparison with that of PTS polymer [10].

The bond lengths of their main chains and the plane-to-plane perpendicular distances between adjacent benzene rings in the same chain are 1.15 (C1 - C1'), 1.37 (C1 - C2), 1.55 (C2 -C2'), and 3.70 Å (ring - ring) for BTFP polymer, 1.19, 1.38, 1.36, and 4.44 Å for DFMP polymer, and 1.19, 1.36, 1.43, and 3.79 Å for PTS polymer, respectively. The substituted phenyl rings of BTFP polymer stack in a similar manner to that of PTS polymer. However, that of DFMP polymer stacks differently because of the bulky trifluoromethyl groups. The dihedral angles between polymer backbone and the phenyl rings of BTFP and DFMP polymers are found to be 58 ^{o} and 67°, respectively.

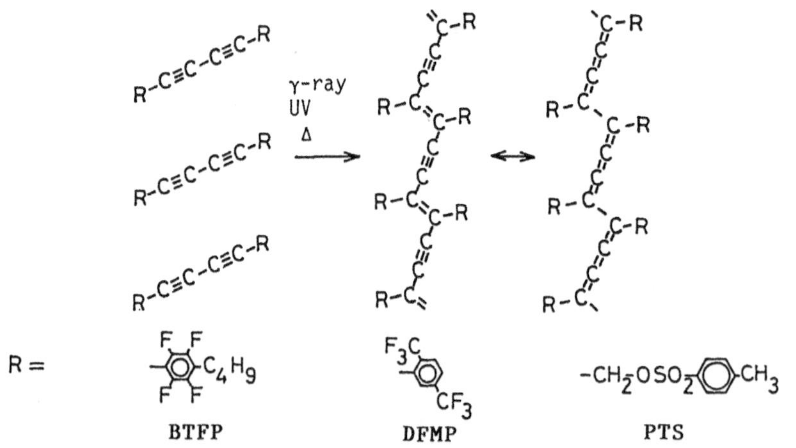

Fig. 2 Solid-state polymerization scheme and the chemical structures of diacetylene derivatives.

Table 1 Crystallographic data of BTFP polymer.

Formula	$C_{24}H_{18}F_8$
Mr	458.4
Crystal System	Monoclinic
Space Group	$P2_1/n$
a (A)	13.627 (11)
b (A)	4.850 (3)
c (A)	15.990 (13)
β (°)	95.61 (10)
V (A^3)	1052 (2)
Z	2
Dx(Mg m^{-3})	1.45
Dm(Mg M^{-3})	1.44

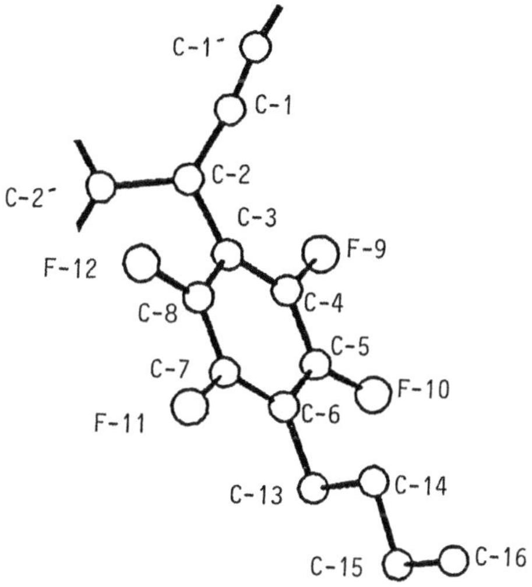

Table 2 Atomic coordinates of BTFP polymer.

Atoms	x	y	z
C-1	0.9844	0.8902	0.4963
C-2	0.9620	0.6151	0.4866
C-3	0.8594	0.5198	0.4532
C-4	0.7807	0.5737	0.4951
C-5	0.6897	0.4562	0.4740
C-6	0.6739	0.2736	0.4084
C-7	0.7523	0.2191	0.3654
C-8	0.8420	0.3440	0.3869
F-9	0.7886	0.7448	0.5634
F-10	0.6159	0.5196	0.5187
F-11	0.7464	0.0472	0.2998
F-12	0.9168	0.2918	0.3404
C-13	0.5748	0.1374	0.3832
C-14	0.5175	0.3024	0.3089
C-15	0.4206	0.1491	0.2808
C-16	0.3643	0.2972	0.2094

Table 3 Crystallographic data of DFMP polymer.

Formula	$C_{20}H_6F_{12}$
Mr	474.2
Crystal System	Triclinic
Space Group	$P\bar{1}$
a (Å)	4.893 (1)
b (Å)	7.401 (3)
c (Å)	12.814 (4)
α (°)	96.73 (3)
β (°)	91.84 (2)
γ (°)	92.27 (3)
V (Å^3)	456.6 (3)
Z	1
Dx(Mg m^{-3})	1.725
Dm(Mg M^{-3})	1.73

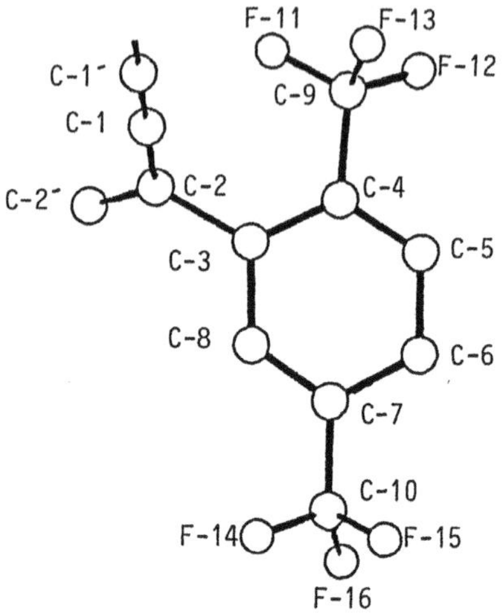

Table 4 Atomic coordinates of DFMP polymer.

Atoms	x	y	z
C-1	0.1183	-0.0023	0.0105
C-2	0.3938	-0.0098	0.0325
C-3	0.4639	-0.0323	0.1471
C-4	0.3950	-0.1935	0.1942
C-5	0.4695	-0.2046	0.2963
C-6	0.6279	-0.0578	0.3593
C-7	0.6967	0.1112	0.3156
C-8	0.6231	0.1141	0.2127
C-9	0.2390	-0.3623	0.1313
C-10	0.8674	0.2676	0.3824
F-11	0.2797	-0.3783	0.0314
F-12	-0.0273	-0.3594	0.1423
F-13	0.3079	-0.5152	0.1649
F-14	0.8258	0.4230	0.3501
F-15	1.1165	0.2470	0.4046
F-16	0.7469	0.3205	0.4682

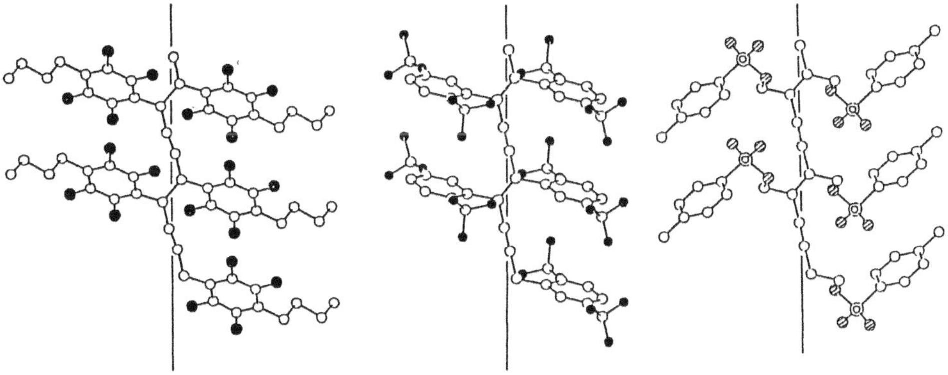

BTFP DFMP PTS

Fig. 3 Comparison of molecular structures of polydiacetylenes.

○ : C, ● : F, ⊘ : O, ◎ : S.

3. PREPARATION OF POLYDIACETYLENE FILMS

The melting points of BTFP and DFMP monomers are 100° and 102 °C, respectively. It was found that both monomers could be sublimated stably. Thus, in order to form thin polymer films on fused quartz plates, the monomers of BTFP and DFMP were vacuum-evaporated under a pressure of 10^{-3} Pa, and then they were subjected to UV-polymerization to give the polycrystalline polymer films. Thin poly-crystalline films of PTS polymer for comparison were made by a different way: The molten monomer (mp. 97 °C) was pressed between two quartz plates and then subjected to thermal polymerization at 60 °C. The thicknesses of these polymer films were about 1 ∿ 0.05 μm (by TALYSTEP) and the sizes of the micro-crystallites were found to be smaller than 100 μm under a polarizing microscope.

4. THG MEASUREMENTS

THG measurements were performed by use of 1.83 ∼ 2.10 μm pulsed (ca. 5 ns) laser light with about 2 mJ/p, which was obtained by wave-mixing of Nd:YAG and dye lasers. The damage thresholds of BTFP and PTS polymer films were greater than about 200 MW/cm^2. However, since that of the DFMP polymer film was about 80 MW/cm^2, the measurements were done with a laser power density of about 50 MW/cm^2; that is, the area of laser focusing was 1.24 x 0.66 = 0.818 mm^2.

For all these polymer films, no fringe patterns were observed in THG measurements by the Maker fringe method as shown in Fig. 4. This is because the sample thickness (L) was much smaller than the coherence length (lc) of these polymers. Therefore, the third-order susceptibility $\chi^{(3)}$ was evaluated by eq. (1), reported by TOMARU et al. [11]:

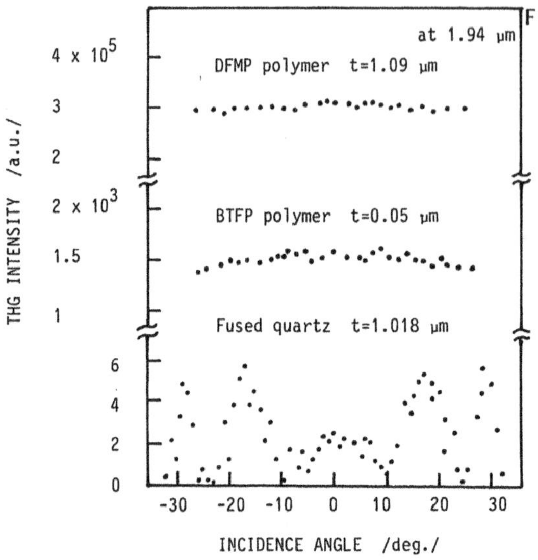

Fig. 4 THG intensity of polydiacetylene thin films and Maker fringes of THG of a fused quartz plate.

$$\chi^{(3)} = \frac{2}{\pi} \chi_s^{(3)} \cdot \frac{\sqrt{P}}{\sqrt{P_s}} \cdot \frac{l_{c,s}}{L} \qquad (1)$$

where P is the THG intensity and suffix (s) indicates the standard sample of fused quartz. The $\chi_s^{(3)}$ value used was 3.0×10^{-14} esu, reported by HERMANN [12]. In the calculation, the difference in refractive indices between samples and fused quartz, and the absorption of TH wave by samples were neglected. In addition, the polydiacetylene films were assumed to be isotropic with respect to the main chain direction, since the area of laser focusing was sufficiently larger than the size of the polymer crystallites as mentioned above. The $\chi^{(3)}$ values obtained are listed in Table 5 and the relation between their absorption spectra and TH

Table 5 $\chi^{(3)}$ values* of polydiacetylene thin films.

Poly-DA	Thickness (μm)	$\chi^{(3)} \times 10^{11}$ (esu)		
		pumping wavelength (μm)		
		1.83	1.94	2.10
BTFP	0.27	15	10	6.8
	0.05	26	15	6.6
DFMP	1.09	2.4	2.6	2.1
	0.15	6.0	3.2	1.7
PTS	0.90	4.7	2.5	1.3

*The $\chi^{(3)}$ values along the chain direction are three times larger than these data.

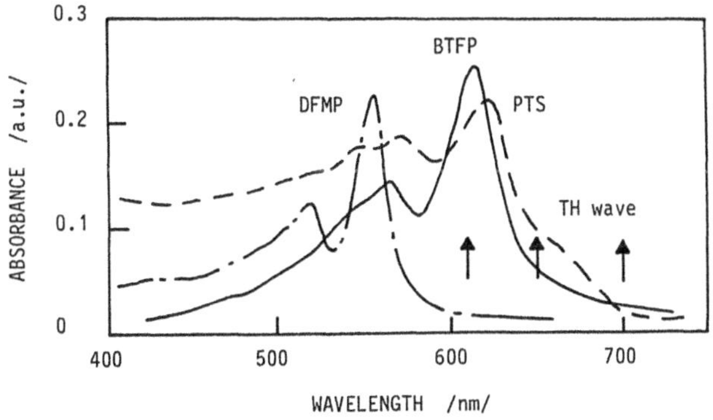

Fig. 5 Absorption spectra of polydiacetylene thin films.

wavelengths are shown in Fig. 5. As seen from Fig. 5, TH wave of 1.83 µm lies in the resonant region for BTFP and PTS polymers, whereas that of 2.10 µm lies in the non-resonant region for all the polymers. It should be noted that at both wavelengths, $\chi^{(3)}$ of BTFP polymer is about 5 times greater than that of the others. There might be several explanations for the enhanced $\chi^{(3)}$, e.g. narrowed band gap, longer dipole, crystallinity and so on. However, it is especially interesting that $\chi^{(3)}$ magnitudes of BTFP and DFMP polymers well reflect the dihedral angles between the polymer backbone and the phenyl substituents (58° for BTFP polymer and 67° for DFMP polymer) as a measure of π-conjugation. CLOUGH et al. [13] have caluclated the band structures of polydiphenyldiacetylene at the various dihedral angles, and found good agreement between the experimental (from the absorption spectra) and the caluclated (using the above-described dihedral angles) values, suggesting that a certain amount of π-conjugation between polymer backbone and phenyl substituents exists in both BTFP and DFMP polymers. Thus, it may be concluded that the directly bound aromatic substituents are surely effective in enlarging the nonlinear optical properties of polydiacetylenes. Further study to decrease the dihedral angles would be interesting.

5. REFERENCES

1. D.S. Chemla, J. Zyss: In Nonlinear Optical Properties of Organic Molecules and Crystals, (Academic Press, 1987), Volume 1 p. 482, Volume 2 p. 276.
2. C. Sauteret, J.-P. Hermann, R. Frey, F. Pradere, J. Ducuing, R.H. Baughman, R.R. Chance: Phys. Rev. Lett. 36, 956 (1976).
3. G.M. Carter, J.V. Hryniewicz, M.K. Thakur, Y.J. Chen, S.E. Meyler: Appl. Phys. Lett. 49, 998 (1986).
4. T. Hattori, T. Kobayashi: Chem. Phys. Lett. 133, 230 (1987).
5. B.J. Orchard, S.K. Tripathy: Macromolecules 19, (1986).
6. G. Wegner: Makromol. Chem. 145, 85 (1971).
7. H. Matsuda, H. Nakanishi, T. Hosomi, M. Kato: Macromolecules 21, 1238 (1988).
8. H. Nakanishi: Papers (Figures and Tables) delivered at the topical workshop on "Organic and Polymeric Nonlinear Optical Materials", the Div. Polym. Chem. ACS, Virginia Beach, May 16 - 19, 1988.

9. K. Kodaira, K. Okuhara: Presented at the 47th annual meeting of the Chemical Society of Japan at Kyoto, April 1-4, 1983.
10. D. Kobelt, E.F.Paulus: Acta Cryst. B30, 232 (1974).
11. S. Tomaru, K. Kubodera, S. Zembutsu, K. Takeda, M. Hasegawa: Electronics Lett. 23, 595 (1987).
12. J.P. Hermann: Opt. Commun. 9, 74 (1973).
13. Presented elsewhere at this meeting.

$\chi^{(3)}$ Evaluation of Organic Nonlinear Optical Materials by THG Measurements

K. Kubodera[1] *and T. Kaino*[2]

[1]NTT Opto-electronics Laboratories, Atsugi-shi, Kanagawa 243-01, Japan
[2]Tokai, Ibaraki 319-11, Japan

$\chi^{(3)}$ evaluation is demonstrated for organic materials by third-harmonic generation (THG) measurements. THG interference effect between film sample and substrate was taken into account for an accurate evaluation. Several kinds of polymer materials were examined; $\chi^{(3)} \simeq 10^{-10}$ esu for vacuum deposited polydiacetylene films, 10^{-11} esu for π-conjugated amorphous polymers, and 10^{-12} esu for monomer-doped polymer materials.

1. INTRODUCTION

Recently, studies on organic materials as third-order nonlinear optical materials have been extensively carried out because some organic materials having conjugated π-electrons show remarkably large nonlinearities /1,2/. As intensity-dependent refractive index materials, organic materials have the advantage of very fast response time and low absorption loss as compared with semiconductor materials. These characteristics make them desirable for use in ultrafast third-order nonlinear optical devices, such as optically gated optical switches and optical bistable devices, which will become key devices in future optical systems /3,4/.

Table 1 shows the calculated optical power densities for p-toluene sulfonate (PTS) polydiacetylene to realize optical bistable devices, compared with a multiple quantum well (MQW) semiconductor material /5/. The n_2 value of 10^{-11} cm^2/W, corresponding to $\chi^{(3)} = 8.5 \times 10^{-10}$ esu for PTS, is seven orders of magnitude less than that for MQW /6/. However, the calculated power density is only four orders of magnitude larger than MQW because of its much smaller absorption loss. If we compare these two materials simply by multiplying the power density and the response

Table 1 Comparison of calculated optical power density for PTS polydiacetylene and MQW semiconductor material to realize optical bistable devices. Medium length l and operation power density I_{th} are calculated from $\alpha_0 l = 1$ %, $n_2 I_{th} l = \lambda / 100$, respectively.

Parameters		PTS	GaAs/AlAs MQW
Nonlinear refractive index	n_2	10^{-11} cm^2/W	10^{-4} cm^2/W
Absorption coefficient	α_0	0.1 cm^{-1}	100 cm^{-1}
Medium length	l	1 mm	1 μm
Operation optical power density	I_{th}	1 MW/cm^2	100 W/cm^2
Response time	τ	10 ps*	10 ns

* restricted by cavity lifetime

Springer Proceedings in Physics, Vol. 36
Nonlinear Optics of Organics and Semiconductors
Editor: T. Kobayashi

time, it can be said that PTS is comparable to MQW from the standpoint of switching energy consumption per unit area.

As a fundamental study on device application, this paper demonstrates the evaluation of third-order nonlinearities of organic materials by third-harmonic generation (THG) measurements. The evaluation was then applied to some polymer films. Polydiacetylenes, monomer-doped polymers, and other amorphous polymers were examined.

2. THG MEASUREMENT AND $\chi^{(3)}$ DETERMINATION

2.1 THG Measurement

A THG measurement system for the film evaluation is shown in Fig.1. The measurements were performed at a pump wavelength around 1.9 μm using $LiNbO_3$ difference-frequency generation between a Q-switched Nd:YAG laser and a tunable dye laser. Since the organic materials in this experiment have absorption edges in the 0.5 - 0.7 μm wavelength region, this wavelength corresponds to the nonresonant or near-resonant region of three-photon resonance. The pump pulse duration was 5.5 ns, the repetition rate was 10 Hz, and the peak pump power density was 80 - 200 MW/cm^2. The film sample was mounted on a goniometer and rotated around an axis perpendicular to the laser beam. The pump light polarization direction was adjusted to be parallel to the rotation axis.

Figure 2 shows typical examples of THG intensity patterns from film samples as a function of incident angle. When film thickness l is much larger than the THG coherence length ($l_c \simeq$ a few μm), a Maker fringe pattern is observed as shown in Fig.2(a), corresponding to the consecutive optical pathlength variation. On the other hand, when l is much less than l_c, a monotonically decreasing pattern is observed, as shown in Fig.2(b). Figure 2(c) is the THG intensity pattern from a 1-mm-thick SiO_2 glass which is used as the standard sample.

Fig.2

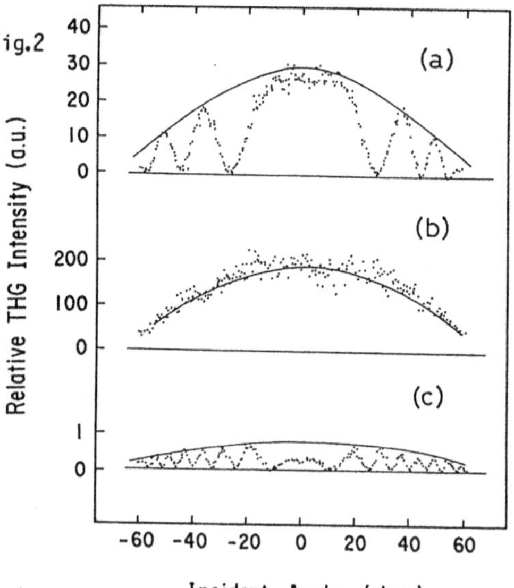

Fig.1

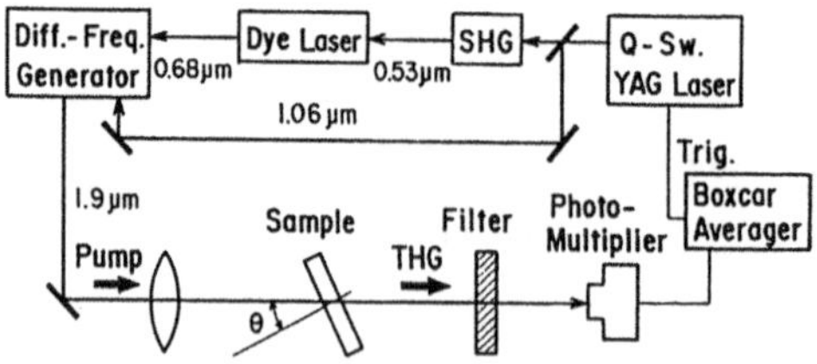

Fig.1 Experimental setup for THG measurement.

Fig.2 Typical THG intensity patterns as a function of incident angle.
(a) a thick film sample: l = 290 μm, $\chi^{(3)} = 3.7 \times 10^{-13}$ esu,
(b) a thin film sample: l = 0.69 μm, $\chi^{(3)} = 7.8 \times 10^{-12}$ esu,
(c) standard SiO_2 glass sample: l = 1 mm.

The $\chi^{(3)}$ value of the film was determined according to the following equations, comparing the THG intensity with the standard sample:

$$\chi^{(3)} = \chi_s^{(3)} (\sqrt{I_{3\omega}} / l_c) / (\sqrt{I_{3\omega,s}} / l_{c,s}) \qquad \text{for } l \gg l_c, \qquad (1)$$

$$\chi^{(3)} = (2/\pi)\, \chi_s^{(3)} (\sqrt{I_{3\omega}} / l) / (\sqrt{I_{3\omega,s}} / l_{c,s}) \qquad \text{for } l \ll l_c, \qquad (2)$$

where $I_{3\omega}$ and $I_{3\omega,s}$ are the THG peak intensities from the film sample and the standard sample, respectively (extrapolated maximum value in the case of Maker fringe), l and l_c are the film thickness and coherence length of the sample, respectively, and $l_{c,s}$ = 18.1 μm and $\chi_s^{(3)} = 2.8 \times 10^{-14}$ esu are the l_c value and $\chi^{(3)}$ value for the standard sample of 1-mm-thick SiO_2 glass at λ = 1.90 μm, respectively /7/. The refractive index difference between the film sample and silica glass was neglected in this calculation. As standard samples, relatively large $\chi^{(3)}$ materials, such as CS_2 and nitrobenzene liquids, were used in addition to the SiO_2 glass /8-10/. Although the air THG effect, which is known to influence the measurement of small $\chi^{(3)}$ materials such as SiO_2, has not been examined in this experiment, it is negligible for these samples.

2.2 Calibration of the Interference Effect Between Film and Substrate

Especially for the very thin film sample on a glass substrate, where the glass THG is comparable to the film THG as shown in Fig.3, the interference effect was theoretically analyzed as follows:

The third-harmonic electric fields from the film and the substrate illustrated in Fig.4(a) are expressed as

$$E_{3\omega,f} = -(i\omega / n_f c)\, \chi^{(3)} E_\omega^3\, l \qquad (l \ll l_c), \qquad (3)$$

$$E_{3\omega,sub} = -(\omega / \pi n_{sub} c)\, \chi_{sub}^{(3)} E_\omega^3\, l_{c,sub}$$

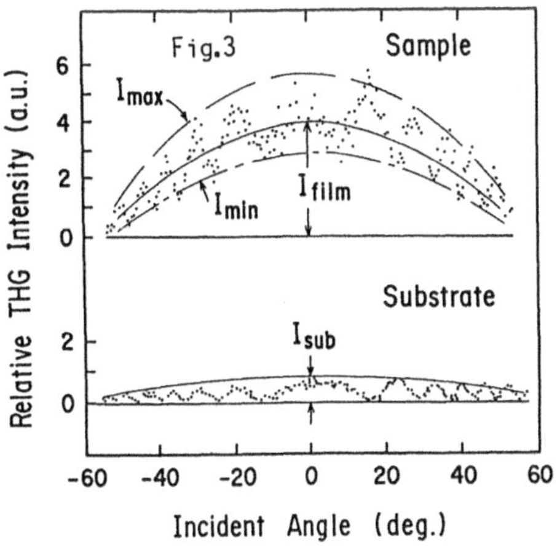

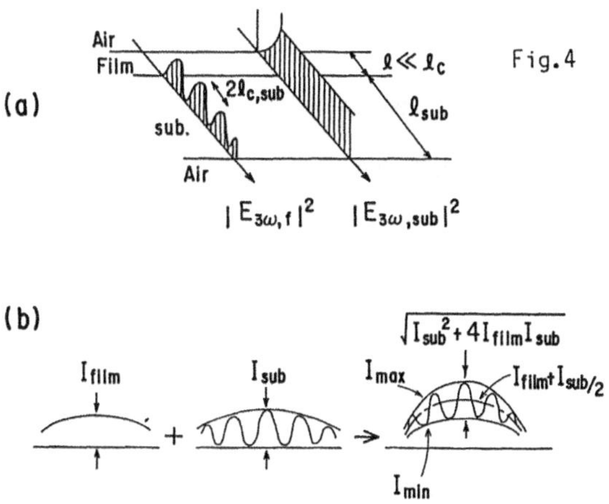

Fig.3 THG intensity pattern from a very thin film sample: l = 0.08 μm, $\chi^{(3)} = 1.4 \times 10^{-11}$ esu.

Fig.4 Analysis of THG interference between film and substrate.

$$\times [\exp (i \pi l_{sub} / l_{c,sub}) - 1] \quad , \tag{4}$$

where E_ω is the electric field of the pump light, n_f, $\chi^{(3)}$, and l are the refractive index, third-order susceptibility, and the sample thickness of the film, respectively, n_{sub}, $\chi_{sub}^{(3)}$, and l_{sub} are those of the substrate, respectively, and $l_{c,sub}$ is the coherence length of the substrate. The sample thickness l was assumed to be much less than the film coherence length ($l \ll l_c$) in this calculation. The total THG intensity $I_{3\omega}$ is given by

$$I_{3\omega} \propto | E_{3\omega,f} + E_{3\omega,sub} |^2 \tag{5}$$

$$\propto I_{film} + I_{sub} \sin^2 (\pi l_{sub} / 2 l_{c,sub}) + \sqrt{I_{film} I_{sub}} \sin (\pi l_{sub} / l_{c,sub}) \tag{6}$$

$$= (I_{film} + I_{sub} / 2) + (\sqrt{I_{sub}^2 + 4 I_{film} I_{sub}} / 2) \sin [(\pi l_{sub} / l_{c,sub}) - \psi] \tag{7}$$

where $I_{film} = [\chi^{(3)}]^2 l^2$ is the THG intensity from the film, and $I_{sub} = (4 / \pi^2) [\chi_{sub}^{(3)}]^2 l_{c,sub}^2$ is the envelope of the THG intensity pattern from the glass substrate. The refractive index difference between the film and the substrate was also neglected in this calculation. The last term in (6) represents the interference effect between the film and the substrate. The result, expressed by (7), is illustrated in Fig.4(b).

From (7) and Fig.4(b), it is easily found that the calibration can be done from the following very simple equation:

$$I_{film} = (I_{max} + I_{min}) / 2 - I_{sub} / 2 \quad , \tag{8}$$

where I_{max} and I_{min} are envelopes of the superimposed THG intensity pattern.

The ultimate accuracy of this measurement is considered to be $\chi^{(3)}_{limit} = 3 \times 10^{-13}$ esu µm, derived from the relation $I_{film} \simeq I_{sub}$.

3. POLYMER FILM EVALUATION

3.1 Polydiacetylene Films

Table 2 summarizes the polydiacetylene films fabricated and evaluated in this work. For more detail on the films, consult the references listed in the table. Single crystals are also listed for comparison with our results. As shown in the table, $\chi^{(3)}$ values of these films range widely depending on the film fabrication method.

Cast films are available by substituting some kinds of long chains as side groups. A solvent casting process was applied to form monomer films, and the polymerization was made by UV-light irradiation. Samples with thicknesses of 10 - 60 µm were prepared for sample No.1 /11/. Polyurethane-type polydiacetylene films are also available by the casting (sample No.2). A spinning technique was applied to form a polyurethane 0.23- µm-thick film on a glass substrate. Although cast films are attractive because of their easy film fabrication and fine-structure patterning, their measured $\chi^{(3)}$ values are relatively small, as shown in the table.

Table 2 $\chi^{(3)}$ values at λ = 1.90 μm for polydiacetylene films fabricated by various methods.

No.	Side groups	$\chi^{(3)}$ ($\times 10^{-12}$ esu)	References
Cast films			
1	$R = C_6H_4NHCOC_{17}H_{35}$	1.4	/11/
2	$R = (CH_2)_3OCONH(CH_2)_3$	5.2	-
Vacuum-deposited films			
3	$R = CH_2OCONHC_4H_9$	14	/12/
4	$R_1 = C_{12}H_{25}$, $R_2 = (CH_2)_8COOH$	50	-
5	$R = (CH_2)_3OH$	11	/13/
6	$R = (CH_2)_4OCONHC_3H_7$	380	/14/
(Single crystals)			
7	$R = CH_2OSO_2C_6H_4CH_3$	850*	/6/
8	$R = (CH_2)_4OCONHC_6H_5$	70*	/6/

* measured at λ = 1.89 μm

Larger $\chi^{(3)}$ values are observed in vacuum-deposited polydiacetylene films. Monomer films were first prepared by vacuum deposition onto glass substrates, and the films were then polymerized by UV-light irradiation. Many types of polydiacetylene films were examined; those having urethane-based alkyl chains (sample No.3 and No.6) /12,14/, those having long alkyl chains (sample No.4), and those having alkyl-hydroxyl chains (No.5) /13/ as side groups. $\chi^{(3)}$ values of these vacuum-deposited films range widely from 1.1 x 10^{-11} esu to 3.8 x 10^{-10} esu, presumably corresponding to such film qualities as π-electron conjugation length and spatial order of polymer chains.

Sample No.6, showing the largest $\chi^{(3)}$ value in our experiment, was prepared at University of Tokyo /15/. The film was fabricated by adding a pre-deposition process of a primary layer and unidirectional rubbing process before the main deposition to realize a highly oriented polydiacetylene film. The high anisotropy of the film due to the high orientation does have an effect on the nonlinear efficiency in this case. The value of $\chi^{(3)}$ = 3.8 x 10^{-10} esu is comparable to that of single crystals, as shown in Table 2. The vacuum-deposition method is very easy compared to the LB film method /16/ because the process is dry and not time-consuming. These features are very important for future applications with nonlinear-optical thin-film devices.

3.2 Monomer-Doped Polymer Films

Polymer films doped with nonlinear optical monomers are examined as sample Nos.1-5 shown in Table 3. Materials were prepared by dissolving the dopant monomer into methylmethacrylate monomer or vinyl-alcohol monomer, and polymerizing the solutions. Films were fabricated by casting. A sample thickness up to about 1 mm is possible in this case.

Table 3 $\chi^{(3)}$ values for many types of nonlinear optical polymer films.

No.	Samples	Concentration	$\chi^{(3)}$ ($\times 10^{-12}$ esu)	Wavelength (μm)	References
Monomer-doped polymer					
1	MNA / PMMA	2 wt%	0.061	1.90	-
2	DANS / PMMA	0.1 wt%	0.058	1.90	-
3	DMSM / PVA	10 wt%	0.37	1.90	-
Monomer-pendant polymer					
4	Azo-benzene / PMMA	25.9 mol%	1.26	2.05	/18/
5	Azo-benzene / PMA	45 mol%	3.2	1.90	-
π -conjugated polymer					
6	PPV		7.8	1.85	/19/
7	PTV		32	1.85	/20/

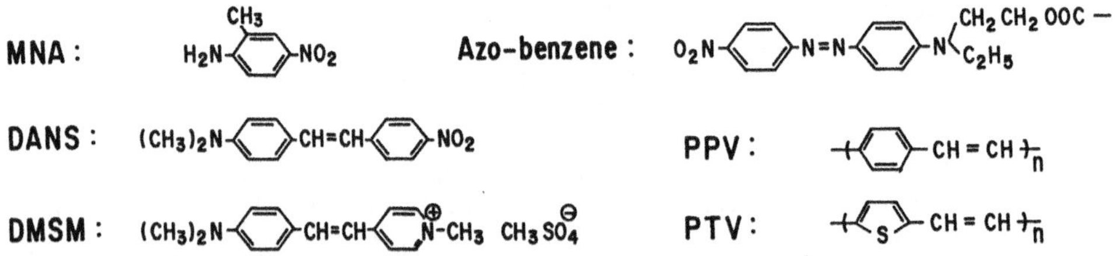

Table 4 Estimated $\chi^{(3)}$ values of nonlinear optical monomers from powder THG and solution THG measurements. λ = 1.90 μm.

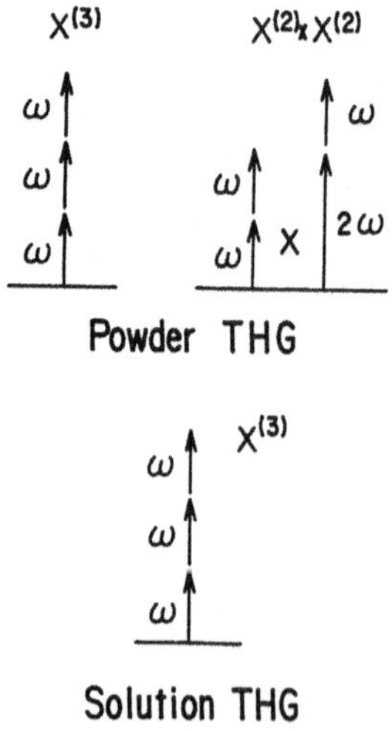

Monomers	Powder THG $\hat{\chi}^{(3)}$ (relative)	Solution THG* $\chi^{(3)}$ (esu)
pNA	1	2.6×10^{-14}
MNA	(15.2)	2.9×10^{-14}
DMSM	(21.9)	11.8×10^{-14}
DANS	3.5	-
Azo-benzene	7.7**	-
PTS	10.7**	-

* 5 wt% / ethanol solvent
** measured at λ = 2.05 μm

As dopant monomers, the following materials were examined in our experiments: 2-methyl-4-nitroaniline (MNA), 4-dimethylamino-4'-nitrostilbene (DANS), and trans-4'-dimethylamino-N-methyl-4-stilbazolium methylsulfate (DMSM) and azo-benzene.

The efficiencies of the above monomers were evaluated in advance, by powder THG and solution THG measurements. Table 4 shows the averaged relative $\chi^{(3)}$ values estimated from the powder THG intensities normalized by para-nitroaniline (pNA). SHG-active materials such as MNA and DMSM were evaluated not by powder THG but by

3.3 Other Amorphous Polymer Films

π-conjugated conducting polymers other than polydiacetylenes also have potential as large $\chi^{(3)}$ materials. Cast film samples of poly(p-phenylenevinylene) (PPV) and poly(2,5-thienylenevinylene) (PTV) were fabricated at University of Kyushu and are shown in Table 3 as samples No.6 and 7 /19,20/. The $\chi^{(3)}$ values were determined to be 7.8×10^{-12} esu and 3.2×10^{-11} esu for PPV and PTV, respectively. Since amorphous polymers have the advantage of easy film fabrication and large film thickness, intensive efforts to develop devices utilizing such materials are expected.

4. CONCLUSIONS

Successful $\chi^{(3)}$ determination by THG measurements was demonstrated using very simple equations. Accurate evaluation was achieved by considering the THG interference between the film sample and the substrate. Several kinds of polymer films were evaluated by this method. $\chi^{(3)}$ value of 10^{-10} esu was determined for vacuum deposited polydiacetylene films. $\chi^{(3)}$ values of 10^{-11} esu and 10^{-12} esu were determined for π-conjugated conducting polymers and monomer-doped polymer materials, respectively. These polymer films are important for future applications because they have the advantage of the ability to form arbitrary shapes and have good optical qualities.

ACKNOWLEDGMENTS

The authors are grateful to Drs. H. Kanbe and H. Hiratsuka for encouragement and discussions. The authors also acknowledge S. Tomaru, S. Matsumoto, T. Kurihara, and H. Kobayashi, Professors T. Koda, K. Ishikawa, and K. Takeda of University of Tokyo, and Professors S. Saito and S. Tokito of University of Kyushu, as coworkers in this work.

References

1. D. J. Williams: Nonlinear Optical Properties of Organic and Polymeric Materials, ACS Symposium Series 233 (American Chemical Society, Washington, D. C. 1983)
2. D. S. Chemla and J. Zyss: Nonlinear Optical Properties of Organic Molecules and Crystals, (Academic Press, Orlando, Florida 1987)
3. S. D. Smith: Appl. Opt., 25, 1550 (1986)
4. P. W. Smith: IEEE Circuits and Devices Magazine, May, 9 (1987)
5. D. S. Chemla, D. A. B. Miller, and P. W. Smith: Opt. Engineering, 24, 556 (1985)
6. C. Sauteret, J. P. Hermann, R. Frey, F. Pradere, J. Ducuing, R. H. Baughman, and R. R. Chance: Phys. Rev. Lett., 36, 956 (1976)
7. G. R. Meredith, B. Buchalter, and C. Hanzlik: J. Chem. Phys., 78 1533 (1983)
8. B. F. Levine and C. G. Bethea: J. Chem. Phys., 63 2666 (1975)
9. G. R. Meredith, B. Buchalter, and C. Hanzlik: J. Chem. Phys., 78 1543 (1983)
10. F. Kajzar and J. Messier: Phys. Rev. A32, 2352 (1985)

11. T. Kurihara, K. Kubodera, S. Matsumoto, and T. Kaino: Polymer Preprints, The Society of Polymer Science, Japan, 36, 1157 (1987) [in Japanese]
12. S. Tomaru, K. Kubodera, S. Zembutsu, K. Takeda, and M. Hasegawa: Electron. Lett., 23, 595 (1987)
13. S. Tomaru, K. Kubodera, T. Kurihara, and S. Zembutsu: Japan. J. Appl. Phys., 26, L1657 (1987)
14. T. Koda, K. Ishikawa, T. Kanetake, T. Nishikawa, Y. Tokura, S. Koshihara, K. Takeda, and K. Kubodera: Third Asia Pacific Physics Conference, Hong Kong, June (1988)
15. T. Kanetake, K. Ishikawa, T. Koda, Y. Tokura, and K. Takeda: Appl, Phys. Lett., 51, 1957 (1987)
16. F. Kajzar and J. Messier: Thin Solid Films, 132, 11 (1985)
17. G. R. Meredith: In Nonlinear Optical Properties of Organic and Polymeric Materials, ed. by D. J. Williams, (American Chemical Society, Washington, D. C. 1983) p.27
18. S. Matsumoto, K. Kubodera, T. Kurihara, and T. Kaino: Appl. Phys. Lett., 51, 1 (1987)
19. T. Kaino, K. Kubodera, S. Tomaru, T. Kurihara, S. Saito, T. Tsutsui, and S. Tokito: Electron. Lett., 23, 1095 (1987)
20. T. Kaino, K. Kubodera, S. Saito, S. Tokito: Micro Symposium on Nonlinear Optical Organic Materials, The Society of Polymer Science, Japan, Tokyo, May, (1988) p.48 [in Japanese]

Resonances of Cubic Nonlinear Optical Susceptibilities in Polydiacetylenes

P.A. Chollet, F. Kajzar, and J. Messier

CEA/IRDI/D.LETI/DEIN, Laboratoire de Physique Electronique des Matériaux, DEIN, CEN Saclay, F-91191 Gif-sur-Yvette Cedex, France

1. INTRODUCTION

Conjugated organic 1D-polymers represent a great potential for practical applications in nonlinear optical devices such as electro and all-optical switches, logic systems, gates, light-modulators, etc. These potentialities are due to the large third-order nonlinear optical susceptibilities of those polymers and their processability and fast response time [1].

Among the different materials with a conjugated 1-D π-electron system polydiacetylenes (PDA) appeared to possess very large cubic susceptibilities ($\chi^{(3)}(-3\omega,\omega,\omega,\omega) = (9.8 \pm 0.5)\ 10^{-10}$ e.s.u. at 1.907 μm [2]) as shown in both theoretical [3,4] and experimental [2] pioneer works. These large nonlinearities are excitonic in origin and the exciton confinement in one dimension gives rise to large dipolar transient moments. The theoretical calculations predict an excitonic length of about 25 Å [5] while the experiments interpreted in terms of the phase filling exciton model give 30 Å from saturation absorption measurements in reflection [6] and 50 ± 25 Å in transmission [7]. Moreover this excitonic origin ensures a subpicosecond response time [6-9].

The possibility of preparing polydiacetylenes as thin films enables THG and EFISHG experiments to be performed in the wavelength absorption range of the harmonic. THG experiments of the blue form of PDA have revealed two resonance enhancements in $\chi^{(3)}(-3\omega,\omega,\omega,\omega)$ at 1.907 μm and 1.35 μm respectively due to three- and two-photon resonances [10]. Consequently it appears that the value reported in [2] was a resonant value. Similarly the EFISHG [11] experiments have confirmed that the two-photon excited state (same parity as the ground state) may generate resonances of $\chi^{(3)}(-2\omega;\omega,\omega,0)$.

In this paper we present a rather thorough study of the wavelength variation of the nonlinear susceptibilities $\chi^{(3)}(-3\omega;\omega,\omega,\omega)$ and $\chi^{(3)}(-2\omega;\omega,\omega,0)$ for several polydiacetylenes in the blue form. The analysis uses the quantum mechanical formulae derived from the time-dependent perturbation method [12, 13]. The simultaneous fits of the absorption spectrum and the two previous nonlinear susceptibilities allow one to determine with precision the energy position and width of the various odd and even excited states.

2. THEORY

The theoretical expressions for the nonlinear susceptibilities are based on the analysis by Orr and Ward [13]. They start from the time-dependent Schrödinger equation :

$$\frac{\delta}{\delta t} \mid \Psi(t) \rangle = (i\hbar)^{-1} H(t) \mid \Psi(t) \rangle$$

Springer Proceedings in Physics, Vol. 36
Nonlinear Optics of Organics and Semiconductors
Editor: T. Kobayashi © Springer-Verlag Berlin, Heidelberg 1989

where $H(t) = H(0) + \epsilon(t)$; $\epsilon(t)$ is the time-dependent perturbation $-eE(t).r$ where $E(t)$ is the oscillating laser light with amplitude $E(t) = E_0 \cos\omega t$. The unperturbed hamiltonian $H(0)$ has eigenvalues $\hbar\Omega$ and normalized eigenfunctions $|n\rangle$: $H(0)|n\rangle = \hbar\Omega_n |n\rangle$. Spontaneous decay is accommodated in a phenomenological way by allowing Ω_n to be complex : $\Omega_n = \omega_n - i\Gamma_n/2$. In the Born-Oppenheimer approximation the wavefunctions are written as a product of electronic and vibronic functions : $|n\rangle = |\Phi_n\rangle \times |\Theta_n^j\rangle$ with energy $\Omega_n^j = \omega_n^j - i\Gamma_n^j/2$ equal to the sum of the electronic and vibronic energy. As the π-electron backbone of the polymers studied (polydiacetylenes) is centrosymmetric the electronic wavefunctions $|\Phi_n\rangle$ are classified into even (g) and odd (u) states.

2.1. Linear susceptibility

Linear electronic polarisability is given by ref. [13], equation 43a. In addition we have taken account of the vibronic functions $|\Theta\rangle$:

$$\alpha_{xy}(-\omega;\omega) = -\left(\frac{e^2}{h}\right) \sum_n{}' \sum_j \left\{ \frac{\langle\Phi_0 | x |\Phi_n\rangle\langle\Phi_n | y |\Phi_0\rangle \, |\langle\Theta_0^0|\Theta_n^j\rangle|^2}{(\Omega_n^j - \Omega_0^0 - \omega)} + \frac{\langle\Phi_0 | x |\Phi_n\rangle\langle\Phi_n | y |\Phi_0\rangle \, |\langle\Theta_0^0|\Theta_n^j\rangle|^2}{(\Omega_n^j - \Omega_0^0 + \omega)} \right\}, \qquad (1)$$

$$\alpha = \alpha' + i\alpha'' \quad .$$

Obviously the sum is restricted to electronic excited states $|\Phi_n\rangle$ of opposite parity to the ground state $|0\rangle$ (one-photon states) and vibronic states $|\Theta_n^j\rangle$ of the same parity as the vibronic fundamental. The resonances of α'', directly connected to the optical absorptions, give the energy position and width (Γ) of the one-photon states. It is straightforward to show from equation (1) that these resonances are lorentzian.

2.2 Nonlinear susceptibilities

In all cases the nonlinear susceptibilities are given by equation 43c of ref [13]. In addition we have introduced the vibronic functions. This has the consequence that the product of the electronic transient moments must be multiplied by the scalar products of vibronic waves :

$\langle\Theta_0^0|\Theta_\ell^i\rangle\langle\Theta_\ell^i|\Theta_m^j\rangle\langle\Theta_m^j|\Theta_n^k\rangle\langle\Theta_n^k|\Theta_0^0\rangle$ for the first Σ', with additional summation over the vibronic states i, j, k ;

$\langle\Theta_0^0|\Theta_m^i\rangle\langle\Theta_m^i|\Theta_0^0\rangle\langle\Theta_0^0|\Theta_n^j\rangle\langle\Theta_n^j|\Theta_0^0\rangle$ for the second Σ', with additional summation over the vibronic states i, j.

In addition in the energy difference Ω_{m0}^i, the vibronic energy of the electronic excited states has been taken into account.

a. In $\chi^{(3)}(-3\omega;\omega,\omega,\omega)$ there are two kinds of resonances :

- one- and three-photon resonances on electronic excited $|n\rangle$ of opposite parity to the ground state (already observed in linear optical absorption spectra)
- two-photon resonances on excited states of the same parity as the ground state (which are not observed by linear absorption).

b. In $\chi^{(3)}(-2\omega;\omega,\omega,0)$ two-photon resonances occur on both kinds of excited states irrespective of their symmetry. This is due to the fact that the centrosymmetry is broken by the applied electric field.

Strictly speaking the theoretical expressions given by Orr and Ward [13] stand for γ rather than for $\chi^{(3)}$. The relations linking γ to $\chi^{(3)}$ are as follows :

$$\chi^{(3)}(-3\omega;\omega,\omega,\omega) = N < f_\omega^3 \, f_{3\omega} \, \gamma(-3\omega;\omega,\omega,\omega)>$$

$$\chi^{(3)}(-2\omega;\omega,\omega,0) = N < f_\omega^2 \, f_{2\omega} \, f_0 \, \gamma(-2\omega;\omega,\omega,0)> \quad .$$

In these expressions N is the number of molecules per volume unit, and the f's are local field factors arising from the difference between the microscopic and macroscopic electric fields. They are related to the molecular frame as γ. The average $< >$ is taken over molecule orientation in the film.

As the polymeric backbone is elongated in one dimension, the factors f for parallel polarization are identically equal to 1. The factors f for perpendicular polarization are functions of the dielectric constants of the π-electron backbone and of the sidegroups. The precise variation is of no great importance as the perpendicular cubic hyperpolarizability is negligible compared to the parallel one. Nevertheless in the vicinity of an absorption band, the mean value of the dielectric film constant may vary sufficiently to affect the Fresnel transmission factor of the generated harmonic light.

3. EXPERIMENTAL PROCEDURES

3.1 Film preparation

Nonlinear susceptibility measurements were performed on thin films (0.1 to 0.3 μm) of various polydiacetylenes :

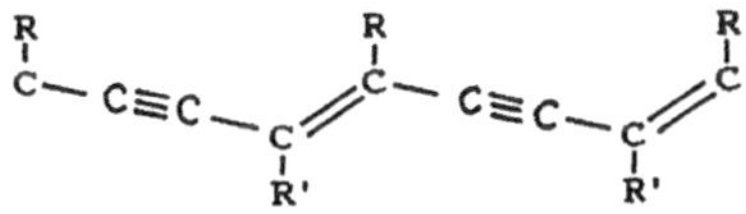

Thin films of monomer are deposited on the silica substrate either by the Langmuir-Blodgett technique (AFA 16/8 ; $R = (CH_2)_{16} - CH_3$; $R' = (CH_2)_8 - COOH$) or by evaporation under vacuum (AFA 16/8, 4BCMU, DCH).

These films are made up of small single crystals. The different lateral groups allow one to choose the film deposition procedure :

(i) Amphiphilic for Langmuir-Blodgett technique. Addition of Cd^{++} ions in the water subphase (10^{-5} N) stabilizes the monomolecular film but may prevent complete polymerization of the transferred film.

(ii) Symmetric diacetylenes (4BCMU : $R = R' = (CH_2)_4 - OCONH-CH_2-COOC_4H_9$) ; DCH : $R = R' = CH_2NC_{12}H_8$) are evaporated under vacuum (10^{-6}T). AFA 16/8 films are also obtained by vacuum evaporation. The absence of Cd^{++} ions allows a quite complete polymerization.

The monomer thin films are polymerized by the UV radiation of a deuterium lamp (AFA 16/8 ; 4BCMU) or thermally for DCH (24h at 150°C). The obtained polydiacetylene thin films are made of microcrystals. Light scattering by these crystals becomes important if their lateral dimensions match either the harmonic or fundamental wavelength. In the case of evaporated films light scattering is strongly reduced by cooling the substrate during evaporation (- 20°C), which results in submicronic lateral dimensions of microcrystals.

3.2 Nonlinear optical measurements

The harmonic generation measurements are performed in the 0.9–1.9 μm fundamental wavelength range using an experimental setup previously described [10,11], see Fig. 1 : a pulsed Nd:YAG laser beam (pulse duration : 13 ns ; energy : 1 J ; wavelength : 1.064 μm) is frequency doubled after crossing a KDP crystal. The 0.532 μm pulses act as pump pulses for a dye laser. The frequency emitted by the dye laser is shifted towards infrared by crossing a high pressure hydrogen cell (stimulated Raman scattering). The desired frequency (1st or 2nd Stokes) is selected by a Pellin-Broca prism. After focusing on the sample the fundamental wavelength is rejected by appropriate filters and the harmonic selected by a monochromator and detected by a photomultiplier. The electric signal delivered by the photomultiplier is integrated by a boxcar triggered in synchronism with the laser. All the experiment is computer controlled.

(i) In the case of THG, care was taken to focus the laser beam on the sample (or silica calibrating slab) under vacuum, in order to avoid air contribution which can give false results [14]. The harmonic signal is averaged by translating the sample. The absolute value of $\chi^{(3)}(-3\omega;\omega,\omega,\omega)$ is obtained by comparing with the Maker frirges emitted by a rotating fused silica slab [10] whose $\chi^{(3)}$ value is 2.79×10^{-14} esu [15].

(ii) In the case of EFISHG, experiments are performed under vacuum, only to avoid dielectric breakdown by the applied high voltage (5 kV/mm). As previously observed [11,16] an internal polarization is set up in the film as a consequence of photoinduced electron-hole pair generation (via multiphoton absorption of the fundamental or direct absorption of the harmonic generated in the film) and separation by the applied electric field reduces the second harmonic signal. This phenomenon may result in underestimation of $\chi^{(3)}(-2\omega;\omega,\omega,0)$. This artefact was avoided by reducing the incident laser power, scanning the sample and alternating the applied electric field. Under these experimental conditions the second harmonic power generated has a perfect quadratic dependence on the fundamental power and electric field, as expected. The absolute value

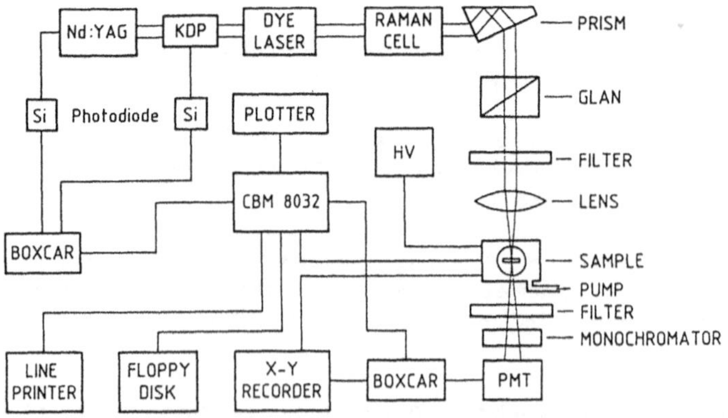

Figure 1 : Experimental setup for nonlinear optical susceptibility measurements

of $\chi^{(3)}(-2\omega;\omega,\omega,0)$ is obtained by calibrating on the second harmonic (Maker fringes) emitted by a wedged α-quartz single crystal whose $\chi^{(2)}(-2\omega;\omega,\omega)$ is 2.25×10^{-9} esu for $\lambda = 1.064$ µm and 2.08×10^{-9} esu for $\lambda = 1.91$ µm [17].

4. RESULTS AND DISCUSSION

4.1 Absorption spectra

Linear optical absorption measurements as well as THG and EFISHG experiments were performed on the blue form of 4BCMU, AFA 16/8 and DCH. Absorption spectra are shown in Fig. 2. It appears that the peak wavelength position depends on sidegroups.
With the use of equation (1) these absorption spectra are split into four lorentzian components using a least squares procedure. The energy difference Ω^i_{u0} and widths Γ^i_u corresponding to the first four even vibronic levels of the one-photon excited state are listed in Table 1. The absorption intensities are proportional to $|\langle\theta^0_g|\theta^i_u\rangle|^2$.

4.2 Nonlinear susceptibilities

The experimental dispersion of these susceptibilities is fitted with a three electronic level model, the ground state $|g\rangle$, the one-photon excited state $|u\rangle$ and a two-photon state $|g'\rangle$ already observed experimentally in AFA 16/8 [10,11]. All the parameters concerning the state $|u\rangle$ are kept without change, as they are listed in Table 1. In the absence of precise knowledge of the vibronic state energy levels inside $|g'\rangle$, the same distribution as inside $|u\rangle$ is assumed (Fig. 3). This implies the same vibronic wavefunction extension and consequently : $\langle\theta^i_u|\theta^j_{g'}\rangle \simeq \langle\theta^i_u|\theta^j_u\rangle = \delta_{ij}$ where δ_{ij} is the usual Kronecker symbol.

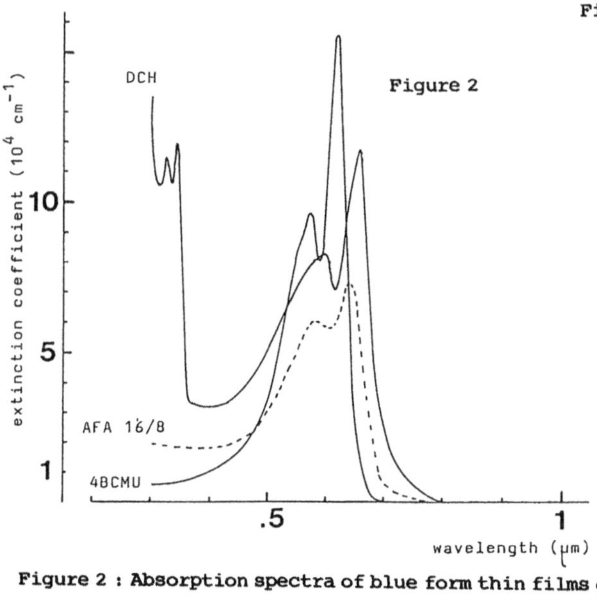

Figure 3

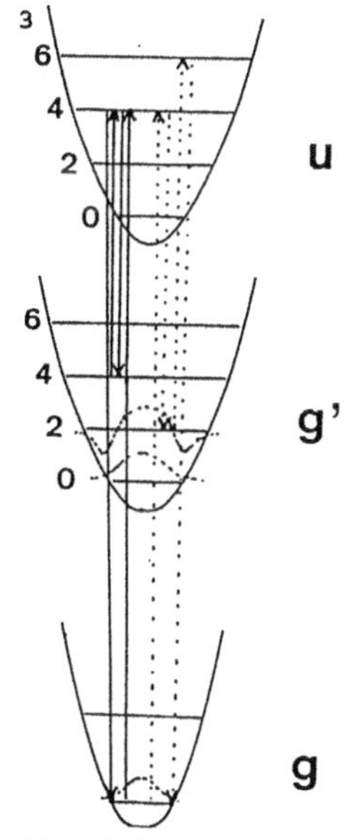

Figure 2 : Absorption spectra of blue form thin films of polydiacetylenes with different sidegroups

Figure 3 : Transient dipolar moments used (continuous lines) for nonlinear susceptibility calculations. The dotted line paths are not taken into consideration

Actually in the calculation of $\chi^{(3)}$, $\langle\theta_u^i|\theta_{g'}^j\rangle$ is taken equal to δ_{ij}. This reduces the terms in the summation to those indicated in Fig. 3. On the other hand the widths $\Gamma_{g'}^j$ are taken as free parameters.
Least squares fits performed simultaneously on $\chi^{(3)}(-3\omega;\omega,\omega,\omega)$ and $\chi^{(3)}(-2\omega,\omega,\omega,0)$ were made by varying the electronic energy of $|g'\rangle$ and the widths $\Gamma_{g'}^j$. The numerical results are gathered in Table 2 and the theoretical dispersion curves compared with the experimental ones in Figs. 4–6.

TABLE 1 : Energies and widths of the four vibronic levels of the electronic excited state $|u\rangle$ (cm^{-1})

Level		4BCMU	AFA 16/8	DCH
0	Ω	16160	15600	15210
	Γ	390	560	600
Intensity		51 %	42 %	43 %
2	Ω	17300	17180	16610
	Γ	820	1220	550
Intensity		25 %	32 %	17 %
4	Ω	18380	18500	17400
	Γ	800	700	550
Intensity		16 %	8 %	13 %
6	Ω	21320	21000	19000
	Γ	1980	2600	2100
Intensity		6 %	16 %	26 %

Table 2 : Energies and widths of the four vibronic levels of the electronic excited state $|g'\rangle$ (cm^{-1})

Level \ Eu-Eg'		4BCMU	AFA 16/8	DCH
Eu-Eg'		500	620	300
0	Ω	15660	14980	14910
	$\Gamma_{g'}^0$	1290	760	720
2	Ω	16800	16560	16310
	$\Gamma_{g'}^2$	440	620	670
4	Ω	17880	17880	17100
	$\Gamma_{g'}^4$	480	300	670
6	Ω	20820	20380	18700
	$\Gamma_{g'}^6$	1100	1100	1800

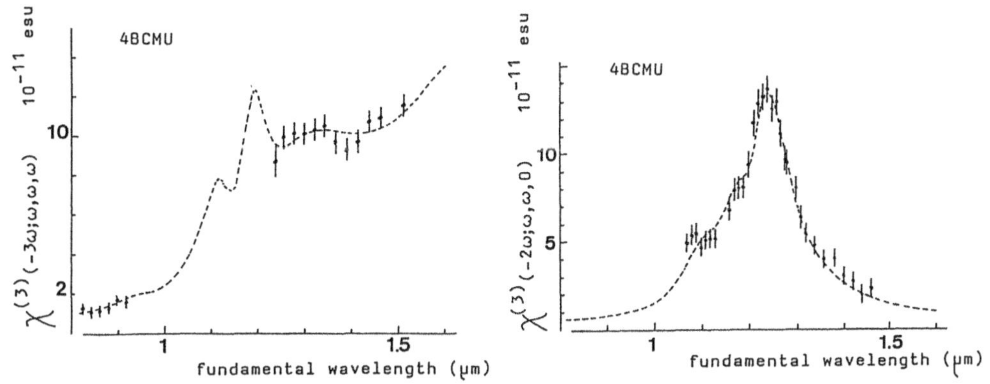

Figure 4 : Cubic susceptibilities of 4BCMU

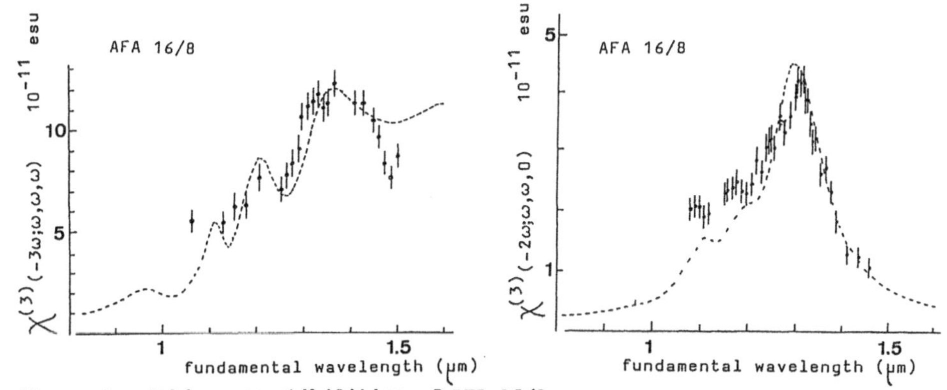

Figure 5 : Cubic susceptibilities of AFA 16/8

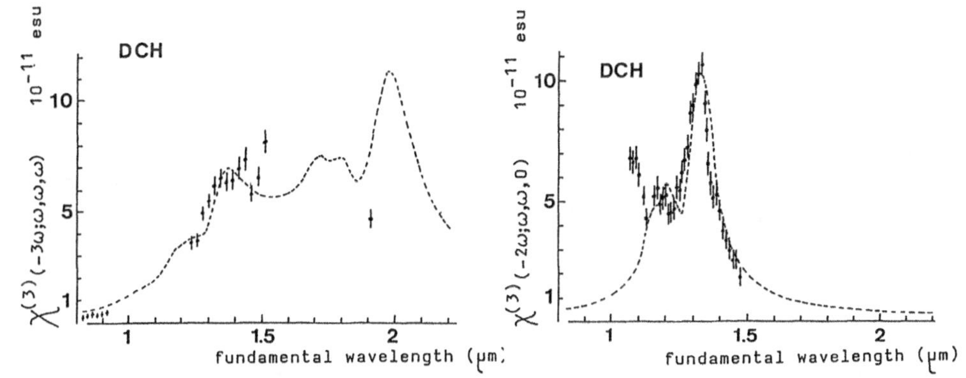

Figure 6 : Cubic susceptibilities of DCH

Figures 4-5-6 : Comparison of the measured and calculated (broken lines) optical cubic susceptibilities for several kinds of polydiacetylene films (blue form). The parameters used in the calculations are listed in Tables 1 and 2

4.3 Discussion

a. Two-photon level energy

As already observed experimentally, the fit of the cubic susceptibility dispersion shows that the excited state |g'> lies under |u> in the blue form of polydiacetylene. Note that the energy difference between these two excited states is small. This is in accordance with the large extension of the exciton [5-7], in which case the distinction between g and u states becomes weak.

b. Two-photon resonance intensity

In THG the three-photon resonance on |u> and two-photon resonance on |g'> are well separated, as they occur for $\omega = \Omega_{ug}/3$ and $\omega = \Omega_{g'g}/2$ respectively. Such is not the case for EFISHG where two-photon resonances of $\chi^{(3)}(-2\omega;\omega,\omega,0)$ occur on |u> as well as on |g>. This is due to the breaking of the centrosymmetry by the electric field. The greatest contribution to the peak resonance around 1.3 μm is due to the cross term of the type

$$\frac{1}{(\Omega_{ug}-2\omega)(\Omega_{g'g}-2\omega)(\Omega_{ug}-\omega)}$$

and was found to be ≃ 70 % of the peak.

c. Influence of the optical quality of the film

Comparison of Figs. 4-6 shows the best fit over the whole wavelength range is obtained for 4BCMU. It is clear from the absorption spectra (Fig. 2), that 4BCMU films have high optical quality. When light scattering is important as in AFA 16/8, and to a lesser extent in DCH, the decreasing transmission of the film towards short wavelengths is artificially taken into account by increasing the vibronic bands intensity of |u> and even shifting them towards higher energy. Owing to the assumption of the same repartition of vibronic levels in |g'>, these two factors together may perturb the fit.

5. CONCLUSION

In this paper, we have made a thorough study of the wavelength dispersion of nonlinear optical cubic susceptibilities in thin polydiacetylene films in the blue form. We have compared the experimental wavelength dispersion of susceptibilities with the theoretical ones obtained from time-dependent perturbation theory including level widths. The calculations are made under the assumption of centrosymmetric isolated molecules. Another possibility would be to consider the polydiacetylene as an infinite molecule described by a band structure [4]. Obviously both descriptions should lead to similar results in the case of very long molecules. The description we used has the advantage of permitting a rather straightforward calculation in the case of $\chi^{(3)}(-2\omega;\omega,\omega,0)$ for instance.

Owing to the centrosymmetry of the polymeric π-electron backbone, the states are classified into even (g) and odd (u) states. We have used a three- level model, the ground state |g>, a one-photon excited state |u> and a two-photon state |g'>. Each electronic excited state is split into four vibronic levels which are connected to the zero-phonon $|g^0\rangle$ state by their overlap, so they possess even symmetry.

By fitting absorption spectra it is possible to determine the energies, widths and overlaps $\langle g^0|u^j\rangle$ of the one-photon state (j denotes the vibronic state). Assuming the same distribution of vibronic levels in $|g'\rangle$ as in $|g\rangle$, it is possible to determine quite accurately the electronic energy level. It is found that $E_{g'}<E_u$, with a difference energy ranging from 300 to 600 cm^{-1} depending on the sidegroups.

The experimental values are available even in the harmonic absorption region because it is possible to obtain films thin enough ($\geqslant 0.3$ µm) to reduce absorption. Simultaneous fits of absorption, $\chi^{(3)}(-3\omega;\omega,\omega,\omega)$ and $\chi^{(3)}(-2\omega;\omega,\omega,0)$ are very good in the whole wavelength region for high optical quality films (4BCMU). A larger discrepancy is found in the short wavelength range for $\chi^{(3)}(-2\omega;\omega,\omega,0)$ in films which scatter light.

Some improvements in the theory may be undertaken such as a better knowledge of the film dielectric variation in the vicinity of an absorption and obtaining the true distribution of vibronic levels in $|g'\rangle$ from photo-induced absorption.

ACKNOWLEDGMENTS

The authors would like to acknowledge Dr D. Grec for synthesis of diacetylene monomers, and A. Lorin and R. Gras for experimental assistance.

REFERENCES

[1] NATO Advanced Research Workshop, Sophia-Antipolis, France, June 19-24, 1988, to be published in NATO ASI Series : Nonlinear optical effects in polymers
[2] C. Sauteret, J.P. Hermann, R. Frey, F. Pradère, J. Ducuing, R.H. Baughman and R.R. Chance, Phys. Rev. Lett. 36, 956, (1976)
[3] K.C. Rustagi and J. Ducuing, Opt. Commun. 10, 258, (1974)
[4] G.P. Agrawal, C. Cojan and C. Flytzanis, Phys. Rev. B17, 776, (1978)
[5] S. Suhai, J. Chem. Phys. 85, 611, (1986)
[6] B.I. Greene, J. Orenstein, R.R. Millard and L.R. Williams, Phys. Rev. Lett. 58, 2750, (1987)
[7] F. Kajzar, L. Rothberg, S. Etemad, P.A. Chollet, D. Grec, A. Boudet and F. Jedju, Opt. Comm. 66, 55, (1988)
[8] G.M. Carter, J.V? Hryniewicz, M.K. Thakur, Y.J. Chen and S.E. Meyler, Appl. Phys. Lett. 49, 998, (1986)
[9] G.M. Carter and E.P. Ippen, Seventh Annual Conference on Lasers and Electro-Optics, (1987)
[10] F. Kajzar and J. Messier, Thin Solid Films 132, 10, (1985)
[11] P.A. Chollet, F. Kajzar, J. Messier, Thin Solid Films 132, 1, (1985)
[12] J.F. Ward, Rev. Mod. Phys. 37, 1, (1965)
[13] B.J. Orr and J.F. Ward, Mol. Phys. 20, 513, (1971)
[14] F. Kajzar and J. Messier, Phys. Rev. A32, 2352, (1985)
[15] G.R. Meredith, B. Buchalter and C. Hanzlik, J. Chem. Phys. 78, 1533, (1983)
[16] P.A. Chollet, F. Kajzar, J. Messier, NATO ASI Ser. n°102 p.317, (1985)
[17] G.R. Meredith, Phys. Rev. B24, 5522, (1981)

Third-Order Nonlinear Susceptibility of Polyalkylthiophene

T. Fukaya[1], *A. Heinämäki*[2], *and H. Stubb*[2]

[1]National Chemical Laboratory for Industry, 1-1, Higashi, Tsukuba-shi, 305 Japan
[2]Semiconductor Laboratory, Technical Research Centre of Finland, SF-02150 Espoo, Finland

We have measured the nonlinear refractive index of poly(3-octylthiophene) (POT) by a surface plasmon resonance (SPR) method. The SPR sample consisting of a 0.4-0.5 μm thick POT layer was prepared by spin coating from a chloroform solution onto a silver evaporated holographic grating. The nonlinear refractive index, $n_{NL} = (9 \pm 3) \cdot 10^{-16} m^2/W$ at 1.06 μm, is about one order of magnitude larger than that of other conjugated polymers.

1. Introduction

The nonlinear optical properties found in the conjugated structure of conducting polymers have been given much attention recently [HEEGER et al. /1/]. The π electron delocalization gives rise to nonresonant optical nonlinearity useful especially in integrated optics, where the powers and pathlengths of light in the waveguides are relatively large and attenuation should be minimized. The immense possibilities of tailoring organic materials nourish such application efforts. These are further facilitated by achievements in processability of conducting polymers e.g. the dissolvable polyalkylthiophenes [RUGHOOPUTH et al./2/].
The nonlinear refractive index is a key property for many applications and various methods are available for measuring it. Time-resolved interferometry [MILAM et al./3/] gives an accurate result directly, but it requires large samples of good optical quality. This is not necessary in three- or four-wave mixing techniques [ADAIR et al./4/, LEVENSON/5/, MILLER et al./6/] but in some cases they require further information on the nonlinear process involved. Measurement by surface plasmons [CHEN et al./7/] is not only simpler than either method, but it also gives the sign of the nonlinearity. The nonlinear refractive index of polydiacetylene was measured using this method [CARTER et al./8/]. In this work we measured the nonlinear refractive index of polyoctylthiophene films by surface plasmons.

2. Experimental

In the surface plasmon resonance (SPR) method the refractive index of a material is measured by coupling light to a plasmon mode at the interface of this material and a metal. The nonlinear refractive index is determined from the dependence of the reflectivity on the angle, where the light couples to surface plasmons. The coupling of light into a plasmon mode causes a dip in the reflection curve, and in the presence of a nonlinear material the corresponding angle becomes intensity dependent.

Springer Proceedings in Physics, Vol. 36
Nonlinear Optics of Organics and Semiconductors
Editor: T. Kobayashi

The main optical set-up includes a laser, goniometer and detector, as described e.g. in [CHEN et al./7/]. Our light source was a p-polarized 1.06 μm Nd^{3+}:YAG laser.

The poly(3-octylthiophene) (POT) used in this work was synthesized chemically by the nickel-catalyzed dehalogenating polycondensation of 3-octyl-2, 5-diiodothiophene. The analogous synthesis of poly(3-hexylthiophene) has been described elsewhere [INGANÄS et al./9/]. The degree of polymerization was about 50 in our samples. In order to couple light into plasmon modes, it is necessary to produce a grating at the boundary between POT and metal [AGRANOVICH et al./10/].
This was performed photolithographically by spin coating a glass substrate with AZ-1470 photoresist and generating the relief type hologram grating with 458 nm Ar laser radiation. A silver layer of 0.3 μm was vacuum evaporated onto the grating surface. The grating space was 1 μm and the relief height was 0.1 μm. Finally, a 0.4-0.5 μm thick POT layer was spun from a chloroform solution onto the metal.

3. Results

We observed a dip in the reflection for high and low incident intensity as shown in Fig.1. The coupling angles are θ_1=20.5 and θ_2=21 and the diffraction order m=1. The corresponding average powers and pulse rates are P_1=210 mW, P_2=35 mW, R_1=500 pulse/s and R_2=5000 pulse/s. For calculation of the peak powers, the pulse width 110 ps (FWHM) and beam diameter 0.9 mm were used. The incident intensity has to be corrected for reflection at the air-POT interface and for field enhancement at the POT-Ag interface. The reflected portion of the incident radiation at the first interface was found theoretically to be 29 %. The field enhancement factor (the ratio between the squared field amplitudes at the metal surface and in POT) was calculated from the theory of HANSEN/11/. The formulas were modified to take into account the absorption in the metal [BORN et al./12/]. In this way we found the value $\Gamma \lesssim 2.1$ at $\theta \approx 21°$. For

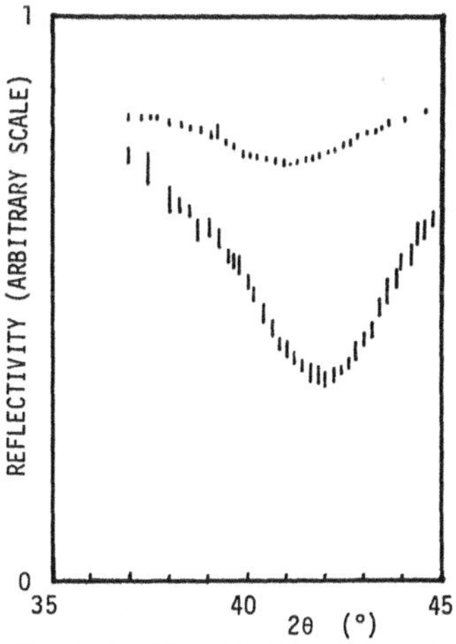

Fig.1 Reflectivity dependence of the angle, where the light couples to surface plasmons. The upper curve corresponds to the incident intensity I_1 = 885 MW/cm^2, the lower curve to I_2 = 0.025·I_1.

the complex refractive index $n=n_{Re}+in_{Im}$ of Ag we used the values $n_{Re}=0.28$, $n_{Im}=7$ [WEAST et al./13/]. Finally, the calculated intensities are $I_1=885$ MW/cm^2, $I_2=0.025\cdot I_1$. According to the theory of surface plasmons [CHEN et al./7/, AGRANOVICH et al./10/], a plasmon is excited when the following coupling condition is fulfilled:

$$\frac{\omega}{c}\left(\frac{\varepsilon_a\,\varepsilon_b}{\varepsilon_a+\varepsilon_b}\right)^{1/2}\left(1+\frac{n_{NL}\,\varepsilon_a\,I}{\varepsilon_b^{1/2}(\varepsilon_a+\varepsilon_b)}\right)=\left|\frac{\omega}{c}\sin\theta+\frac{2\pi m}{\Lambda}\right| \qquad (1)$$

where

ω = angular frequency of laser light
c = speed of light
ε_a, ε_b = first order permittivities of metal and nonlinear material, respectively
I = light intensity
θ = angle of incidence of the radiation
Λ = grating constant
m = diffraction order
n_{NL} = proportionality constant to be determined in intensity-dependent refractive index $n_b=(\varepsilon_b)^{1/2}+n_{NL}\,I$

From this formula, we can calculate

$$\left|n_{NL}\right|=\left|\left(\frac{\varepsilon_a+\varepsilon_b}{\varepsilon_a}\right)^{3/2}\cos\theta_1\frac{\Delta\theta}{\Delta I}\right|\approx(9\pm3)\cdot10^{-16}\ m^2/W \qquad (2)$$

where the values $\varepsilon_a=-49$, $\Lambda=1\ \mu m$ and $\varepsilon_b^{1/2}=n_{POT}=1.7$ were used. The last is an average value for conjugated polymers [FLYTZANIS /14/], and the possible error made with this assumption is not significant since n_{NL} depends only weakly on n_{POT}. The error limits are determined by the inaccuracy in n_{POT} and especially in the coupling angles,

4. Discussion

The measured value for the nonlinearity in POT is somewhat larger than those reported for other polymers. We believe that the nonlinearity is of electronic origin, although thermal origin is hard to exclude completely when using the present method. If the temperature is increased considerably, changes in the material occur. Except for very high temperatures, the changes are reversible but the relaxation time is very long, over days or weeks [INGANÄS et al./9/].
Our results were tested in the following way. We made measurements at high power levels close to the threshold of optical damage. The temperature effects were visible in the asymmetric shape of the reflection curve around the plasmon coupling angle. The powers used to determine the nonlinearity were considerably smaller and the corresponding reflection curves were symmetric. Another test was made by comparing two low-power measurements made before and after a high-power one. The results were found to be alike. Furthermore, the effective powers were of the same magnitude as those used by CHEN et al./8/. They concluded that the non linearity is of electronic origin. The values of the optical nonlinearity now measured in POT should be sufficient for e.g. optical switching applications with a power of some hundred mW [INGANÄS et al./9/].

However, for these applications other material parameters such as relaxation time, absorption and maximum change in refractive index are also important and should be measured.

Acknowledgment

The financing by the Technology Development Centre of Finland for T. Fukaya's visit to Finland is gratefully acknowledged. The financial support by Emil Aaltonen Foundation and Swiss Federal Institute of Technology is acknowledged by A. Heinämäki. The polymer material was synthesized by Dr. J.-E. Österholm's group at Neste Ltd.

References

1. A.J.Heeger, D.Moses, M.Sinclair, Synth.Met. 15,95(1986)
2. S.D.D.V.Rughooputh, M.Nowak, S.Hotta, A.J.Heeger, F.Wudl, Synth.Met. 21, 41 (1987)
3. D.Milam, M.J.Weber, J.Appl.Phys. 47, 2497 (1976)
4. R.Adair, L.L.Chase, S.A.Payne, J.Opt.Soc.Am. B 4, 875 (1987)
5. M.D.Levenson, IEEE J.Quant.El.QE-10, 110 (1974)
6. D.A.B.Miller, D.S.Chemla, D.J.Eilenberger, P.W.Smith, A.C.Gossard, W.Wiegmann, Appl.Phys.Lett. 42, 925 (1983)
7. Y.J.Chen, G.M.Carter, Appl.Phys.Lett. 41, 307 (1982)
8. G.M.Carter, Y.J.Chen, S.K.Tripathy, Appl.Phys.Lett. 43, 891 (1983)
9. O.Inganäs, W.R.Salaneck, J-E.Österholm, J.Laakso, Synth.Met. 22, 395 (1988)
10. Surface Polaritons, ed. by V.M.Agranovich and D.L.Mills, Ch.3, North-Holland, Amsterdam (1982)
11. W.N.Hansen, J.Opt.Soc.Am. 58, 380 (1968)
12. M.Born, E.Wolf, Principles of Optics, Pergamon Press, Inc., New York (1959)
13. CRC Handbook of Chemistry and Physics, ed. by R.C.Weast and M.J. Astle, 60th Edition, CRC Press (1981)
14. C.Flytzanis "Optical Nonlinearities and Photoinduced Solitons in Conjugated Polymer Crystals" in Nonlinear Optical Properties of Organic and Polymeric Materials, ed. by D.J.Williams, ACS Symposium Series 233 (1983)

Highly Oriented Polydiacetylene Films for Possible Nonlinear Optical (NLO) Application: Method of Fabrication, Morphology and Some NLO Properties

T. Kanetake[1], *K. Ishikawa*[1], *T. Koda*[1], *K. Takeda*[2], *K. Kubodera*[3], *and H. Kobayashi*[3]

[1]Department of Applied Physics, University of Tokyo, Tokyo 113, Japan
[2]Department of Synthetic Chemistry, University of Tokyo, Tokyo 113, Japan
[3]NTT Opto-electronics Laboratories, Atsugi-shi, Kanagawa 243-01, Japan

1 Introduction

Polydiacetylenes (PDAs) have been attracting considerable interest for the feasibility of application to NLO devices. The notable features of these PDAs are the large third-order electric susceptibility $\chi^{(3)}$, the very fast response time, and the fairly high threshold for irradiation damage. These specific features are closely associated with the conjugated π-electron system in the polymer backbone chains. To take the best advantage of these features, it is necessary to fabricate high quality PDA films, in which individual polymer chains are spatially aligned along a specific direction. Such a highly-oriented PDA film is useful, not only to optimize anisotropic NLO response of PDA, but also to minimize the light scattering loss which is supposed to be a serious problem in an optical wave-guide device with a PDA clad layer, for example. Here, we report on our recent study on highly-oriented PDA films which have been prepared by the vacuum-deposition technique / 1,2/. Some details of the fabrication method and the morphological examination are described, together with preliminary NLO results.

2 Method of Film Fabrication

Some diacetylene monomers can be vaporized without being thermally polymerized or decomposed in a heating crucible, and can be deposited upon a suitable substrate to form a uniform monomer film. Subsequent irradiation with UV light converts these monomer films to PDA films /1/. The feasibility of fabrication of high quality PDA films is strongly dependent on the chemical and thermal properties of the diacetylene monomer, R-C≡C-C≡C-R. Most of monomers which are useful for vacuum-deposition contain in R an urethane group (-OCONH-) which is linked to the conjugated carbons by an alkyl chains. The hydrogen bonds which are formed between the urethane groups in adjacent side-groups are supposed to stabilize a plane structure of the polymer crystal when the monomer crystal is photo-polymerized.

In order to discover the optimum condition required for the chemical structure of diacetylene monomers suitable for vacuum-deposition, we synthesized a series of urethane-based diacetylene monomers with the side-group $[(CH_2)_m OCONH\text{-}C_nH_{2n+1}]$, m and n determining the lengths of alkyl chains on both sides of the urethane group. The melting point, thermal and optical polymerization activities, which are important factors in the evaporation process, can be systematically changed by m and n, thus allowing us to choose appropriate monomers for vacuum deposition. As a result, we have found that those monomers with m=4 are useful. (We designate the corresponding PDAs as PDA-C_mUC_n.) Furthermore, these PDA films , which are otherwise composed of randomly oriented microcrystalline grains, can be made to have a fairly high degree of structural as well as optical anisotropy by means of a suitable treatment /2/ (especially for PDA with m=4 and n=3,4,5). The procedure of highly-oriented PDA films consists of the following sequential steps: First, the surface of the substrate, usually a glass plate or a fused silica plate (other

Springer Proceedings in Physics, Vol. 36
Nonlinear Optics of Organics and Semiconductors
Editor: T. Kobayashi

materials are also applicable), is coated by a primary thin monomer layer (about 10 nm thick) by vacuum-deposition. The evaporation temperature is about 390K and the deposition speed is about several 0.1 nm per min. Then the monomer layer is photo-polymerized in air by UV light from a high pressure Hg lamp. In the next step, this primary PDA layer is softly rubbed by a silicone cloth in one direction, until the blue-colored polymer layer is almost wiped off the substrate surface. According to optical microscopic observations, however, there remains a very thin polymer layer in which individual polymer grains are preferentially aligned along the rubbing direction. The final step is the vacuum-deposition of a thicker monomer layer, usually of 100-200 nm thickness, upon the same substrate, and the subsequent photo-polymerization of this secondary film. It was found that the optical absorption spectra of PDA films prepared in this way are highly anisotropic, as seen in Fig.1. The absorbance ratio for the polarized light parallel and perpendicular to the rubbing direction is as large as 30 or more at the exciton absorption peak at about 2.0 eV in the blue PDA films.

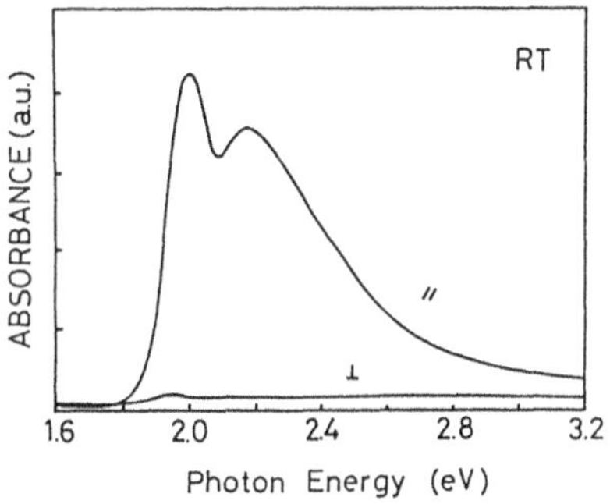

Fig.1
Polarized absorption spectra of oriented PDA-C_4UC_4 film at room temperature. Polarized components in the direction parallel and perpendicular to the rubbing axis are indicated by // and ⊥.

3 Morphology of Oriented PDA Films

The mechanical treatment described above is rather empirical, being based on our previous experience to fabricate an oriented film of conducting polymer $(SN)_x$ /3/. However, there are several good reasons for anticipating a successful result when, in particular, the same method is applied to PDAs. Namely, the conjugated backbone chains of PDA have a very strong tensile strength, so that the randomly oriented polymers will be stretched out by rubbing without being torn off, leading to preferential orientation of polymer axis along the rubbing direction. In addition, since the polymer crystal is obtained by the so-called topochemical reaction of the monomer crystal, the lattice constants of monomer and polymer crystals must be almost identical. Then, one may expect that the vacuum-deposition of monomers upon the oriented polymer layer will take place in a quasi-epitaxial manner. Various morphological examinations, including optical microscopic observations, SEM study, small-angle X-ray scattering /2/, and electro-absorption measurements, indicate that above mentioned mechanisms are playing an important role.

An example of SEM image of an oriented PDA film is shown in Fig.2. From these observations, it was found that the film was composed of elongated microcrystals of polymer, with the average diameter of about 0.3 µm and that long axis was directed preferentially along the rubbing direction. The conjugated backbone chains of polymers are parallel to the long axis of elongated grains, whose (1,0,0) plane is parallel to the substrate surface. However, since the film has a microcrystalline structure, the spatial orientation of individual polymer chains will be slightly distributed in the film. In order to make a quantitative estimation of the spatial distribution of polymers, we have measured electro-absorption spectra by applying an external electric field which is directed in the following three geometries; the electric field F parallel to the rubbing direction, perpendicular to the rubbing direction but parallel to the film plane, and perpendicular to both the rubbing

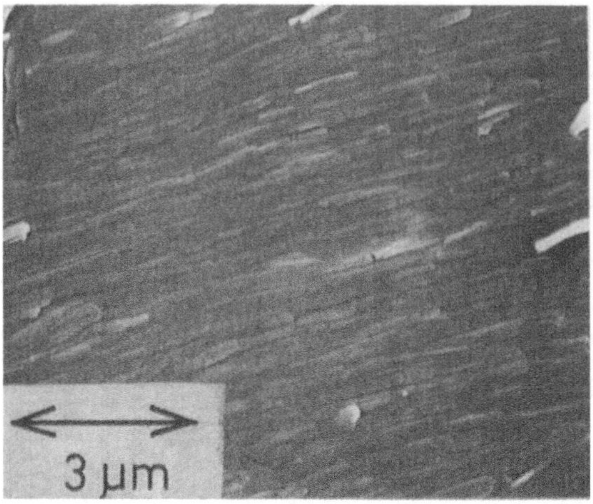

Fig.2
SEM photograph of an oriented PDA-C_4UC_4 film (80nm thick). The elongated microcrystalline grains of polymers preferentially oriented along the rubbing direction of the substrate.

direction and the film surface. Since the electro-absorption of an exciton is related to the component of the exciton transition moment projected to the direction of applied field, we can estimate the relative magnitude of the squared transition dipole moment of the π-π^* exciton projected to the three directions defined before. The result we have obtained from an analysis of the electro-absorption spectra is 1200:40:1 for the respective directions. The optical anisotropy in the same exciton absorption is in a quantitative agreement with this result.

4 Nonlinear Optical Properties of Oriented PDA Films

The absorption spectra of vacuum-deposited PDA films show a prominent peak around 2.0 eV (about 620 nm) due to the lowest singlet π-π^*(^1B_u) exciton in the conjugated polymer backbone chains. Because of the one-dimensional nature of the exciton, this absorption is predominantly polarized along the polymer axis, as seen in Fig.1. From the centrosymmetric structure of the present PDAs, these π-π^* excitations respond nonlinearly to a strong laser light via the third-order electric susceptibility $\chi^{(3)}$. The dispersion of $\chi^{(3)}$ is then expected to show a significant resonance behavior to the relevant electronic excited states. In particular, a three photon transition to the lowest singlet exciton will give rise to a strong resonance enhancement effect in the $\chi^{(3)}$ of PDA.

In order to evaluate the $\chi^{(3)}$ value of vacuum-deposited PDA films, we have measured the third harmonic generation (THG) in oriented and unoriented PDA films. The excitation was at 1.9 μm (0.65 eV), which is nearly three-photon resonant to the ^1B_u exciton. The pulsed pump beam was obtained from a Q-switched Nd:YAG laser via a frequency conversion system composed of an SHG converter, a dye laser and a $LiNbO_3$ difference-frequency generator /4/. The pulse duration of laser light was 5.5 ns with a repetition rate of 10 Hz, and the peak power density of about 80 MW/cm^2. The THG measurements were taken, for the moment, in the air at room temperature, but the effect of the environmental air has been proved to be small because we used the short focusing lens for the pumping beam. The intensities of the THG light emitted from samples were measured as a function of the rotating angle θ, as is usually done in the Maker fringe measurements.

Typical examples of the THG intensities versus the rotation angle are shown in Fig.3; Fig.3(a) and (b) for an oriented PDA-C_4UC_4 film, Fig.3(c) for the same unoriented PDA film, and Fig.3(d) for a glass substrate. The respective THG intensities are plotted on different vertical scales depending on their relative THG intensities. The results shown in Fig.3(a) and (b) were obtained when the

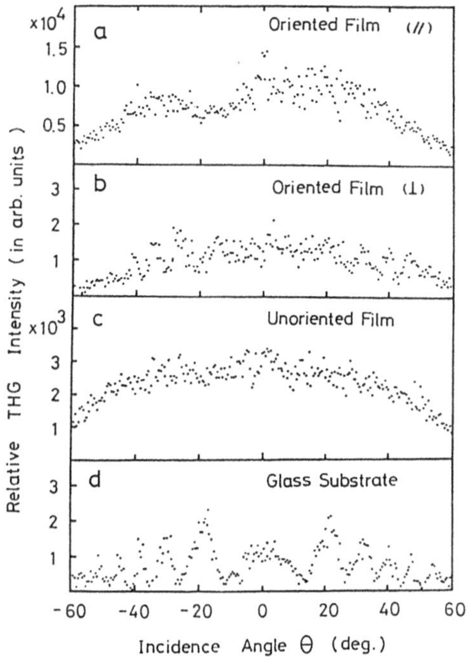

Fig.3
THG intensities plotted against the rotation angle θ, (a) for an oriented PDA-C_4UC_4 film in the parallel polarization, (b) for the perpendicular polarization, (c) for an unoriented PDA-C_4UC_4 film, and (d) for a 1mm thickness glass substrate. The thickness of PDA films is 108 nm. Note the different vertical scales in the respective plots.

electric vector of pumping laser light was polarized parallel and perpendicular, respectively, to the rubbing direction of the oriented film. The THG intensities in the former case are nearly four orders of magnitude larger than the latter, indicating a high degree of anisotropy in the nonlinear optical property of this film. In the unoriented film, the nonlinear responses of individual polymers are averaged over the randomly distributed polymer chains in the film. The observed THG intensities in the unoriented film (Fig.3(c)) can be quantitatively accounted for by assuming a random distribution of polymers in the plane parallel to the substrate surface, as concluded from the morphology study.

A crude estimation of the $\chi^{(3)}$ values was made by using the reported $\chi^{(3)}$ value of fused silica plate (2.8×10^{-14} esu at 1.9 μm) for the standard. The result was $\chi^{(3)}=2.9\times10^{-10}$ esu for the parallel polarization and $\chi^{(3)}=3\times10^{-12}$ esu for the perpendicular polarization. Although these values are not viewed with 100% confidence as yet possibly due to the correction for the air effect, the observed THG intensities in the oriented PDA films are nevertheless sufficiently large and highly anisotropic to make us confident in the usefulness of the PDA films fabricated by the present method for the future NLO applications. To achieve this goal, however, greater effort is required on the better molecular design of diacetylene monomers, further improvement of the fabrication process, and more detailed characterization of the linear as well as nonlinear optical properties of these films.

References

1. T. Kanetake, Y. Tokura and T. Koda : Solid State Commun. 56, 803 (1985).
2. T. Kanetake, K. Ishikawa, T. Koda, Y.Tokura and K. Takeda : Appl. Phys. Lett. 51, 1957 (1987).
3. T. Mitani, H. Mori, S. Suga, T. Koda, S. Shin, K. Inoue, I. Nakada and H. Kanzaki : J. Phys. Soc. Jpn. 47, 679 (1979).
4. S. Tomaru, K. Kubodera, T. Kurihara and S. Zembutsu : Jpn J. Appl. Phys. L1657 (1987), also K. Kubodera and T. Kaino, in this book.

Nonlinear Optical Properties of Polymer Ion–Dye Complexes

H. Matsuda, S. Okada, T. Nishiyama, H. Nakanishi, and M. Kato

Research Institute for Polymers and Textiles,
1-1-4, Higashi, Tsukuba, Ibaraki 305, Japan

1. INTRODUCTION

Polymers so-far studied for nonlinear optics may be divided into the following three groups. The first group is for the conjugated polymers which themselves inherently show high nonlinear optical responses such as poly-diacetylenes [1], [2], polyacetylene [3], and so on. The second one is for those which play the role of matrix to disperse nonlinear optically active chromophores, e.g. poly-methymethacrylate [4] and a liquid crystalline polymer [5] with azodyes. The third one is the same as the second except that the chromophores are covalently bonded to polymers. Compared with these polymers, polymer ion – dye complexes are considered to have several advantages as listed in Table 1. For example, polymer ion complexes must be superior to the first group in the processability, and to the second group in both the time-temperature stability and the polymer-chromophore compatibility, and even to the third group in that the numbers of ion pairs can be easily prepared by simple ion exchange reaction, leading to the tailorability of structures. To confirm these possibilities, poly-styrenesulfonic acid complex with the hemicyanine has been investigated [6].

Table 1 Advantages of utilizing polymer ion – dye complexes in the field of nonlinear optics.

1. Processability
2. Time and Temperature Stability
3. Polymer-Chromophore Compatibility
4. Multiplicity of Ion-Pairing
5. Tailorability of Structures

2. PREPARATION OF POLYMER ION COMPLEX

The preparative pathway of the polymer ion – dye complex of poly-styrenesulfonic acid and hemicyanine (PSS-HC) is shown in Fig. 1. To a 5 mM solution of poly-styreneslufonic acid lithium salt in methanol-water (10:1), the equi-molar solution of hemicyanine iodide in the same mixed solvent was added, and the homogeneous solution was kept quiet at room temperature. After sevral hours, red-

Springer Proceedings in Physics, Vol. 36
Nonlinear Optics of Organics and Semiconductors
Editor: T. Kobayashi

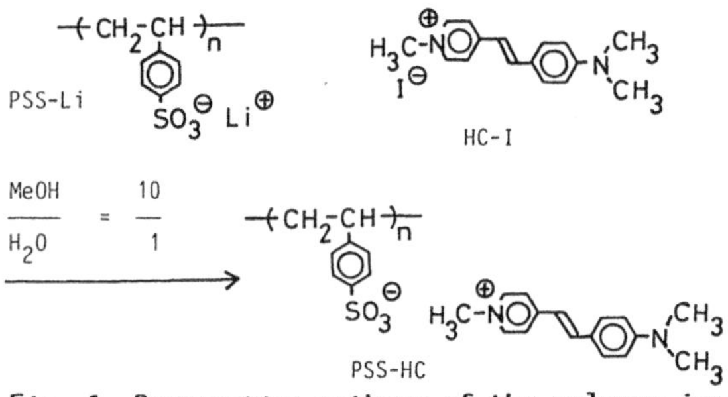

Fig. 1 Preparative pathway of the polymer ion complex of PSS-HC.

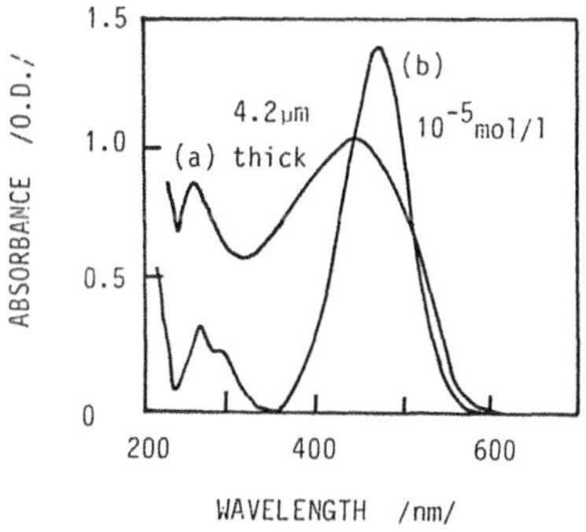

Fig. 2 Absorption spectra of (a) PSS-HC film and (b)hemicyanine iodide in methanol solution.

colored transparent films precipitated on the bottom of flask. The precipitates were washed with methanol to remove the unreacted hemicyanine iodide, and then resolved in the methanol-water (10:1) mixed solvent around 50 ^{o}C. Homogeneous, transparent, and red-colored thin films of 4.2 and 0.55 μm in thickness were obtained by conventional casting of the solution on fused quartz substrates.

In the absorption spectrum shown in Fig. 2, the absorption maximum is located at 450 nm, which is 20 nm shorter than that of hemicyanine iodide in methanol solution. Elementary analysis of the cast-films revealed that as much as 0.7 molar ratio of the hemicyanine with respect to the styrenesulfonate unit were contained in the polymer complex. Stable dispersion of hemicyanine molecules in the polymer were confirmed at room temperature; that is, no segregation of crystalline hemicyanine was observed under polarizing microscope and by x-ray diffraction.

3. THG MEASUREMENTS

THG measurements of the films were done by the Maker fringe method using 1.94 and 1.50 μm pulsed laser light with about 3 mJ/p, which was obtained by wave-mixing of Nd:YAG and dye lasers. The fringe patterns of the films were shown in Fig. 3 together with that of a fused quartz plate. The fringe patterns of the 0.55 μm thick film at 1.94 μm correspond to the coherence length (lc) of fused quartz. It means that the lc of this complex film is longer than 0.55 μm. On the other hand, the fringe patterns of the 4.2 μm thick film at the same wavelength had two maxima near $\pm$ 10 degree and decreased monotonously with inclination angles. At the shorter wavelength of 1.50 μm the patterns were recognized to have numbers of maxima and minima. Of the Maker fringe method [7], lc depends on incident angle θ as is given in eq. (1):

$$lc(\theta)= \frac{\lambda}{6\,|n_{\omega}\cos\theta'_{\omega} - n_{3\omega}\cos\theta'_{3\omega}|} \simeq \frac{lc}{\cos\theta'_{\omega}} \qquad (1)$$

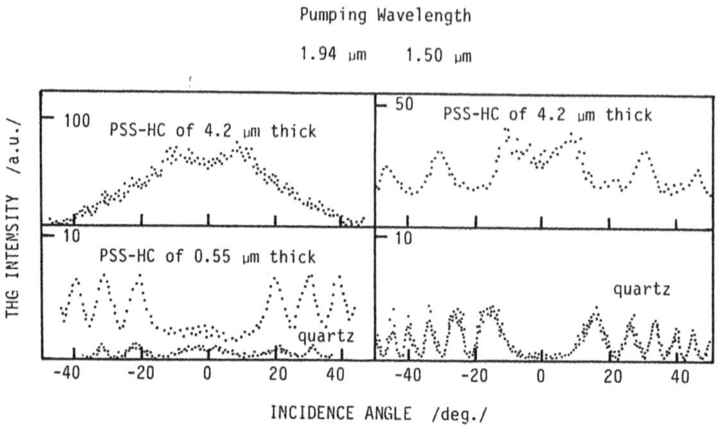

Fig. 3 Maker fringes of THG of PSS-HC films and a fused quartz plate.

where λ is the pumping wavelength and θ' is refractive angle. Using eq. (1), fringe patterns $F(\theta)$ can be expressed by eq. (2):

$$F(\theta)= \sin^2\left(\frac{\pi}{2}\frac{L}{lc(\theta)}\right) = \sin^2\left(\frac{\pi}{2}\frac{L}{lc}\cos\theta'\right) \approx \sin^2\left(\frac{\pi}{2}\frac{L}{lc}\cos\theta\right) \tag{2}$$

where L is the thickness of the sample and for the simplification, $\cos\theta'$ is assumed to be equal to the $\cos\theta$.

From the simulations of the positions of maxima and minima by use of eq. (2), lc's at 1.94 and 1.50 μm were derived to be 1.4 and 0.32 μm, respectively. The refractive index of the film ($n_{0.63}$), measured at 0.63 μm by ellipsometry, was 1.67. From the definition of lc and by assuming $n_{0.63} \approx n_{0.65}$ and $n_{1.94} \approx n_{1.50}$, $n_{1.94}$ and $n_{0.50}$ were caluculated to be 1.44 and 2.22.

The third order susceptibilities $\chi^{(3)}$ of PSS-HC film were evaluated in comparison with fused quartz. The $\chi^{(3)}$ of fused quartz used was 3.0×10^{-14} esu [8]. Since the TH wave was partly absorbed by PSS-HC film itself, $F(\theta)$ should be modified to be eq. (3) [9], [10]:

$$F(\theta)= \exp(-\alpha L)\cdot\left[\left\{1 - \exp\left(-\frac{\alpha L}{2}\right)\right\}^2 + \exp\left(-\frac{\alpha L}{2}\right)\cdot\sin^2\left(\frac{\pi}{2}\frac{L}{lc(\theta)}\right)\right] \tag{3}$$

The absorption coefficient α of the film was 4,100 cm^{-1} at 0.50 μm and those at longer wavelengths were taken as zero. The $\chi^{(3)}$ values thus obtained were 4.6×10^{-12} esu at 1.94 μm and 1.7×10^{-11} esu at 1.50 μm, respectively, indicating a remarkable resonant effect. The damage threshold was greater than 100 MW/cm^2. It should be noted that by this simple method of ion complex formation transparent and isotropic polymer films of the same orders of $\chi^{(3)}$ as that of PBT [11] can be obtained with dispersion stability.

In conclusion, some of the advantages of polymer ion - dye complexes in Table 1 have been demonstrated. Further study to demonstrate the remaining terms is currently in progress.

4. REFERENCES

1. C. Sauteret, J.-P. Hermann, R. Frey, F. Pradere, J. Ducuring, R.H. Baughman: Phys. Rev. Lett. 36, 956 (1976).
2. H. Matsuda, H. Nakanishi, T. Hosomi, M. Kato: Macromolecules 21, 1238 (1988).
3. K.D. Singer, J.E. Sohn, S.J. Lalama: In Nonlinear Optical Materials, edited by D.A.B. Miller (Extended Abstracts 1985 Fall Meeting of the Material Research Society, Boston, Massachusetts) pp. 7-9
4. A.J. Heeger, D. Moses, M. Sinclair: Synthetic Metals 17, 343 (1987).
5. G.R. Meredith, J.G. Van Dusen, D.J. Willams: In Nonlinear Optical Properties of Organic and Polymeric Materials, edited by D.J. Williams (ACS Symposium Series 233, Washington, D.C., 1983) pp.109-133.
6. Preliminary results of this study, together with those of cyanine dye ion complexes, was presented at the MRS International Meeting, Tokyo, May 30 - June 3 (1988); H. Nakanishi, H. Matsuda, S. Okada, M. Kato: In MRS Conference Proceedings, in press.
7. J. Jerfhagnon, S.K. Kurtz: J. Appl. Phys. 41, 1667 (1970).
8. J.P. Hermann: Opt. Commun. 9, 74 (1973).
9. J.L. Oudar: J. Chem. Phys. 67, 446 (1977).
10. J. Zyss, D.S. Chemla, J.F. Nicoud: J. Chem. Phys. 74, 4800 (1981).
11. P.N. Prasad: Proc. SPIE 682, 120 (1986).

Nonlinear Polysilane Thin Films

F.M. Schellenberg [1,*], *R.L. Byer* [1], *J. Zavislan* [2], *and R.D. Miller* [2]

[1]Edward L. Ginzton Laboratory, Stanford University, Stanford, CA 94305, USA

[2]IBM Almaden Research Center, 650 Harry Rd, San Jose, CA 95120, USA

*on leave of absence from IBM Almaden Research Center

The new class of organo-silane polymers consist of long (>1000 atoms) σ-bonded silicon backbones with two organic substituent groups attached to each Si atom. As nonlinear optical materials, these have several advantages over π-bonded carbon chain polymers. First, the σ-σ* absorption edge is typically ≈3.9 eV, so the material is transparent at all visible wavelengths. Second, this absorption band edge is related to the unperturbed length of the silicon backbone, and can be further shifted by adding various substituent groups making resonant enhancement of nonlinear effects at visible wavelengths possible. Finally, these materials are soluble in hydrocarbon solvents such as toluene and isooctane, making fabrication of thin film waveguides with photoresist spinning technology possible.

Of special interest is the compound poly-(di-n-hexylsilane), a polymer with two hexyl chains attached to each Si atom. At temperatures below 42°C, the hexyl side chains crystallize forcing the Si backbone (normally helical) into planar zigzag conformation [1]. The larger transition matrix element associated with this new orientation can be seen in the formation of a strong absorption shifted from 313 nm (the peak for the amorphous region) to 375 nm and an enhancement of $\chi^{(3)}$ by an order of magnitude [2]. A plot of a typical absorption spectrum of a 250 nm thick film of poly-(di-n-hexylsilane) is seen in fig. 1. This sidechain crystallization is observed for n-hexyl and larger sidegroups (e.g. n-heptyl and n-octyl), but at room temperature the Si backbone of polymers with shorter sidechains (e.g. n-butyl and n-pentyl) remains helical, and polymer films of these materials are amorphous.

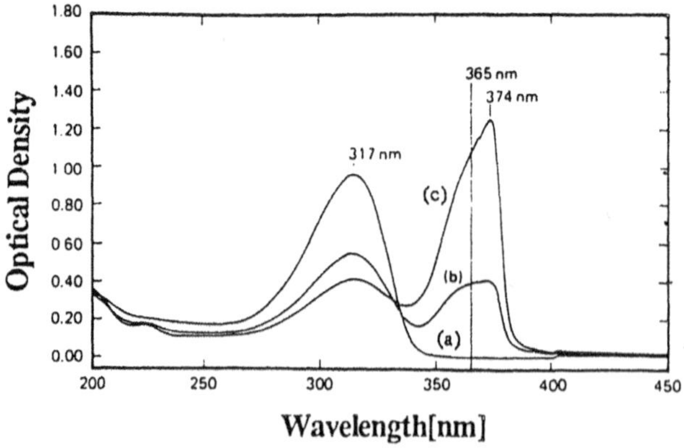

Fig. 1: Poly-(di-n-hexylsilane) UV spectra a) after baking at 100°C. b) cooled for 90 min at 30°C, and c) after 120 min at 30°C

Springer Proceedings in Physics, Vol. 36
Nonlinear Optics of Organics and Semiconductors
Editor: T. Kobayashi

Polysilanes have been extensively investigated as photoresists[3], and films can be easily spun onto quartz wafers using a standard photoresist spinner. These films are optically dense in the UV, and so typically only very thin films (about ≈250 nm thick) have been used. However, for optical waveguide applications, thicker films must be fabricated which are optically homogeneous, with uniform thickness and minimal scattering due to surface roughness. The initial films spun were far from ideal. Attempts to spin films thicker than 4 μm resulted in problems with nonuniform solvent evaporation, which caused gross surface cracking. Somewhat thinner films (≈0.8 μm) did not exhibit surface cracking, but thickness variations as large as 30% could be seen across the film. These thickness variations could be controlled by choosing the appropriate polymer solution concentration and application procedure, but scattering from surface roughness remained a problem. Examination of these films with a transmission polarizing microscope indicated they were highly poly-crystalline, as can be seen in fig. 2a, with crystal grain sizes on the order of 200 μm. These crystalline domains were observed to form under the microscope as the film was cooled through the phase transition at 42° C. As they formed, the surface roughness, as seen with the Nomarski interference contrast attachment to the microscope, dramatically increased.

These properties are obviously unacceptable for the fabrication of waveguides. However, if the crystal grain size can be made small compared to optical wavelengths, the material would appear to be optically homogeneous while retaining the enhanced spectral and nonlinear properties associated with the crystalline state. This is indeed possible if the molecular weight of the polymer is significantly increased. Figures 3a and 3b show photographs taken under the same conditions as those in Fig 2, for a film with molecular weight two orders of magnitude larger (M_W ≈$6x10^6$ measured by GPC). The absorption spectrum for these films still exhibits the strong crystalline peak at 375 nm, but the film is black when placed between crossed polarizers, indicating a pseudo-glassy state. Surface roughness was also significantly reduced for the high molecular weight films.

Fig. 2: 720 nm thick film of poly-(di-n-hexylsilane) on quartz, prepared from a 15% solution in isooctane. Polymer molecular weight was 42,000 by GPC. a) Transmission through crossed polarizers, showing grains ≈200 μm across. b) Nomarski image, showing surface roughness.

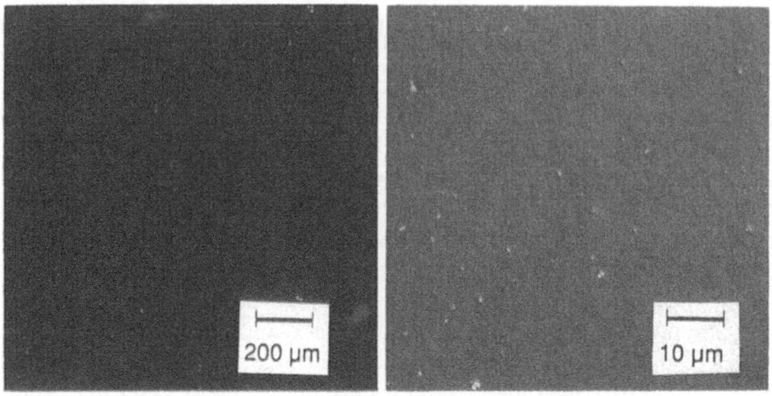

Fig. 3: 560 nm thick film of poly-(di-n-hexylsilane) on quartz, prepared from a 2% solution in isooctane. Polymer molecular weight is 6,000,000 by GPC. a) Transmission through crossed polarizers. b) Nomarski image showing surface roughness.

For comparison, solutions of low and high molecular weight poly-(di-n-pentylsilane) and poly-(di-n-octylsilane) were prepared and spun to make films of comparable thickness. Both of the pentyl films were found to be amorphous with relatively smooth surfaces, regardless of molecular weight, while both octyl films were found to be crystalline, with extremely rough surfaces. Hexyl was the only polymer found to exhibit both amorphous surface properties and crystalline spectral properties. Prism waveguide coupling measurements for both TE and TM modes were made for all films, to evaluate their quality as waveguides and measure the index of refraction. The results are tabulated in table 1. All materials were found to be birefringent, with Δn on the order of $3.5x10^{-2}$.

Table 1: Indices of refraction for poly-(di-x-silane), where x represents the sidegroup. Uncertainty of index of refraction values is ±.005.

Sidegroup	n-butyl	n-pentyl	n-hexyl	n-octyl
n_{TE}	1.573	1.562	1.578	1.545
n_{TM}	1.543	1.528	1.538	1.522
$\Delta n x 10^2$	3.1	3.4	4.0	2.3

When polysilanes are exposed to UV light, chain scission occurs, shifting the crystalline band absorption to shorter wavelengths. Even though polysilanes are normally transparent throughout the visible spectrum, exposure to intense visible light from a pulsed dye laser causes the same kind of chain scission for the polymer chains oriented parallel to the laser polarization. This anisotropic change in absorption has an associated change in index of refraction, and maximum induced birefringence can be large enough ($\Delta n \approx 2.6x10\text{-}2$ at λ=632.8 nm) to allow fabrication of channel waveguides by photoexposure. Note that this is is approximately the same magnitude as the

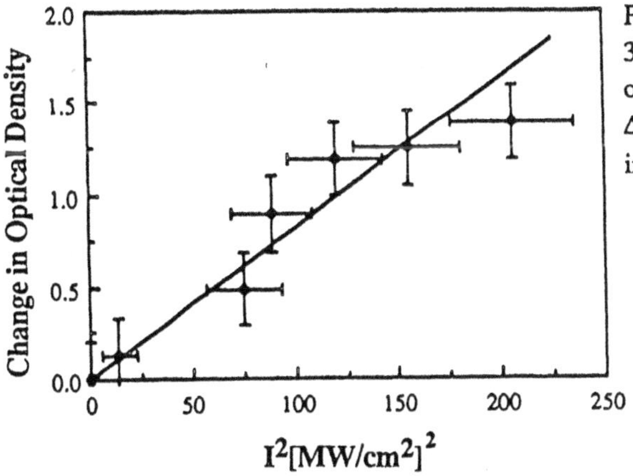

Fig. 4: Change in absorption at 375 nm for a 13.4 J exposure of 560 nm laser light. Fit is $\Delta(OD)=8.2\times10^{-3}I^2$, where I is in MW/cm^2.

birefringence between TE and TM modes, suggesting that the orientation of the polymer chains is parallel to the substrate.

The induced changes in absorption depend quadratically on laser power, as shown in Fig. 4. This suggests a two-photon absorption process. Normalizing the change in optical density with nonlinear exposure to the change induced by one photon absorption, the number of absorbed photons can be determined, and $\chi^{(3)}$ for the process can be estimated to be

$$\text{Im}\,\chi^{(3)}(-\omega;\ \omega,\omega,-\omega)\approx3.5\times10^{-10}\ \text{esu.}\ [4]$$

References

1. J.F. Rabolt, D. Hofer, R.D Miller, and G. N. Fickes, *Macromolecules* 19, 611 (1986), and H. Kuzmany, J.F. Rabolt, B. L. Farmer, and R.D Miller, *J. Chem. Phys* 85, 7413 (1986)
2. J.C. Baumert, G.C. Bjorklund, D.H. Jundt, M.C. Jurich, H. Looser, R.D. Miller, J. Rabolt, J. D. Swalen, R.J. Twieg, in *Conference on Lasers and Electro-Optics Technical Digest Series 1988, Vol Cleo'88* (Optical Society of America, Washington, D.C 1988) paper WW5.
3. R.D. Miller, D. Hofer, G.N. Fickes, C.G. Willson, E. Marinero, P. Trefonas III, and R. West, *Polym. Eng. Sci* 26, 1129 (1986)
4. F.M. Schellenberg, R.L. Byer, R.D. Miller, and R. Sooriyakumaran, in *XVI International Conference on Quantum Electronics Technical Digest,* (Japan Society of Applied Physics, Tokyo, Japan 1988) pp.702-703

Alternating Copolymer of Vinylidene Cyanide and Vinyl Acetate: A New Nonlinear Optical Material

I. Seo [1], *M. Kishimoto* [1], *H. Sato* [2], *and H. Gamo* [3]

[1] Advanced Materials Laboratory, Mitsubishi Petrochemical Co. Ltd., 8-3-1 Chuo, Ami, Ibaraki 300-03, Japan

[2] Department of Electrical Engineering, National Defense Academy, Yokosuka 239, Japan

[3] Department of Electrical Engineering, University of California, Irvine, CA 92717, USA

Poled polar polymers such as vinylidene fluoride (VDF) and trifluoroethylene (TrFE) copolymer, vinylidene cyanide (VDCN) and vinyl acetate (VAc) copolymer have been recently recognized as new nonlinear optical materials [1,2], in addition to the applications of their piezoelectric, pyroelectric and ferroelectric properties. In particular, the VDCN/VAc copolymer is noncrystalline or poorly crystalline according to its X-ray diffraction pattern; nevertheless it has strong piezoelectric and pyroelectric activities [3,4] and exhibits ferroelectriclike dielectric behavior [5]. These strong activities and the dielectric anomalies are considered due to the existence of large C≡N dipole moments in the molecule and to the oriented polar molecules in polymer films poled by high dc field. The VDCN/VAc copolymer is an alternating copolymer in a head-to-tail arrangement with a VDCN-VAc unit. Under poling at high dc electric field, the VAc unit located between VDCN would facilitate the rotation of main chains, which ultimately causes the dipole orientation of cyanide group.

We have observed the second harmonic generation (SHG) in this copolymer film by using Nd:YAG laser. Since its nonlinear dipole moment should be greatly enhanced by poling and stretching, we experimentally investigated the dependence of SHG upon poling and stretching processes for an alternating VDCN/VAc copolymer using a pulsed Nd:YAG laser. The piezoelectric, pyroelectric and dielectric properties of poled and stretched VDCN/VAc copolymer thin films will also be discussed.

The sample used here was an alternating copolymer of VDCN and VAc, the same as used in previous investigations [1,4]. Gold was evaporated onto each side of the sample to serve as electrodes. All samples were poled by applying dc electric fields.

The complex permitivitty was measured over the frequency range of 100Hz to 100KHz at temperatures between 30° C and 200° C using a computer controlled apparatus. Figure 1 shows the temperature spectra of the complex dielectric constant of a VDCN/VAc copolymer measured for four different frequencies. At the temperature below 160° C, both ε' and tan δ curves are nearly independent of temperature and frequency. As the temperature exceeds 170° C, however, both ε' and tan δ increase sharply at all frequencies. This is the first noncrystalline or poorly crystalline polymer with such a high dielectric constant at 100Hz. As the temperature is raised further, the values of ε' and tan δ either continue to decrease or reverse the direction and increase sharply, apparently as a result of dc ionic conduction. The polarization peak, which decreases in height and increases in temperature position increasing frequency, is attributed to the strong cooperative interaction of C≡N dipoles in the copolymer.

Figure 2 shows the pyroelectric coefficient Py and the remanent polarization Pr as a function of poling field for uniaxially stretched VDCN/VAc copolymer films. The pyroelectric coefficient and the remanent polarization of the VDCN/VAc

Springer Proceedings in Physics, Vol. 36
Nonlinear Optics of Organics and Semiconductors
Editor: T. Kobayashi

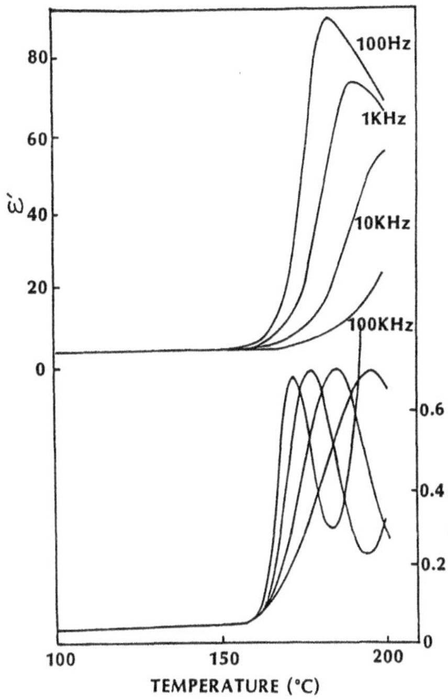

Fig.1 Temperature spectra of ε' and tanδ for a VDCN/VAc copolymer measured at four frequencies.

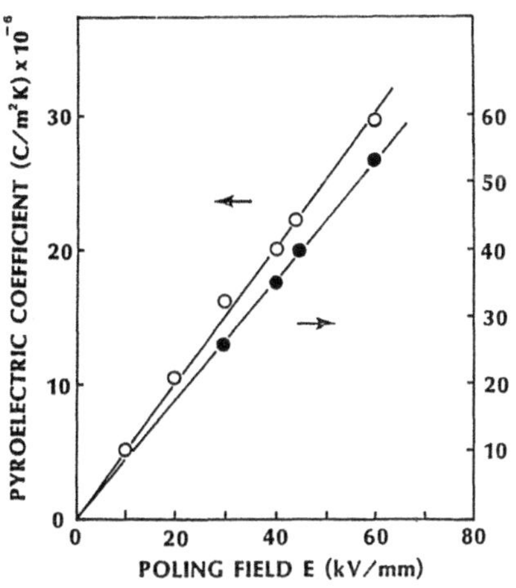

Fig.2 Effect of poling field E of pyroelectric coefficient Py and remanent polarization Pr. (stretching : 1 x 3)

copolymer films were measured by the charge integration technique. The remanent polarization and the pyroelectric coefficient increase linearly with the poling field E. It shows a definite correlation between increasing pyroelectricity and increasing remanent polarization. Tasaka et al.[3] have found in this copolymer a strong piezoelectric activity which is proportional to the remanent polarization.

In measuring the piezoelectric parameters in the direction of the thickness K_{33}, the piezoelectric resonance method developed by Ohigashi was employed. The piezoelectric strain coefficient D_{33} were calculated from the following equation:

$$D_{33} = K_{33}\sqrt{\varepsilon_{33}/ C_{33}}$$

where C_{33} is the elastic constant, and ε_{33} is the dielectric constant. A Cartesian coordinate system was defined by the axis 1, 2 and 3 for the stretched film, with 1 parallel to the direction, 2 normal to the stretched direction in the film plane and 3 normal to the film plane.

The apparatus employed for SHG measurements is shown in Fig.3(a). A pulsed Nd:YAG laser was used as a fundamental optical source, having the peak power of about 1kW and the pulse width 100 μs. The film was typically cut with dimensions of 30 mm in the x direction and tapered from 3 mm to zero in the y direction, as shown in Fig.3(b). The direction of stretching was chosen so as to accord to the x, and the x and y directions, respectively, for uni- and biaxially stretched specimens. The SHG signal was detected by the photomultiplier through the interference filter at 0.53 μm.

The SHG power is plotted as a function of the poling field E, together with the relation of piezoelectric activity D_{33} and E for various stretched films in Fig.4. The piezoelectric activity and the SHG power increase with poling field and stretching ratio. The SHG power tends to increase proportionally with poling field and stretching ratio. In Fig.5 the obtained SHG power was depicted as a function of the piezoelectric coefficient, showing its quadratic dependence. This strong piezoelectricity is considered due to the introduction of large dipole moments as C≡N group in the molecule and to the orientation of this large dipole moment in the film. In the VDCN/VAc copolymer, the C≡N groups are assumed to contribute to the nonlinear polarization. Translating the tapered film along the x direction, i.e., changing the path length for the fundamental and SHG signal, we could observe the phase matching conditions, as illustrated in Fig.6. From the

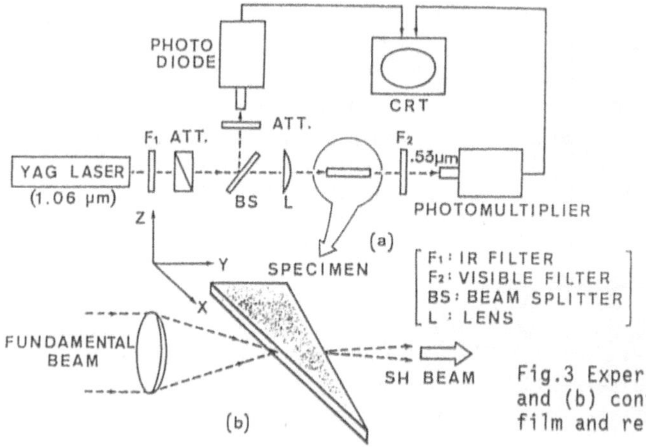

Fig.3 Experimental setup: (a) apparatus and (b) configuration of the copolymer film and related coordinates.

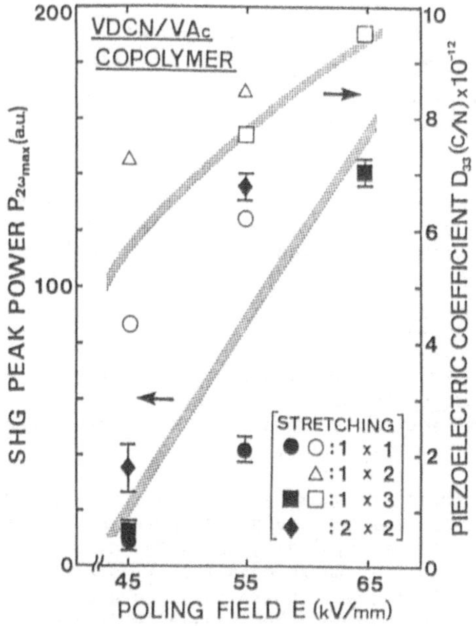

Fig.4 Effect of poling field E on SHG peak power $P_{2\omega}$ and piezoelectric coefficient D_{33} .

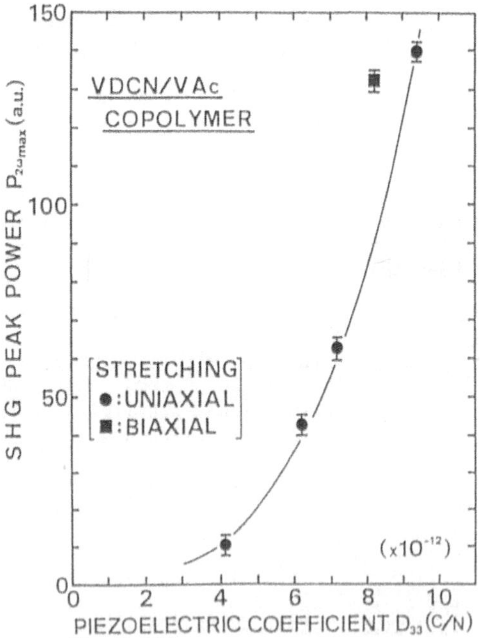

Fig.5 SHG power P vs piezoelectric coefficient D .

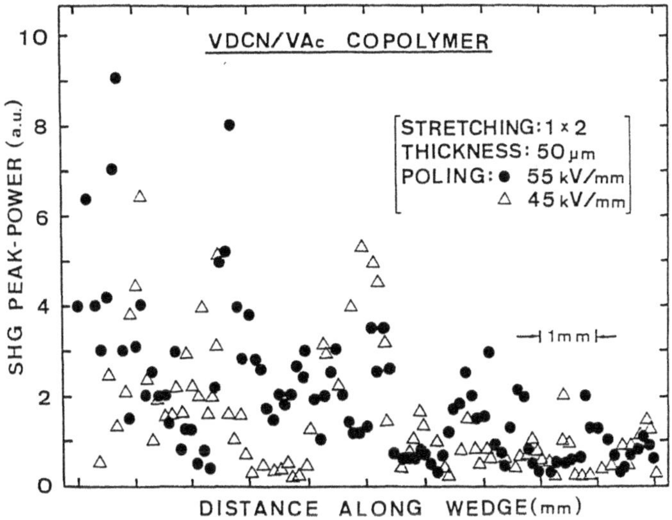

Fig.6 SHG peak power as a function of the distance along the wedge of film (the x direction).

interval between peak SHG signal and the tapered configuration of the film we can estimate the coherence length ℓc to be about 200 μm. SHG peak power should decay exponentially with increasing wedge distance, i.e., path length y (from the left to the right in the figure), owing to the absorption effects for both the fundamental and the SHG beams.

In conclusion, we observed SHG of VDCN/VAc copolymer even though it is noncrystalline or poorly crystalline polymer. According to the SHG measurement, the nonlinear polarization mainly consisting of C≡N dipoles will be increased by the electric field and stretching ratio. This copolymer also shows strong piezo- and pyroelectric activities and ferroelectriclike dielectric anomalies. This flexible and excellent processable nonliear optical materials will be useful in the electro-optical devices, optical signal processors, or visual sensors.

References

1. H.Sato and H.Gamo: Jpn. J. Appl. Phys. 25, L990(1986)
2. H.Sato, T.Yamamoto, I.Seo and H.Gamo: Optics Letters, 12, 579(1987)
3. S.Tasaka, K.Miyasato, M.Yoshikawa,S.Miyata and M.Ko: Ferroelectrics, 57, 267(1984)
4. I.Seo and K.Nakajima: Nikkei New Materials, 50(1986.2.24)
5. T.Furukawa, M.Date, K.Nakajima, T.Kosaka and I.Seo: Jpn. J. Appl. Phys. 25, 1178(1986)

Part VI

Molecular Crystals – Single Crystals

Dispersion of the Second-Order Nonlinear Optical Coefficient d_{11} of Organic Crystal 2-Methyl-4-Nitroaniline

R. Morita[1], *T. Kondo*[1], *N. Ogasawara*[1], *S. Umegaki*[2], *and R. Ito*[1]

[1]Department of Applied Physics, Faculty of Engineering, University of Tokyo, 7-3-1 Hongo, Bunkyo-ku, Tokyo 113, Japan
[2]Department of Electronics, Tokyo Engineering University, 1404-1 Katakura-cho, Hachioji-shi, Tokyo 192, Japan

Some organic materials have been shown to have large optical nonlinearities [1]. This has generated considerable interest in their potential applicability to nonlinear optical devices. However, data on nonlinear optical coefficients are surprisingly limited. For example, although 2-methyl-4-nitroaniline (MNA) is one of the best-studied materials, the value of d_{11}, its largest second-order nonlinear coefficient, is known only at 1.064 μm, the wavelength of the Nd^{3+}:YAG laser [2]. In this paper, we report on the results of the measurement of d_{11} of an MNA crystal over a wavelength range of 1.20 to 1.54 μm. This has been made possible by the use of laser diodes as light sources. First observation of multiple-reflection effects in optical second-harmonic generation using a single longitudinal and transverse mode laser diode will be also reported.

Measurement of the nonlinear optical coefficient of an MNA crystal was done by observing Maker fringe patterns of optical second-harmonic powers [3]. Maker fringes were obtained by translating a wedge-shaped sample in the direction perpendicular to the incident fundamental beam. A $LiNbO_3$ crystal, which is well known as a nonlinear inorganic material, was used as a reference. The wedge-shaped MNA sample was prepared from melt in the gap between two glass plates. A film spacer was placed on one side of the gap to control the apex angle of the wedge. Both surfaces of the wedge-shaped $LiNbO_3$ were polished optically-flat.

The light sources were laser diodes operating continuously. The wavelengths used for the measurement were 1.20, 1.30, 1.47 and 1.54 μm. All the laser diodes were multi-longitudinal-mode devices except for the one that emitted at 1.54 μm. The nonlinear optical coefficients involved were d_{11}($\simeq$ 250 pm/V) [2,4] of MNA and d_{33} ($\simeq$ 40 pm/V) [4] of $LiNbO_3$, the largest components of these materials.

The change of the optical second-harmonic intensity was measured as the samples were displaced perpendicularly to the direction of the incident fundamental beam, that is, as the optical pathlength in the samples was varied. Typical experimental results are shown in Fig. 1. Here, the optical second-harmonic power is plotted against the wedge displacement; a displacement of 1mm corresponds to a pathlength change of 2.2 μm (MNA, Fig. 1(a)) and 2.0 μm ($LiNbO_3$, Fig. 1(b)). The solid curves are Maker fringe patterns calculated according to a theoretical analysis [5]. The clearly resolved Maker fringe of MNA is considered to testify its crystal quality.

Springer Proceedings in Physics, Vol. 36
Nonlinear Optics of Organics and Semiconductors
Editor: T. Kobayashi

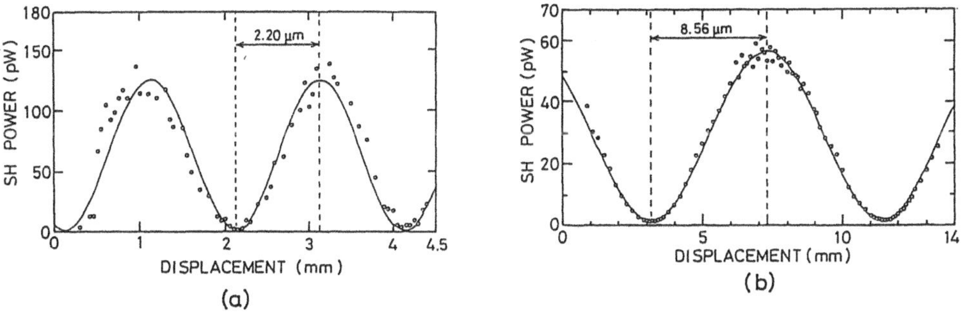

Fig. 1. Maker fringes from (a) MNA and (b) $LiNbO_3$. The open circles show experimental data. The solid curves are theoretical results. The horizontal bars with arrows denote the coherence lengths. (λ =1.47μm)

The sum of the maximum and the minimum values of the Maker fringe pattern gives the maximum optical second-harmonic power $P_0^{2\omega}$ that would be obtained from a parallel plate (apex angle θ=0). From the ratio of $P_0^{2\omega}$ of MNA to that of $LiNbO_3$, we can deduce the ratio of the nonlinear optical coefficients of these materials. For the MNA sample, since there are the glass plates sandwiching the MNA crystal, a small correction for the glass transmission coefficient was made. The open circles in Fig. 2 denote the ratio of the nonlinear coefficients of MNA and $LiNbO_3$, d_{11}(MNA)/d_{33}($LiNbO_3$), which has been determined in this manner. The closed circle in Fig. 2 denotes a value calculated from published data [2,4]. Since the dispersion of d_{33}($LiNbO_3$) is negligible over this wavelength range [6], the ratio d_{11}(MNA)/d_{33}($LiNbO_3$) directly gives the value of d_{11}(MNA), as indicated by the ordinate on the right. The solid curve in Fig. 2 represents a theoretical fit to the formula based on a simple two-level model [7,8]. The pole at λ=0.33 μm agrees with the UV absorption data [9].

During the course of the measurement on $LiNbO_3$ using the 1.54 μm laser diode, which operated in a single longitudinal and transverse mode, we observed that the optical second-harmonic intensity was modulated by short-period oscillations superimposed on the well-known sine-squared-function dependence (Fig. 3(a)) [10]. The short-period oscillations are more clearly seen in Fig. 3(b), which shows part of the curve in Fig. 3(a), plotted in an expanded horizontal scale. It turns out that the short-period oscillations are decomposed into two components, one corresponding to the Fabry-Perot resonance of the fundamental wave, and the other corresponding to that of the second-harmonic wave. The solid curves in Fig. 3 are the results of a theoretical analysis taking account of multiple reflections within a parallel plate. Here, the values of refractive indices have been calculated using empirical formula [11]. We can see that the agreement between theory and experiment is excellent. Conversely, when the values of the refractive indices are unknown, they can be determined from the coherence lengths and the intervals of the Fabry-Perot resonances of both the fundamental and second-harmonic waves. (The apex angle of the wedge must be determined beforehand.) Fabry-Perot resonances were not observed when multi-mode laser diodes were used as light sources; it is considered that they were smeared out because of the overlap of many resonances.

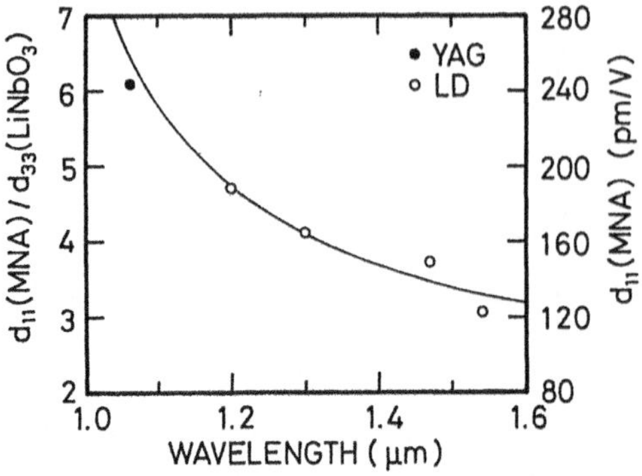

Fig. 2. Dispersion of the nonlinear coefficient d_{11} of MNA. The abscissa represents the fundamental wavelength

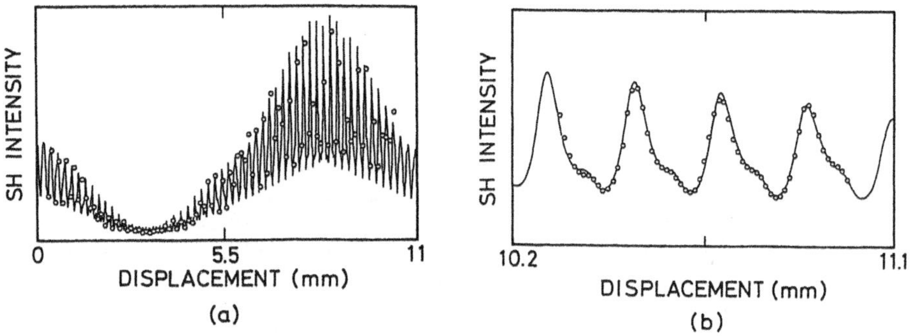

Fig. 3. Multiple-reflection effects on Maker fringes. (b) is magnified from (a). The open circles are experimental data and the solid curves are calculated results. (λ =1.54 μm)

In conclusion, we have measured the dispersion of the second-order nonlinear optical coefficient d_{11} of the organic crystal MNA over a wavelength range 1.20 to 1.54 μm. The nonlinear optical coefficient strongly depends on wavelength, exhibiting a resonance enhancement. The measurement has been made possible by the use of laser diodes as light sources. The compactness, low noise and wide range of wavelengths available make laser diodes very attractive light sources in studying nonlinear optical properties of materials. We have also presented the first observation of multiple-reflection effects in optical second-harmonic generation using a single-mode laser diode (λ =1.54 μm) as a light source.

The authors would like to thank Dr. K. Aiki of the Takasaki Works, Hitachi, Ltd. and Dr. M. Sakuta of the Oki Electric Company for supplying the laser samples. Thanks are also due to Mr. M. Ooasa of Sumitomo Metal Mining Co., Ltd. for supplying the $LiNbO_3$ samples. This work was partially supported by the Grant-in-Aid for Scientific Research, #63850012, from the Ministry of Education, Science and Culture of Japan.

References

1. D.S. Chemla and J. Zyss (eds.): *Nonlinear Optical Properties of Organic Molecules and Crystals*, Vol. 1 and 2 (Academic Press, Orlando, 1987)

2. B.F. Levine, C.G. Bethea, C.D. Thurmond, R.T. Lynch and J. L. Bernstein: J. Appl. Phys. **50**, 2523 (1979)
3. R. Morita, T. Kondo, Y. Kaneda, A. Sugihashi, N. Ogasawara, S. Umegaki and R. Ito: Jpn. J. Appl. Phys. **27**, L1131 (1988)
4. S.K. Kurtz, J. Jerphagnon and M.M. Choy: in *Landolt- Börnstein New Series,* Group III, Vol. 11, ed. by K.H. Hellweg and A.M. Hellweg (Springer-Verlag, Berlin, 1979)
5. G.D. Boyd, H. Kasper and J.H. McFee: IEEE J. Quantum Electron. **QE-7**, 563 (1971)
6. L.M. Dorozhkin, V.A. Kizel', V.D. Shigorin and G. N. Shipulo: Pis'ma Zh. Teor. Fiz. **24**, 363 (1976) (JETP Lett. **24**, 329 (1976))
7. J.L. Oudar and D.S. Chemla: J. Chem. Phys. **66**, 2664 (1977)
8. J. Zyss and J.L. Oudar: Phys. Rev. **A26**, 2028 (1982)
9. C.C. Teng and A.F. Garito: Phys. Rev. **28**, 6766 (1983)
10. R. Morita, T. Kondo, Y. Kaneda, A. Sugihashi, N. Ogasawara, S. Umegaki and R. Ito: Jpn. J. Appl. Phys. **27**, L1134 (1988)
11. M.V. Hobden and J. Warner: Phys. Lett. **22**, 243 (1966)

Synthesis and Second Harmonic Generation of Deuterated L-Arginine Phosphate Monohydrate Crystals

T. Sasaki[1], *A. Yokotani*[3], *K. Fujioka*[2], *T. Yamanaka*[2], *S. Nakai*[2], *and C. Yamanaka*[3]

[1]Department of Electrical Engineering, Osaka University, Suita, Osaka 565, Japan
[2]Institute of Laser Engineering, Osaka University, Suita, Osaka 565, Japan
[3]Institute for Laser Technology, Osaka, Osaka 550, Japan

1. Introduction

We are investigating an improved harmonic generator for laser fusion experiments in place of the conventional KDP crystals. The necessary conditions for the crystals are 1) higher conversion efficiency, perhaps over 70 % to 80 % from the fundamental (ω_1 = 1.06 μm) to the third harmonic (ω_3 = 0.35 μm) for an input power density of 1 - 2 GW/cm^2; 2) higher laser damage threshold, perhaps over 10 J/cm^2 for a 1ns pulse of ω_1; 3) low loss at frequencies ω_1 and ω_3; 4) possibility of growth to large dimensions, perhaps over 10 × 10 × 10 cm^3; and 5) chemical stability (they must be nonhygroscopic).

Organic L-Arginine Phosphate Monohydrate (LAP) is a new nonlinear optical material which was presented in 1983 by Shangdon University, China [1] and is a promising material for laser fusion experiments [2]. The structural formula is shown in Fig. 1. LAP crystal has a monoclinic structure and is optically biaxial.

The problem is that LAP has considerable absorption at 1.06 μm due to overtones of molecular vibration of the hydrophilic functional groups such as -OH, $-NH_3^+$, =NH, $-NH_2$, $=NH_2^+$, etc., which are encircled by dashed lines in Fig. 1. The absorption can be reduced by deuterizing these functional groups. In this paper we describe the synthesis of deuterated LAP (DLAP) crystals and the resultant properties of absorption and second harmonic generation.

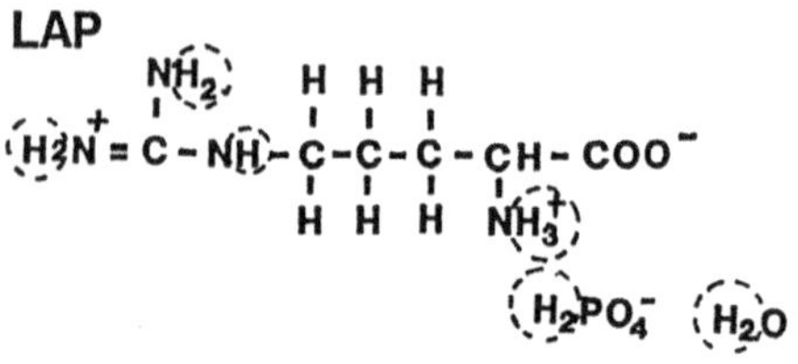

Fig. 1. Structual formula of LAP. Functional groups encircled by dashed lines are the origins of absorption.

2. Growth of DLAP Crystals

Deuterated ortho-phosphoric acid (D_3PO_4), which was made from the reaction of phosphorus pentoxide (P_2O_5, Guaranteed Reagent:GR) and 99.8 % deuterated water (D_2O), was added to L-arginine which was dissolved in 99.8 % D_2O so that the mole ratio of L-arginine and D_3PO_4 was 1 to 1. The solution was maintained at 60°C for two hours after mixing to react completely. The reaction was performed in a glove box filled with dry nitrogen gas to prevent the deuterium in the solution from being replaced with the hydrogen from the atmospheric moisture. All instruments for the reaction were heated at 200°C for two hours before use to remove the moisture adsorbed on the walls.

The growth vessel made of acrylic resin had a volume of 1.5 *l*. Magnetic sealing was adopted to isolate the solution in the vessel from the atmosphere and to rotate the seed crystal and the stirrer. The crystal was grown by the temperature reduction method (temperature range 45°- 20°C). The reduction rate was 0.1 °C/day

Springer Proceedings in Physics, Vol. 36
Nonlinear Optics of Organics and Semiconductors
Editor: T. Kobayashi

on average. The crystals, whose size was 1.5 × 5.0 × 4.5 cm^3 (a × b × c axes), were obtained after twenty days. The maximum growth speed, without degradation of crystal quality, of the b axis, which was the largest of the three axes, was 1.2 mm/day.

DLAP crystals were usually colorless and transparent, but we occasionally obtained yellowish crystals, despite the same growth conditions. The investigation of inorganic impurities by ICP quantometer and fluorescent x-ray spectroscopy showed no difference between these crystals. We found that when the solution was irradiated by ultra-violet (UV) light, both the solution itself and the resultant grown crystal become yellowish. It is considered that the phenomenon is induced by the decomposition of constituent organic substances. The same thing occurred in LAP crystals. The cut-off wavelength was 0.26 μm for both LAP and DLAP. We could use these crystals for fourth harmonic (ω_4=0.265 μm) generation, but the yellowish crystals showed bad transmittance (cut-off wavelength 0.52 μm).

We estimated the deutrium concentration rate σ by NMR (Nuclear Magnetic Resonance) [3] and an IR (infrared) method [4]. By NMR, two kinds of σ can be measured separately for hydrophilic ($-NH_2$, $=H_2N^+$, -NH, $-NH_3^+$, -OH) and hydrophobic (-CH) groups of LAP in Fig. 1. As it was observed that the hydrogens in the hydrophobic groups were not substituted with deuterium, we defined the degree of deuterium substitution for hydrogen in hydrophilic groups as σ. The measured σ was 89 % ± 1 %.

By the IR method, σ could be estimated by measuring the absorption due to OH vibration and σ of 92 % was obtained. But in this method, there was a problem because the absorption spectrum due to OH was not completely isolated from the peaks from other molecules.

The precise absorption coefficient at 1.06 μm was estimated by measuring the temperature rise with a thermoviewer by means of CW YAG laser irradiation on the crystals. The estimated absorption coefficients were 0.09 cm^{-1} and 0.02 cm^{-1} for LAP and DLAP, respectively.

3. Second Harmonic Generation

DLAP crystals are biaxial. There exist many combinations of angles θ and ϕ which satisfy the phase matching condition. We calculated the phase matching angles of Type I and Type II for θ and ϕ [5] by using measured refractive indices n_x, n_y and n_z at ω_1 (λ=1.06 μm) and ω_2 (λ=0.53 μm). Table 1 shows these refractive indices. The calculated results are shown with the experimental values in Fig. 2. O.A. illustrates an optical axis.

The Q-switched YAG laser was used for the SHG experiment. For a given angle ϕ, the angle θ was determined to obtain maximum SHG. The experimental data are shown by the dots in Fig. 2. They agreed fairly well with the calculation.

The most intense SHG outputs were observed at θ=60°,ϕ=16° for Type I and θ=43°, ϕ=0 for Type II within the range of measured angles. The tuning curves were obtained by tilting the crystal surface in the arrow directions (normal to the calculation curves) in Fig. 2. The crystal length was 6 mm. Fig. 3 shows the tuning curves. The curve of KDP (Type II) is also illustrated as a reference. The acceptance angles were 0.52 (mrad·cm) and 1.1 (mrad·cm) for Type I and Type II, respectively and approximately a half that of KDP. The maximum intensities of SHG were 2.3 and 3.2 times that of KDP for Type I and Type II, respectively.

Table 1. Refractive indices of DLAP Crystal.

Wavelength	1.064 μm	0.532 μm
n_x	1.4977	1.5086
n_y	1.5595	1.5767
n_z	1.5686	1.5850

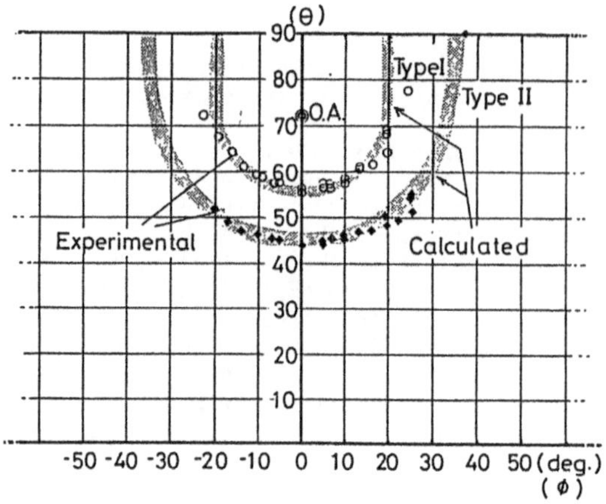

Fig. 2. Phase matching angle of second harmonic generation for DLAP. Curves show calculations and dots illustrate experimental data.

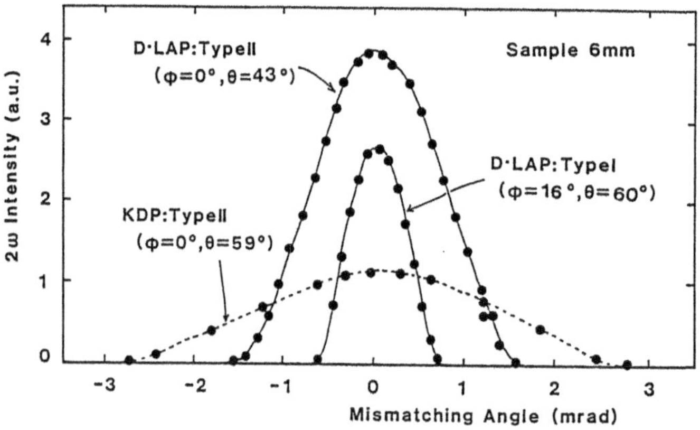

Fig. 3. Angular tuning curves for DLAP crystals. (Thickness of samples : 6mm)

4. Conclusions

The above-mentioned results are summarized in Table 2. The absorption coefficients of LAP crystal at 1.06 μm were reduced to 0.02 cm^{-1} from 0.09 cm^{-1} by deuterization. The cut-off wavelength of LAP and DLAP for UV region was 0.26 μm and they can be used for the FHG. d_{eff}/d_{eff}(KDP) in Table 2 was obtained from the square root of the ratio of the SHG peak intensity of DLAP and LAP divided by that of KDP (Type II).

The acceptance angle is almost a half that of KDP, on the other hand, d_{eff} is approximately double. A preliminary study of laser damage showed a higher durability than a conventionally grown KDP crystal for 1 ns pulses at 1.06 μm (over 30 J/cm^2). It is as easy to grow large crystals of LAP or DLAP as KDP, and they are more non-deliquescent than KDP. From these facts, we can say that LAP or DLAP can be used in place of KDP as a frequency converter for fusion lasers or higher average power lasers.

Table 2. Laser properties of DLAP, LAP and KDP crystals.

		DLAP	LAP	DKDP	KDP	BBO
Absorbance at 1.06 μm (cm^{-1})		0.02 (a-axis)	0.09 (a-axis)	0.003 (c-axis)	0.05 (c-axis)	0.001
UV Cut-off, 95% for $5mm^t$ (μm)		0.26	0.26	0.2	0.2	0.198
Angular Acceptance of 2ω(mrad·cm)	Type I	0.52	-	-	1.1	0.52
	Type II	1.1	0.84	2.2	2.1	-
Change of Matching Angle against Temperature(mrad/℃)	Type I	-	-	-	-	0.023
	Type II	0.04	0.06	0.11	0.12	-
Temperature Acceptance of 2ω(℃·cm)	Type I	-	-	-	-	23
	Type II	27	14	20	18	-
d_{eff}/ d_{eff}(KDP Type II)	Type I	1.5 θ= 60° φ= 16°	-	-	0.75 θ=42° φ=45°	4.4 θ= 22° φ= 0°
	Type II	1.8 θ= 43° φ= 0°	1.9 θ= 43° φ= 0°	0.92 θ= 57° φ= 0°	1 θ= 59° φ= 0°	-

References

1. D.Xu, M.Jiang and Z.Tan: Acta Chemica Sinica 41 570 (1983)
2. S.B.Monaco, L.E.Davis, S.P.Velsko, F.T.Wang and D.Eimerl: J. Crystal Growth 85 252 (1987)
3. R.M.Silverstein and G.C.Bassler, T.C.Monill (eds.): Spectrometric Indentification of Organic Compounds (Wiley, New York 1981)
4. C.Belouet and M.Monnier: J. Crystal Growth 30 151 (1975)
5. J.Q.Yao and T.S.Fahlen: J. Appl. Phys. 55 65 (1984)

Crystal Structures of a New Organic Second Order Nonlinear Optical Material [Cyano(ethoxycarbonyl)methylene]-2-ylidene-4,5-dimethyl-1,3-dithiole

Y. Shimizu [1], *T. Uemiya* [1], *N. Yoshie* [2], *H. Yoshioka* [2], *K. Nakatsu* [2], *T. Nogami* [3], *and Y. Shirota* [3]

[1] Basic High Technology Laboratories, Sumitomo Electric Industries, Ltd., Shimaya 1-2, Konohana-ku, Osaka 554, Japan
[2] Department of Chemistry, School of Science, Kwansei Gakuin University, Uegahara 1-1-155, Nishinomiya, Hyogo 662, Japan
[3] Department of Applied Chemistry, Faculty of Engineering, Osaka University, Yamadaoka, Suita, Osaka 565, Japan

1. Introduction

Organic compounds with conjugated π-electron systems are noticed because of their large molecular hyperpolarizability, high speed response to electric field, high optical damage threshold, etc. The 2nd order nonlinear optical materials demand a noncentrosymmetric molecular arrangement in their crystals, which means that some molecular dipoles are maintained without cancelling each other. As for SHG (Second Harmonic Generation), one of the ideal crystal structures is a non-centrosymmetric one with space group P1, in which all molecular dipoles are kept remained with no cancelling among dipoles of neighboring molecules.

In this work, we found the title compound **1** exhibits polymorphs from a viewpoint of SHG activity (SHG active and inactive), and X-ray structure analyses were carried out to find the correlation between the SHG activity and the molecular packing in both crystals.

The title compound was prepared according to the procedure of Jensen and Hendriksen with minor modifications [1]. Slow evaporation of an acetone solution of the compound gave almost colorless, transparent, plate-like crystals, which were named the α-form. On the other hand, crystallization from an ethanol solution produced colorless, transparent, columnar crystals, named the β-form, which exhibited highly efficient SHG (24 times as large as that of the urea in powder method) with irradiation of YAG laser-light [2].

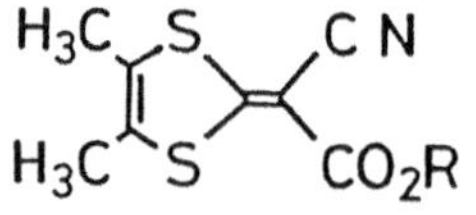

R= CH_2CH_3

1

	Recryst. solvent	SHG
α- form	acetone	inactive
β- form	ethanol	active

2. Structure Determination

The crystal data are given in Table 1. The SHG active β-form belongs to the non-centric space group P1, while the inactive α-form belongs to centric P$\bar{1}$. These findings are quite consistent with those of the general principle. Intensities of the reflections were collected on a four-circle diffractometer. The independent reflections within the range of 2θ<55 were measured using MoKα radiation. Altogether, 2247 and 1337 independent reflections were obtained for

Springer Proceedings in Physics, Vol. 36
Nonlinear Optics of Organics and Semiconductors
Editor: T. Kobayashi

Table 1 Crystal data

	α	β
Crystal System	Triclinic	Triclinic
Space Group	$P\bar{1}$	P1
Cell dimensions		
a/Å	8.711(2)	8.043(9)
b/Å	9.748(2)	9.018(8)
c/Å	7.800(1)	4.088(7)
α/°	95.50(2)	97.81(13)
β/°	107.67(2)	93.71(12)
γ/°	109.14(2)	98.13(9)
$V/Å^3$	582.0(2)	289.7(6)
Z	2	1
$D_x/g.cm^{-3}$	1.3771(5)	1.383(3)
$D_m/g.cm^{-3}$ (by floatation method)	1.37	1.37
Final R	0.044	0.099

α- and β-crystals respectively, which had intensities greater than three times their standard deviations.

The structures were solved by the direct method using MULTAN 78 [3], and refined by the least-squares method. Anisotropic thermal vibrations were assumed for the non-hydrogen atoms. All the hydrogen atoms in the α-crystal were found from a difference Fourier map and refined isotropically, while in the β-crystal the hydrogen atoms were not found, probably due to imperfection of the crystal. The final residual indices were 0.044 and 0.099 for α- and β-crystal respectively.

3. Results and Discussions

Figure 1 shows both molecular structures, which are similar to each other except for the difference in orientations of the ethyl groups. No unusual bond lengths and angles can be seen.

In the crystal structure of α-form, shown in Fig.2, the molecules are related to each other by the center of symmetry at (1/2,1/2,1/2). The donor part of a molecule is stacked on top of the acceptor part of the adjacent molecule, with the interplanar separation of 3.52 Å. This is slightly less than the expected value.

Fig.3 shows the crystal structure of the β-form. The molecules are all nearly on the (011) plane, with the same orientation. Thus the molecular dipoles are arranged towards the same direction without cancelling each other. This type of

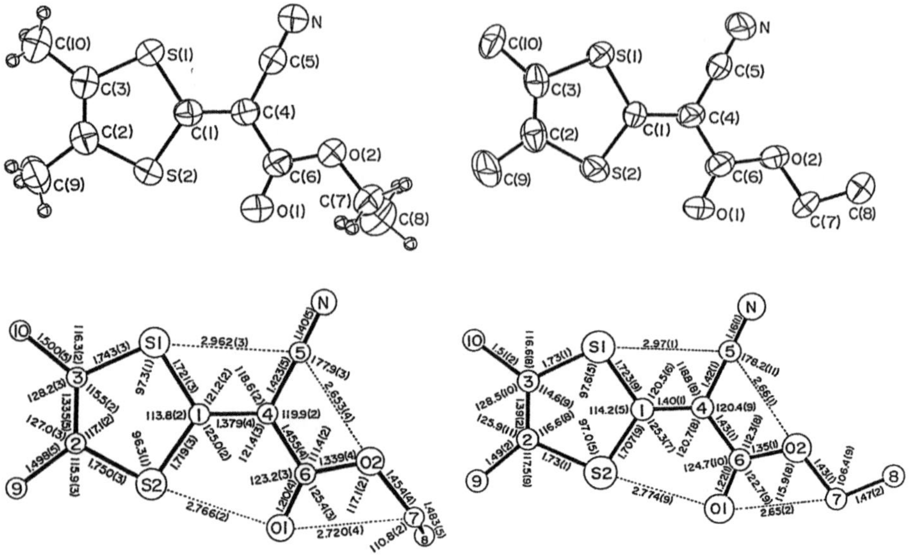

Fig.1 Molecular structures in α-form (left) and β-form (right).

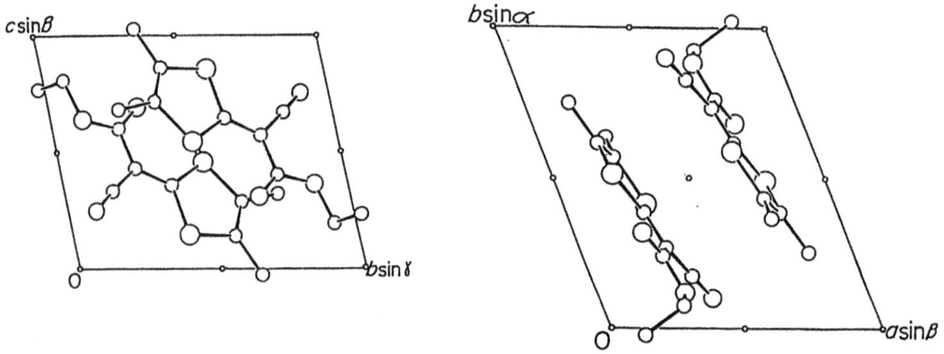

Fig.2 Crystal structure of α-form (SHG inactive). Small circles indicate a center of symmetry.

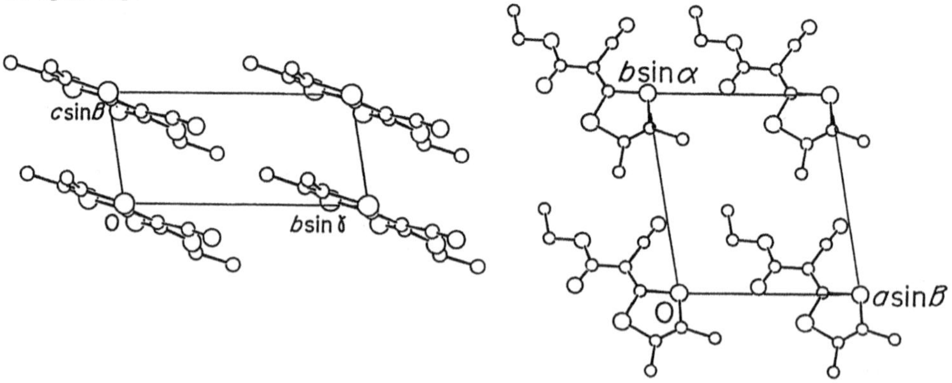

Fig.3 Crystal structure of β-form (SHG active).

orientation of the dipoles should contribute to the observed SHG. The intermolecular separation in the β-form is 3.62 Å, which is larger than that found in the α-form.

4. References

1. K. A. Jensen and L. Hendriksen, Acta Chem. Scand., 22, 1107 (1968).
2. T. Nogami, H. Nakano, Y. Shirota, S. Umegaki, Y. Shimizu, T. Uemiya and ,N. Yasuda, in contribution.
3. P .Main, S. E. Hull, L. Lessinger, G. Germain, J. P. Declercq, and M. M. Woolfson, "MULTAN 78", Univs. York, England, and Louvain, Belgium(1978).

Maker Fringe and Phase Matched SHG Studies of Crystalline Dicyano Substituted Benzene Compounds

*C.H. Grossman, T. Wada, S. Yamada, A. Yamada, H. Sasabe, and A.F. Garito**

Frontier Research Program, RIKEN,
(The Institute of Chemical and Physical Research),
Hirosawa 2-1, Wako-shi, Saitama 351-01, Japan

In searches for new organic structures that possess nonlinear optical properties, dicyano substituted benzene compounds were found to have a microscopic second order susceptibility, β, comparable to the nitroaniline based molecules [1,2,3]. In order to find a dicyano compound that crystallizes into a noncentrosymmetric structure, many derivatives were synthesized that include methoxy, methyl and aromatic rings as substituent groups. Each compound was screened for phase matched SHG (PMSHG) properties by the Kurtz-Perry powder SHG method [4]. Thus, by such combined synthesis and measurements, many compounds can be studied to determine which ones are worthy of in-depth study. One such compound, 2-dicyanovinylanisole (DIVA), produced an intense SHG signal under irradiation from the fundamental output of a Q-switched Nd:YAG laser. This was checked as a function of particle size and showed increasing PMSHG with larger and larger microcrystals. A large quantity of purified DIVA was then synthesized for the growth of single crystals.

Very high quality crystals were grown from vapor by low pressure sublimation. In this method, pure DIVA was carefully placed in the bottom of a glass tube such that no material was accidently placed on the walls. The tube was pumped down to 1 millitorr and back-filled with argon gas. This process was repeated several times to remove oxygen from the tube, and in the final step the pressure was set at 2 Torr of argon. The tube was sealed with a stopcock and removed from the vacuum system so it could be placed in a temperature gradient. The bottom of the tube was inserted into a copper block inside a heating mantle and heated to 70° C while the top was allowed to remain at room temperature. The region in between the top and the bottom was surrounded with insulation, thus forming a temperature gradient along the axis of the tube. In this region, high quality crystals formed in 1-2 days and were easily removed from the walls of the glass tube. Typical size is $1x1x10\ mm^3$. Larger crystals were grown from solution by solvent evaporation. Approximately 36 grams of DIVA was dissolved in 100 ml of acetone at room temperature. The solution was left to slowly evaporate over 2-3 days. Medium quality crystals formed although some of the samples have highly transparent regions that are suitable for studies of bulk PMSHG. Typical size is $10x10x5\ mm^3$.

The x-ray crystal structure was determined by standard diffraction methods, and importantly, no evidence of polymorphism was found. DIVA crystallizes into a monoclinic unit cell with a point group symmetry $P2_1$. The orientation of the molecules within the unit cell is shown in Figure 1. The methoxy groups ($O\text{-}C\text{-}H_3$) are nearly aligned along the crystallographic B axis, shown in the horizontal axis, while the

* Permanent address; Department of Physics, University of Pennsylvania, Philadelphia, PA 19104, U. S. A.

Springer Proceedings in Physics, Vol. 36
Nonlinear Optics of Organics and Semiconductors
Editor: T. Kobayashi

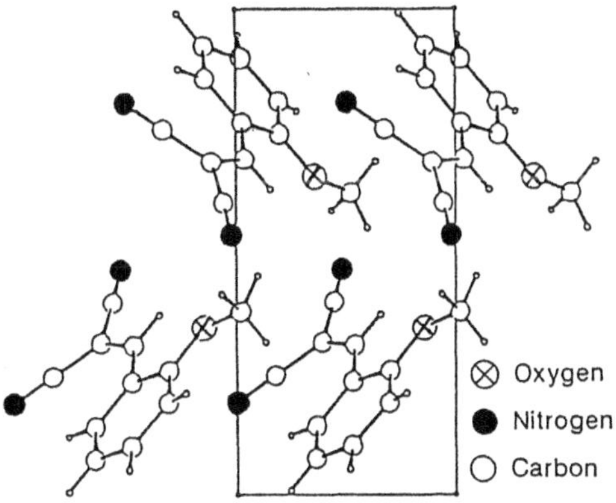

Fig. 1. The DIVA crystal structure shows a net dipole alignment along the crystallographic B axis, shown here in the horizontal direction. The unit cell is monoclinic with a $P2_1$ point group symmetry. This structure is suited for phase matched SHG where large off-diagonal SHG tensor components are favorable.

dicyano groups have approximately a 55° direction cosine to this axis. Therefore, the molecular dipole moments have a common polar axis in the B direction. In the A-C plane the dicyano groups point in opposite directions, and the projection in this plane is centrosymmetric. With this type of molecular orientation we would expect to find large off-diagonal SHG tensor components like d_{21}, d_{23}, and d_{25}, but smaller diagonal components like d_{22}.

In studies of the solution grown crystals, we observed both Type I and Type II PMSHG for laser light incident into the naturally grown crystal faces. This eliminates the difficult and time consuming task of crystal cutting and polishing. Type I phase matching was observed from an unfocused Q-switched Nd:YAG laser incident at 30° from the normal of the (001) face. In this case, the fundamental was polarized in the A-C plane and the second harmonic along the B axis. Type II PMSHG was observed near normal incidence in the (100) face with the fundamental polarized at 45° from the B axis and the second harmonic polarized along the B direction. The Type I PMSHG was stronger than the Type II and, the conversion efficiency was measured using an unfocused incident beam. The incident beam had a peak power of 150 KW, and the measured SHG 300 Watts.

High precision measurements were performed on the vapor grown crystals to determine the d_{22} SHG tensor component and the effective Type I PMSHG coefficient. The experimental setup was described earlier [5], but a rotation stage was used for the sample instead of a translation stage. Both Maker fringe and PMSHG experiments measured the SHG intensity polarized along the B axis as a function of crystal rotation about the same B axis. The d_{22} Maker fringes were generated when the fundamental was incident in the crystal A-C plane and polarized along the B axis. The results are shown in Figure 2 along with the calculated pattern [6] (solid line). The derived value of d_{22} is 2.6×10^{-8} esu ≈ 20 x d_{11} of quartz.

PMSHG was generated at the specific PM angle, same angle as in the solution grown crystal case, with the fundamental incident in the A-C plane and also polarized in the A-C plane. Figure 3 shows the measured and calculated SHG intensity as a function of angle around the PMSHG position. From the amplitude of the fitted curve, we calculated the effective PMSHG coefficient (a combination of d_{21}, d_{23}, d_{25}, and directioncosines) with the result that $d_{eff} = 4.9 \times 10^{-8}$ esu ≈ 40 x d_{11}

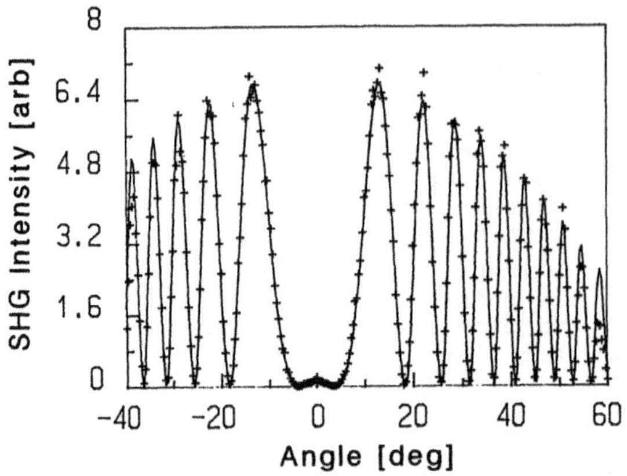

Fig. 2. Maker Fringe SHG at 1.064 μm was measured in DIVA with the fundamental and second harmonic polarized along the principal Y axis. Similar Maker Fringe data were measured from quartz and the fringe amplitudes were compared in order to calculate the d_{22} SHG coefficient.

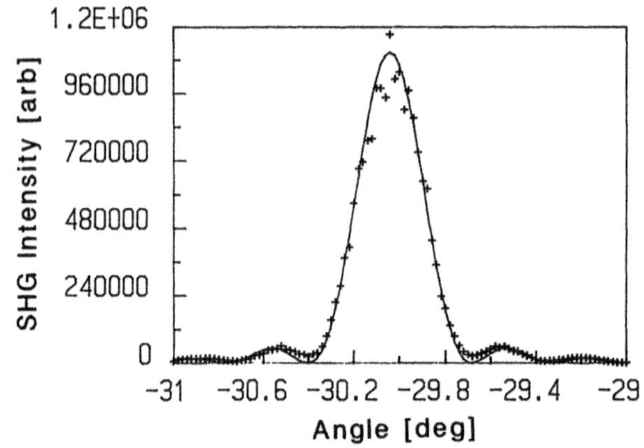

Fig.3. Angular dependence of Type I PMSHG was measured with the fundamental polarized in the A-C plane while the crystal was rotated about the B axis. The SHG was polarized along the B axis. The phase matchable SHG coefficient d_{eff} (a combination of d_{21}, d_{23}, and d_{25}) was calculated from the amplitude and its value is approximately 40 times d_{11} of quartz.

of quartz. In the limit $n^{2\omega}$ approaches n^{ω}, the PMSHG power is given by [6]

$$P^{2\omega} = \frac{(8\pi)^5 \, (d_{eff})^2 \, L^2 \, (P^{\omega})^2}{\pi \, c \, \lambda^2 \, w^2 \, (n+1)^6} \tag{1}$$

where L is the crystal length, λ the fundamental wavelength, w the beam radius, n the refractive index, and P^{ω} the fundamental power. With L = 0.5 mm, n = 1.65, w = 3 mm, and P^{ω} = 150 KW the calculated PMSHG power is 407 Watts which is in good agreement with the measured value, especially considering the relatively poor quality doughnut mode of the fundamental beam.

In conclusion, single crystals of DIVA have been shown to be phase matchable for SHG in the near infrared range. Crystals are easily grown and do not require cutting and polishing for both Type I and Type II PMSHG. Continued development of crystal growth experiments can easily yield large high optical quality DIVA crystals for linear and nonlinear optical studies.

[1]Y. Mori, M. Okamoto, T. Wada, and H. Sasabe, Materials Res. Soc. Symp. Proc., 109, 345 (1988).

[2]H. E. Katz, K. D. Singer, J. E. Sohn, C. W. Dirk, L. A. King, and H. M. Gordon, J. Am. Chem. Soc., 109, 6561 (1987).

[3]J. B. Stamatoff, A. Buckley, G. Calundann, E. W. Choe, R. DiMartino, G. Khanarian, T. Leslie,G. Nelson, D. Stuetz, C. C. Teng, and H. N. Yoon, Proc. SPIE **1986**, 682, 85 - 92.

[4]S. K. Kurtz and T. T. Perry, J. Appl. Phys. 39, 3798 (1968)

[5]K. D. Singer and A. F. Garito, J. Chem. Phys.,75, 3572 (1981).

[6]J. Jerphagnon and S. K. Kurtz, J. Appl. Phys., 41 , 1667 (1970).

Nonlinear Optical Properties of 2-(α-Methylbenzylamino)-5-nitropyridine

T. Kondo[1], *N. Ogasawara*[1], *S. Umegaki*[2], *and R. Ito*[1]

[1]Department of Applied Physics, Faculty of Engineering, University of Tokyo, 7-3-1 Hongo, Bunkyo-ku, Tokyo 113, Japan

[2]Department of Electronics, Tokyo Engineering University, 1404-1, Katakura-cho, Hachiohji-shi, Tokyo 192, Japan

Some organic materials have attracted considerable attention as preferable materials in the field of nonlinear optics owing to their extremely large optical nonlinearities compared with most inorganic materials which are transparent in the visible range [1]. In particular, those organic materials are expected to be used for frequency doubling of semiconductor lasers. In order to realize efficient second-harmonic generation (SHG) of such low-power light sources, we have to find materials which have large second-order nonlinear optical coefficients and short absorption cut-offs. TWIEG *et al.* pointed out that one of the donor-acceptor substituted pyridine derivatives, 2-(α-methylbenzylamino)-5-nitropyridine (MBANP), has high SHG efficiency in the powder form and the absorption cut-off measured in an ethanol solution is in a relatively short-wavelength region [2,3]. Recently, BAILEY *et al.* have reported its single crystal growth and optical characterization including the observation of type I phase-matched SHG of a Nd:YAG laser [4,5].

We have developed a comprehensive study of MBANP independently, motivated by the results of TWIEG *et al* . In this paper, we describe an experimental study of the molecular hyperpolarizability measured in a solution, the crystal structure determined by x-ray diffraction analysis, and several important optical properties, such as the refractive index, optical transmission, and the nonlinear optical coefficient.

MBANP was synthesized according to a method previously reported [2,6]. MBANP displays an intense absorption band due to strong intramolecular charge-transfer in the near ultraviolet. However, the charge-transfer absorption band is appreciably shifted to shorter wavelengths relative to typical donor-acceptor substituted benzene derivatives (see Table 1). The molecular hyperpolarizability β of MBANP was measured in a p-dioxane solution using the modified electric-field-induced second-harmonic generation (EFISH) technique [7] with a Nd:YAG laser operating at $\lambda = 1.064$ μm. As summarized in Table 1, β of MBANP is comparable with those of para-nitroaniline-like molecules.

Table 1. λ_{max} in ethanol and β for nitrobenzene (NB), para-nitroaniline (p-NA), 2-methyl-4-nitroaniline (MNA), and MBANP

material	λ_{max} [μm]	β [10^{-30}esu]
NB	0.255	2.0
p-NA	0.368	12.6
MNA	0.373	14.2
MBANP	0.359	15.0

Springer Proceedings in Physics, Vol. 36
Nonlinear Optics of Organics and Semiconductors
Editor: T. Kobayashi

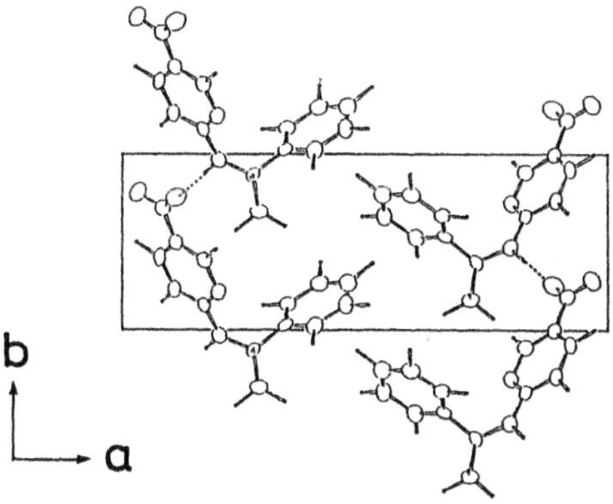

Fig. 1. The molecular packing projected along the c axis. The hydrogen bonds are represented by dotted lines

The crystal structure of MBANP [monoclinic, $P2_1$: $a = 17.945(4)$, $b = 6.372(1)$, $c = 5.401(2)$ Å , $\beta = 94.70(2)°$] was determined by x-ray diffraction analysis [8,9]. Crystal packing is shown in Fig. 1. As is readily apparent, molecules are aligned so that the intramolecular charge-transfer axis in the pyridine ring of each molecule makes a relatively small angle (33.2°) with the twofold screw axis ($\equiv b$ axis). Therefore, the nonlinear optical coefficient d_{22} is expected to be very large.

Single crystals were grown by slow cooling of a seeded toluene solution. Samples with optically-flat cleaved (100) surfaces were used in the optical measurements. The optical transmission spectra for light propagating perpendicular to bc plane and polarized along b axis were measured. The absorption cut-off ("cut-off " refers to a wavelength of "onset" of the transmission decrease) is around 0.45μm.

The refractive index n_Y (principal dielectric axis Y is parallel to the crystallographic b axis in the case of crystals belonging to point group 2) for $\lambda = 0.532$ μm and for $\lambda = 0.589$ μm was measured by means of the Becke line method [10] and n_Y for $\lambda = 1.064$ μm was determined from the Fabry-Perot resonances in the transmission through thin plates :

$$n_Y(0.532\mu\text{m}) = 1.91 \pm 0.01, \tag{1}$$

$$n_Y(0.589\mu\text{m}) = 1.84 \pm 0.01, \tag{2}$$

$$n_Y(1.064\mu\text{m}) = 1.77 \pm 0.02. \tag{3}$$

The second-order nonlinear optical coefficient d_{22} was measured using the Maker fringe method with a Nd:YAG laser operated at $\lambda = 1.064$ μm. The observed Maker fringes for d_{22} of a 595-μm-thick plate are shown in Fig. 2. To determine the relative value of d_{22} of MBANP, we used quartz (d_{11}) and $LiNbO_3$ (d_{33}) as references. The relative envelope intensities of these fringes for d_{22} of MBANP was found to be 71.2 times larger than that for d_{11} of quartz and 1.38 times larger than that for d_{33} of $LiNbO_3$. These relative values and the measured refractive indices yield the following results : [11]

$$\begin{aligned} d_{22}^{\text{MBANP}} &= (126 \pm 20) d_{11}^{\text{Quartz}} \\ &= (1.46 \pm 0.21) d_{33}^{\text{LiNbO}_3}. \end{aligned} \tag{4}$$

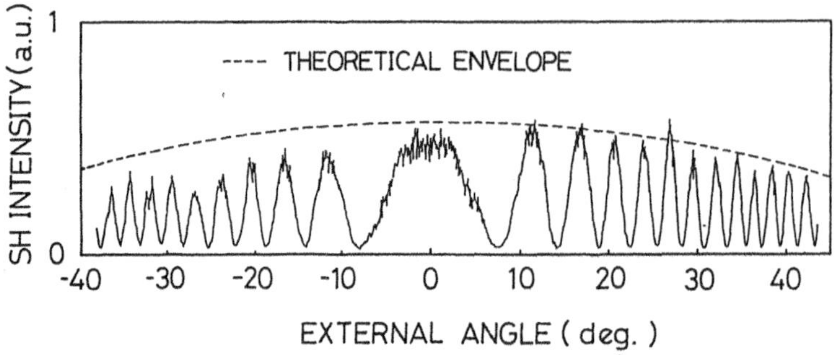

Fig. 2. Maker fringes and theoretical envelope for d_{22} of MBANP. Sample thickness is 0.595mm

In conclusion, we have shown that MBANP is a promising nonlinear optical material. The molecular hyperpolarizability of MBANP is as large as those of para-nitroaniline-like molecules, while the charge-transfer absorption band is appreciably shifted to shorter wavelengths. MBANP has an absorption cut-off around 0.45μm in the crystalline state. As is expected from the molecular alignment in the crystal, the nonlinear optical coefficient d_{22} was revealed to be very large : d_{22} of MBANP is 126 times larger than d_{11} of quartz and 1.46 times larger than d_{33} of $LiNbO_3$. The large optical nonlinearity and the transparency in the relatively short-wavelength region make MBANP attractive as a material for frequency doubling of semiconductor lasers.

We would like to thank Professor M. Hidai, Mr. T. Murata, Dr. K. Ishida, and Dr. T. Tanase of the Department of Synthetic Chemistry of the University of Tokyo for their help in synthesis and x-ray analysis. We also acknowledge Mr. A. Nagai of Kureha Chemical Industry Co., Ltd. for providing equipment for crystal growth. This work was partially supported by the Grant-in-Aid for Scientific Research, #63850012, from the Ministry of Education, Science and Culture of Japan.

References

1. D.S. Chemla, J. Zyss (eds.): *Nonlinear Optical Properties of Organic Molecules and Crystals* (Academic Press, Orlando 1987)
2. R.J. Twieg, K. Jain, Y.Y. Cheng, J.I. Crowley, A. Azema: Polym. Prepr. **23**, 147 (1982)
3. R. Twieg, A. Azema, K. Jain, Y.Y. Cheng: Chem. Phys. Lett. **92**, 208 (1982)
4. R.T. Bailey, F.R. Cruickshank, S.M.G. Guthrie, B.J. McArdle, H. Morrison, D. Pugh, E.A. Shepherd, J.N. Sherwood, C.S. Yoon, R. Kashyap, B.K. Nayar, K.I. White: Opt. Commun. **65**, 229 (1988)
5. R.T. Bailey, F.R. Cruickshank, S.M.G. Guthrie, B.J. McArdle, H. Morrison, D. Pugh, E.A. Shepherd, J.N. Sherwood, C.S. Yoon, R. Kashyap, B.K. Nayar, K.I. White: J. Mod. Opt. **35**, 511 (1988)
6. G. Bettoni, S. Catsiotis, R. Perrone, V. Tortorella: Gazz. Chim. Ital. **107**, 111 (1977)
7. S. Umegaki: In preparation
8. T. Kondo, N. Ogasawara, R. Ito, K. Ishida, T. Tanase, T. Murata, M. Hidai: Acta Cryst. **C 44**, 102 (1988)
9. Twieg *et al.* also determined the crystal structure (R.J. Twieg , C.W. Dirk: IBM report RJ5237(54077) (1986)). Their results are the almost same as ours, except that the crystallographic *a* and *c* axis are defined in a different way.

10. N.H. Hartshorne, A. Stuart: *Crystals and the Polarising Microscope* (Edward Arnold, London 1970)
11. From these results, the relation,

$$d_{33}^{\mathrm{LiNbO_3}} = (86\pm10) d_{11}^{\mathrm{Quartz}},$$

is deduced. This is in good agreement with the reported results $d_{33}^{\mathrm{LiNbO_3}} = (64\pm16) d_{36}^{\mathrm{KDP}}$ [D.A. Kleinman, R.C. Miller: Phys. Rev. **148**, 302 (1966)] and $d_{11}^{\mathrm{Quartz}} = (0.77\pm0.04) d_{36}^{\mathrm{KDP}}$ [J. Jerphagnon, S.K. Kurtz: Phys. Rev. **B 1**, 1739 (1970)], which give the relation

$$d_{33}^{\mathrm{LiNbO_3}} = (83\pm20) d_{11}^{\mathrm{Quartz}}.$$

Pockels Effect in Organic Thin Films

T. Yoshimura and Y. Kubota

Fujitsu Laboratories Ltd., Atsugi, Japan

I. Introduction

Organic materials with polarized conjugated systems have received much attention because of the large second harmonic generation (SHG) activity. This implies that organic solids are also expected to show a large Pockels effect which, along with SHG, is involved in second-order nonlinear optical phenomena. In Fig.1, the figures of merit F for electro-optic phase retardation are plotted for various organic nonlinear optical materials. For 2-methyl-4-nitroaniline (MNA) and $LiNbO_3$, F was measured by LIPSCOMB et al. [1]. For the other materials, F was estimated from SHG efficiencies [2]. It is expected that F is about 10 times larger in styrylpyridinium cyanine dye (SPCD) and 3 times larger in MNA, when compared with $LiNbO_3$. We examined the Pockels effect in the SPCD and MNA thin-film crystals by a new ac modulation method at 633 nm. We also studied the influence of the interaction between the molecules in crystals on the second-order nonlinearity by simulation.

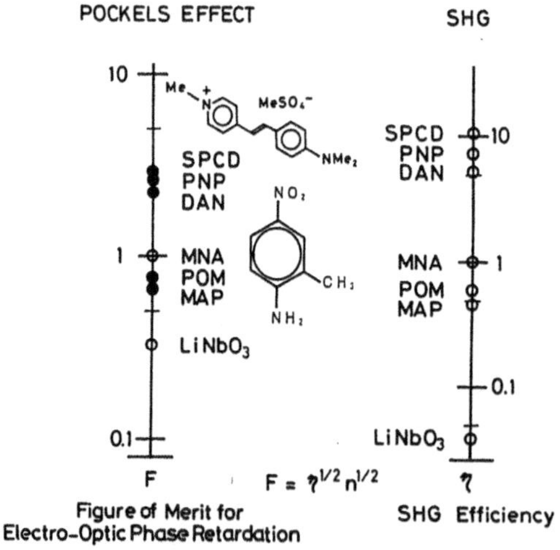

Fig.1 Expected Pockels effect in organic materials

II. Procedure for Pockels effect characterization [3]

Figure 2 diagrams the experimental setup for the new ac modulation method. To apply an electric field to crystals, a slit-type NiCr electrode with a 5 µm gap was pressed against the thin-film crystal. Linearly polarized light at 633 nm from a He-Ne laser passed through the crystal and the light transmitted through the analyzer was detected

Springer Proceedings in Physics, Vol. 36
Nonlinear Optics of Organics and Semiconductors
Editor: T. Kobayashi

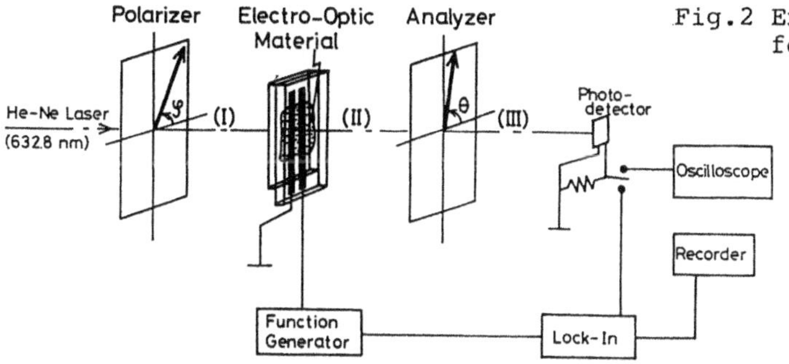

Fig.2 Experimental setup for ac modulation

by a solar cell. When ac voltage is applied to the crystal, the detector output is modulated by electro-optic phase retardation $\Delta\delta$ induced in the crystal. The relationship between the detector output change V_{AC} and the 0 V detector output V_{DC} is given by Eq.(1):

$$V_{AC}/V_{DC} = [E_{0\theta}(\delta_0)^2 - E_{0\theta}(\delta_0+\Delta\delta)^2]/E_{0\theta}(\delta_0)^2 \tag{1}$$

$$E_{0\theta}(\delta)^2 = E_0^2[(\cos\phi\cos\theta + \sin\phi\sin\theta\cos\delta)^2 + (\sin\phi\sin\theta\sin\delta)^2]$$

δ_0: phase retardation at 0 V.

Substituting V_{AC}, V_{DC}, δ_0, ϕ, and θ into Eq.(1) yields $\Delta\delta$. F can then be estimated with the following formula:

$$F = (\lambda_0\Delta\delta D)/(2\pi VL) \tag{2}$$

λ_0: wavelength D: electrode gap L: film thickness
V: applied voltage.

III. Pockels effect in organic thin film crystals

(A) SPCD [3]

SPCD thin-film crystal was grown between two paired quartz substrates from a methanol solution. Figure 3 shows the typical crystal shape and the absorption spectra. The refractive index of the crystal was estimated to be $n_Z=1.55$ and $n_Y=1.3$ using the prism coupler method. Figure 4

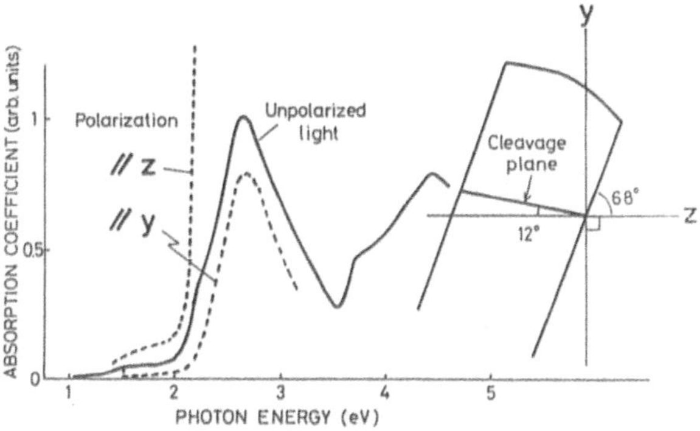

Fig.3 Schematic shape and absorption spectra of SPCD thin-film crystal

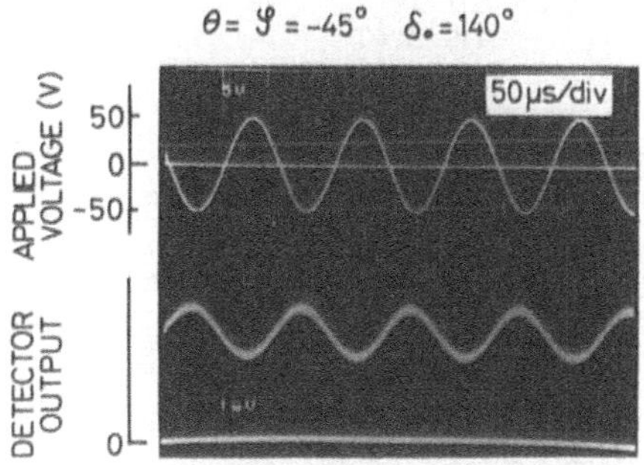

Fig.4 Light modulation induced by ac voltage in SPCD

shows a typical example of the light modulation introduced by an 8-kHz sine wave. Figure 5 graphs the dependence of $\Delta\delta$ on the applied voltage. F and the electro-optic coefficient r_{33} of SPCD are compared with $LiNbO_3$ in Fig.6. For SPCD, F is 6 times and r_{33} 14 times larger than in $LiNbO_3$.

(B) MNA

MNA thin-film crystal was grown between two paired quartz substrates from the melt using the setup diagramed in Fig.7. Figure 8 plots the time chart for temperature and the time derivative of temperature in the crystal growth process. Initially, MNA powder is heated at a constant rate. When the temperature approaches the melting point T_M of MNA, the time derivative of temperature decreases due to the endothermic reaction of MNA powder. By detecting the temperature T_0 at which the rate of the rise starts to decrease, a feedback control keeps the melt temperature at $T_P = T_0 + \Delta T$. The maximum melt temperature can thus be controlled within an accuracy of 0.1°C. Figure 9 plots the T_P dependence of F. As to increases T_P, F decreases drastically. This indicates that T_P must be kept as close as possible to T_M to get a large F. The MNA crystal grown under optimum conditions exhibits an F 3 times larger than $LiNbO_3$. Figure 10 graphs the influence of the purity of MNA powder on F. To get a large F high-purity MNA is necessary.

We expected that one of the following three reasons was responsible for the decrease in F : 1) a change in molecular structure, 2) a change in crystal structure, or 3) a change in interaction between the neighboring molecules. Differential thermal analysis and X-ray diffraction analysis reveal that obvious changes in molecular and crystal structure do not occur. Consequently, the decrease in F is attributed to the change in interaction between molecules in the crystal.

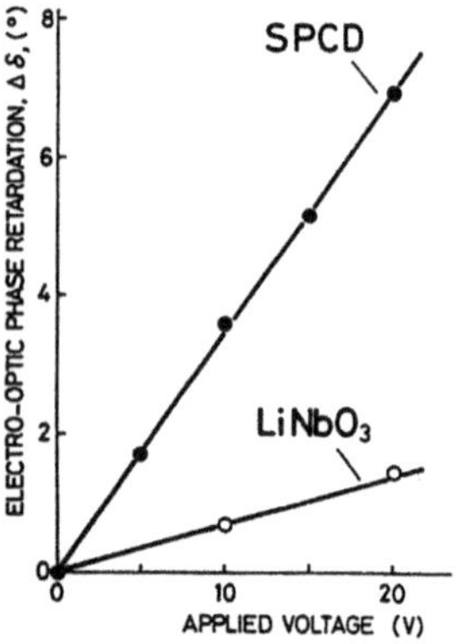

Fig.5 $\Delta\delta$ vs V characteristics in SPCD

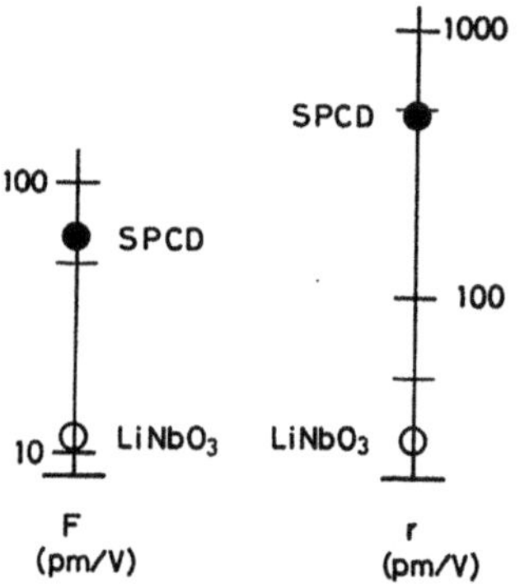

Fig.6 F and r_{33} in SPCD

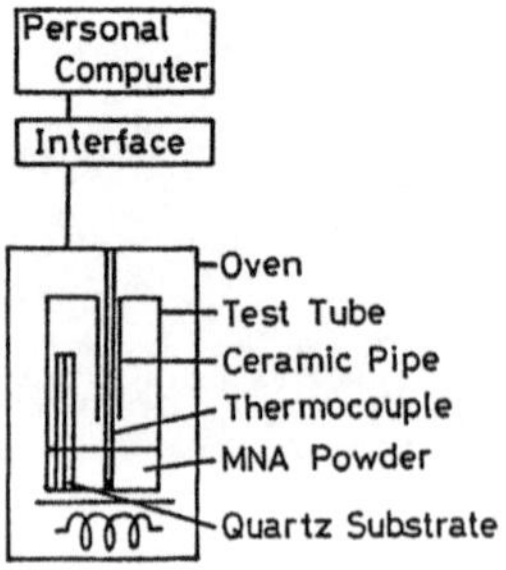

Fig.7 Setup for MNA crystal growth

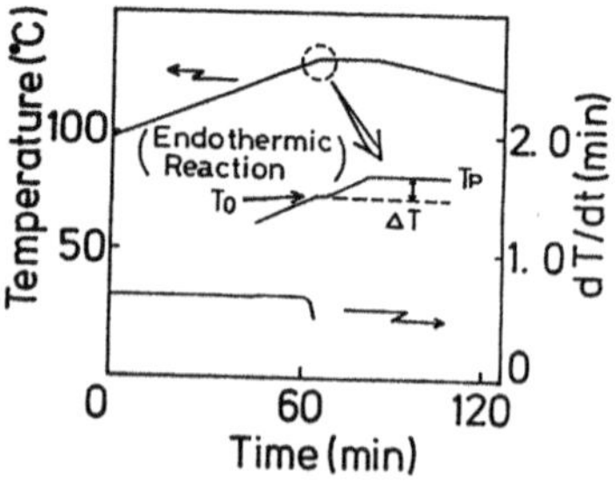

Fig.8 Temperature in the crystal growth process

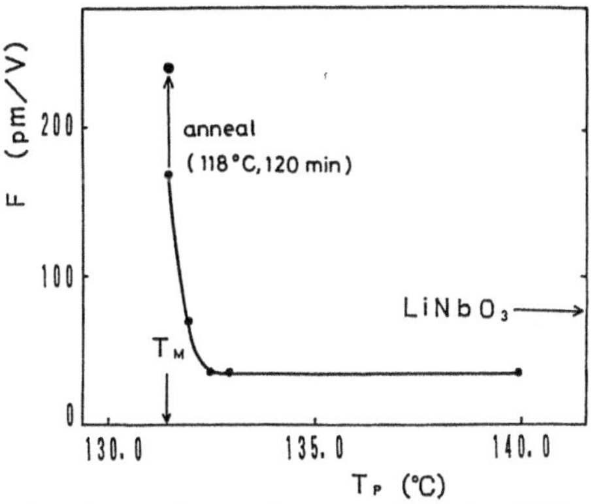

Fig.9 T_P dependence of F in MNA

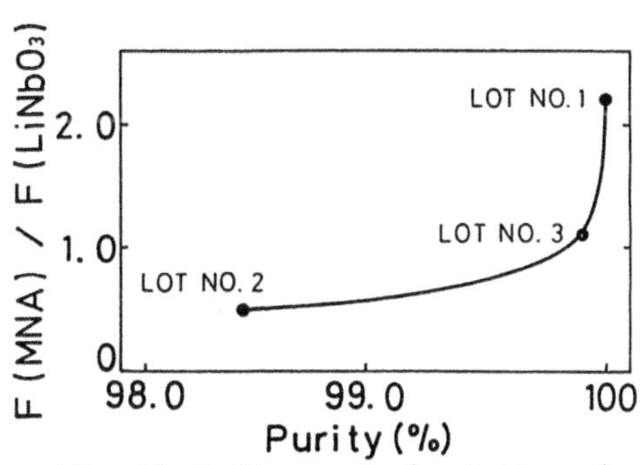

Fig.10 Influence of purity of MNA powder on F

IV. Influence of interaction between molecules on the second-order nonlinear susceptibility β in the crystal

Figure 11 diagrams the four atomic arrangements of MNA molecules considered. Each molecule is rotated around the lattice point by $\theta = 0°$, 15°, 35°, and 65°. The molecular orbitals for each arrangement were calculated by ANCHOR and AM1. From Eq.(3) [4], β of the central molecule was estimated:

$$\beta = 3\mu_{ge}^2 (\mu_e - \mu_g) E_G^2 / (E_G^2 - 4E^2)(E_G^2 - E^2) \quad (3)$$

μ_g : ground-state dipole moment μ_e : excited-state dipole moment
μ_{ge} : transition dipole moment E_G : energy gap   E: photon energy.

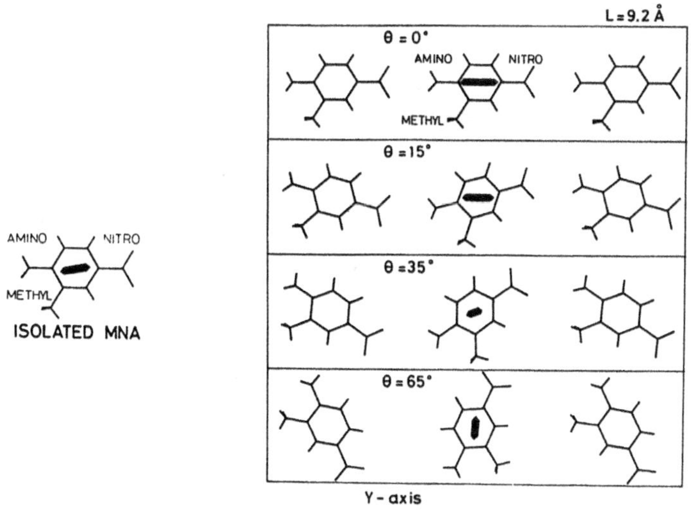

Fig.11 Influence of arrangement of MNA molecules on β

The values of β are represented by bars in Fig.11. When θ increases from 0°to 35°, β decreases. With further increases in θ to 65°, β again increases. This result suggests that the interaction between molecules in solids greatly affects the second-order nonlinearity, and confirms the possibility that the T_P dependence of F in the MNA crystal is attributed to the change in the interaction between MNA molecules. Also note that, in the configuration of $\theta=0°$, β is larger than in the isolated molecule. This indicates that enhancement of β occurs by interaction between neighboring molecules depending on the arrangement of molecules.

V. Summary

1. In SPCD, the figure of merit for electro-optic phase retardation is 6 times and for MNA 3 times as large as that in $LiNbO_3$ crystal. The result suggests that organic nonlinear optical materials are promising for electro-optic materials.
2. In MNA, the Pockels effect decreases with increases in the maximum melt temperature during the crystal growth process and with decreasing MNA powder purity. These changes of the Pockels effect are attributed to interactive changes between MNA molecules in the crystal.
3. The effect of interaction between neighboring molecules on the second-order susceptibility β in MNA crystal was simulated. Decreases and enhancements in β occur depending on the arrangement of molecules.

Acknowledgments

We thank Associate Professor S. Umegaki of Tokyo Engineering University, and Professor Y. Ohkatsu of Kogakuin University for their helpful discussions, Messrs. K. Nishida. and I. Yamamoto of Fujitsu Limited for their kind instructions and support in the molecular orbital calculation by ANCHOR, and Drs. H. Nakajima and T. Akamatsu of Fujitsu Laboratories Ltd. for their encouragement.

References

1) G.F. Lipscomb, A.F. Garito, and R.S. Narang, J. Chem. Phys.75, 1509 (1981).
2) R.J. Twieg and K. Jain, ACS Symposium Series 233, 57 (1983).
3) T. Yoshimura, J. Appl. Phys. 62, 2028 (1987).
4) J.L. Ouder, J. Chem. Phys. 67, 446 (1977).

Part VII

Molecular Crystals – Powder

Synthetic Design of Organic Nonlinear Optical Materials: Cyanovinylheteroaromatics and Related Compounds

M. Matsuoka[1], *T. Kitao*[1], *and K. Nakatsu*[2]

[1]Department of Applied Chemistry, University of Osaka Prefecture, Sakai, Osaka 591, Japan

[2]Department of Chemistry, Kwansei Gakuin University, Nishinomiya, Hyogo 662, Japan

1. INTRODUCTION

The laser as the artificial light has opened up enormous opportunities to develop many electronic systems by using dye chromophores as a light absorber. Nonlinear optical properties of organic molecules developed a new field in dye chemistry in which dye chromophores can be used as a material to produce a new laser light. Design and synthesis of organic molecules for efficient second-harmonic generation (SHG) and third-harmonic generation (THG) are now studied widely and the introductory works are summarized in the book[1].

The practical applications of nonlinear properties of dye chromophores are focused on developing materials to get a new laser light (390-410 nm) from the diode laser by their SHG response.

In this paper, a series of cyanovinylheteroaromatics and related compounds were synthesized and their SHG activities were evaluated by the powder test method, and correlated to chemical structures on the basis of the calculated β values and the X-ray analysis data.

2. PPP MO CALCULATION

Molecular design of a SHG active compound can be conducted by the PPP MO method. The molecular polarizability β of intramolecular charge transfer chromophores is defined by the Oudar's equation[2]. The β value is a function of the laser fundamental photon energy $h\nu$, the electron charge e and the mass m, the energy gap ΔE, the oscillator strength f of the CT transition, and the difference in the dipole moment $\Delta\mu_{ge}$ involved in the first transition. To maximize β it is favorable to minimize ΔE, maximize $\Delta\mu$ and f at a constant $h\nu$. The larger β value is expected to get larger SHG response. There are two main requirements to get a large β value for a molecule, namely, it does not have a molecular symmetry and has π-electron conjugated chromophores with intramolecular charge-transfer character.

Springer Proceedings in Physics, Vol. 36
Nonlinear Optics of Organics and Semiconductors
Editor: T. Kobayashi

From the point of using the diode laser as a light source which emits laser at 780-830 nm, there is a strong restriction to minimize ΔE, that is, the chromophores don't absorb at 400 nm. The λ_{max}, $\Delta\mu$ and f values of chromophores can be calculated quantitatively by the PPP MO method and then appropriate chromophores can be designed for effective SHG response. Some calculation results are shown in Table 1.

Table 1. PPP MO calculation of β values for cyanostyryls

	μ_e [D]	μ_g [D]	$\Delta\mu_{ge}$ [D]	ΔE [eV]	f	$\beta^{a)}$ [$\times 10^{-30}$esu]	λ_{max} [nm]
1a	17.07	8.31	8.76	3.029	0.937	88	428
1c	21.02	8.90	12.11	2.605	0.731	326	512
2	22.20	9.58	12.62	2.741	1.319	368	480

a) Calculated at 1064 nm (YAG laser).

3. X-RAY ANALYSIS OF CRYSTAL

The SHG efficiency can be evaluated by the nonlinear susceptibility (X^2) of the material. To maximize the X^2, the crystal should not have a symmetric center in a unit cell of the crystals. It is desirable to postulate the crystal structures of compounds from their molecular structure but it is impossible to know the real crystal structures without the X-ray diffraction analysis. The molecular mechanics calculation methods have been applied to know the crystal structures in a unit cell but desirable results have not been obtained yet. The empirical rules to control the packing modes of molecules in crystals are now practically used[1]. The relationship between an effective packing in crystals and the chemical structures of molecules must be rationalized practically by the X-ray analysis.

4. CYANOVINYLHETEROAROMATICS AND RELATED COMPOUNDS

Cyanovinylaromatics have generally a good crystallinity, which is required for the efficient SHG crystals for optical devices. Their SHG efficiency was evaluated by the powder test method, and the powder efficiency (P.E) was denoted against urea. The syntheses of the related compounds were carried out by the condensation reaction of the corresponding arylaldehyde with cyanomethylene compounds in the presence of base as catalyst. They have good crystallinity in general and can be purified by recrystallization from organic solvent such as ethanol. The SHG efficiency and other properties are summarized in Table 2. The β values are calculated by the PPP MO method.

Table 2. Some properties of cyanovinylaromatics

No.	R	X	Z	λ_{max} a)	β b)	SHG	PE c)
1a	$N(Me)_2$	H	CN	428	88	○	
1b	$N(Me)_2$	H	COOMe	421	38	×	
1c	$N(Me)_2$	CN	CN	512	326	×	
1d	$N(Et)_2$	H	CN	434	88	×	
1e	$N(Et)_2$	H	COOMe	427	38	○	9
1f	$N(Et)_2$	H	COOEt	426	38	○	16
2			CN	480	409	×	
3a		NH	COOEt	363	9	×	
4a		O		296	1.6	○	1.5
4b		N	OH	265	—	○	— d)
5a	H		COOMe	382	13	×	
5b	H		COOEt	383	13	○	25
5c	Me		COOEt	387	—	×	
6a	Me	O		329	13	○	5.5
6b	Et	O		330	13	×	
6c	Me	N	OH	283	—	○	— d)
7a	Me		CN	407	82	○	— d)
7b	Me		COOEt	398	50	○	— d)
7c	Et		COOEt	398	50	○	— d)

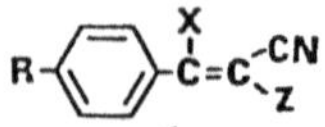

1

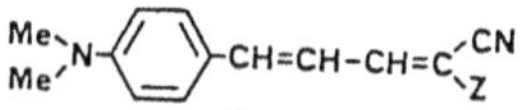

2

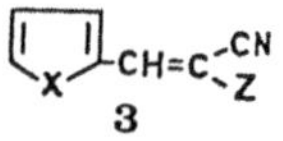

3

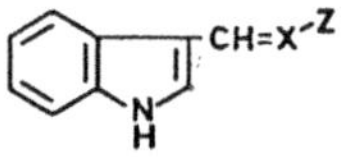

4

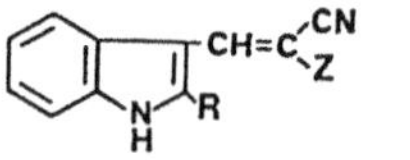

5

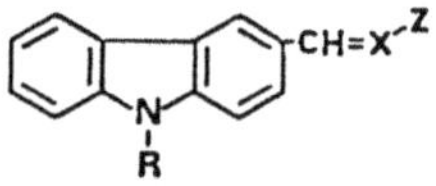

6

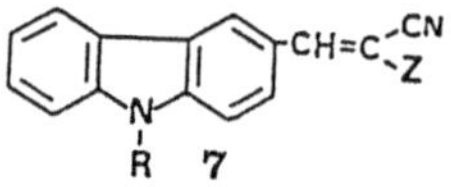

7

a. Measured in EtOH

b. Molecular polarizability calculated by the PPP MO method ($\times 10^{-30}$esu)

c. Powder efficiency against urea

d. Not determined

In a series of cyanostyryl **1**, **1a** has good efficiency (PE=44 U) but **1b** and **1f** are inactive. The substitution of a cyano group in **1a** for methoxycarbonyl group **1b** decreased the β value from 88 to 38, while the λ_{max} only produced a hypsochromic shift of 7 nm. The introduction of a cyano group is effective to get a large β value. An additional cyano group in **1c** increased the β value from 88 of **1a** to 328 in **1c**, but made **1c** inactive. The substitution of a dimethylamino group for a diethylamino group as the donor moiety changed their SHG efficiency. Compounds **1e** and **1f** are active but **1d** is inactive. The extension of a conjugating unit in **2a** also made it inactive. The SHG efficiency is largely influenced by the substituents which control the packing mode of molecules in crystals.

The effects of heteroaromatics, pyrrole (6π system), indole (10π) and carbazole (14π), on the SHG efficiency of cyanovinyl compounds were studied. Pyrrole derivative **3a** has a small β value and is inactive. 3-Formylindole **4a** is active (PE = 1.5 U) in spite of its small β value. The oxime **4b** also has the SHG efficiency. Cyanovinyl derivatives of indole **5** show remarkable substituent effects, that is, **5b** shows good efficiency (PE=25) but its methyl ester **5a** and 2-methyl analogue **5c** are inactive. Carbazole derivatives have generally good SHG efficiency. 3-Formylcarbazole **6a** has a larger β value than indole **4a**, and the efficiency of **6a** (PE = 5.5) is larger than that of **4a** (PE=1.6). N-Ethyl derivative **6b** is inactive but its oxime has the SHG efficiency. Cyanovinyl derivatives of carbazole **7** have quite large β values and show a good SHG efficiency in many types of derivatives.

The crystal data of **1a** obtained by the X-ray analysis are as follows. The yellow needle crystals obtained from methanol solution belong to the monoclinic space group of $P2_1$ with 2 molecules in a unit cell. The crystal structure viewed along the b axis is shown in A, and the molecular overlap viewed perpendicular to the molecular plane is shown in B of Fig. 1, respectively. The dipole moment to induce the effective SHG response of the crystal is directed to b axis, as shown in C of Fig. 1.

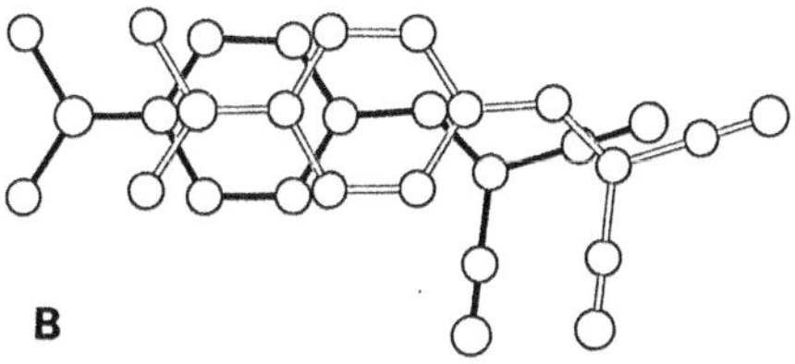

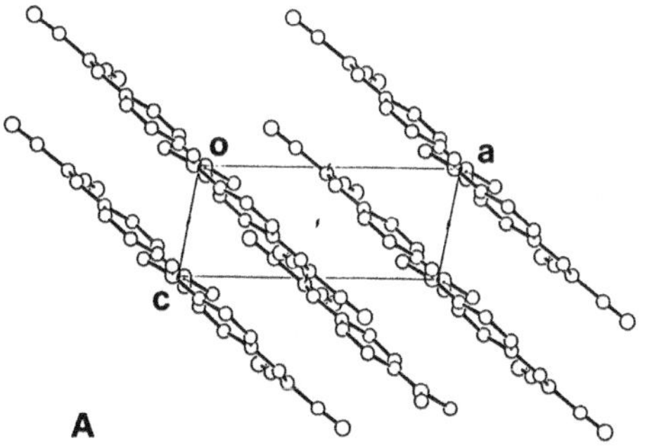

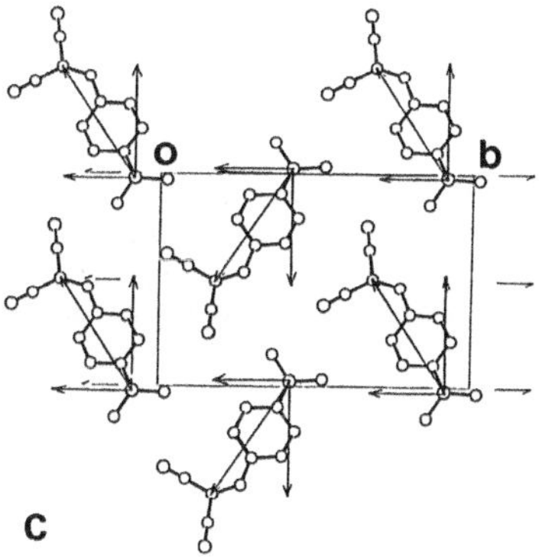

Fig. 1. The X-ray diffraction data of **1a**

5. REFERENCES

1. D. S. Chemla, J. Zyss: Nonlinear Optical Properties of Organic Molecules and Crystals, Vol.1, 2 (Academic Press, London, New York 1987)
2. J. L. Oudar: J. Chem. Phys. **67**, 446 (1977)

Optical Second-Harmonic Generation from a Series of [Cyano(alkyloxycarbonyl)-methylene]-2-ylidene-1,3-dithioles and Their Methyl and Dimethyl Derivatives

T. Nogami[1], *H. Nakano*[1], *Y. Shirota*[1], *S. Umegaki*[2], *Y. Shimizu*[3], *T. Uemiya*[3], and *N. Yasuda*[3]

[1]Department of Applied Chemistry, Faculty of Engineering, Osaka University, Yamadaoka, Suita, Osaka 565, Japan
[2]Department of Electronics, Faculty of Engineering, Tokyo Engineering University, Katakura, Hachioji, Tokyo 192, Japan
[3]Basic High Technology Laboratories, Sumitomo Electric Industries, Ltd., Shimaya, Konohana-ku, Osaka 554, Japan

Second-harmonic generation (SHG) of a series of title molecules was investigated mainly by a powder technique. A Nd:YAG laser was used as a fundamental wave. The material 3 (R = Et) recrystallized from methyl alcohol gave large SHG (24 times that of urea), but that recrystallized from acetone gave no SHG. The material 2 (R = Et) also gave large SHG (10 times that of urea). However, other materials gave weak or no SHG. These materials are almost colorless. They are attractive materials for frequency doubling of infrared semiconductor lasers in blue wavelength region.

1. Introduction

Conjugated organic molecules have drawn much attention because of their large optical nonlinearities [1]. Intramolecular charge-transfer (CT) interaction has been found to play a crucial role especially to the second-order nonlinearity [2]. A variety of intramolecular CT systems have been investigated. Most of the materials contain amino- and dialkylamino-groups or their modified structure as the donor sites and nitro- or cyano-group as the acceptor sites. 2-Methyl-4-nitroaniline (MNA) is the most typical material studied extensively [1]. For frequency doubling, the materials should be transparent not only at a fundamental wavelength but also at the second-harmonic (SH) wavelength. Considering practical applications, infrared semiconductor laser will be used as the fundamental wave. In this case, the materials should be transparent in the visible range. Generally speaking, second-order hyperpolarizability becomes larger with the increase of the conjugation length of the organic molecules. However, it concomitantly results in a bathochromic shift of an absorption edge. Thus, larger hyperpolarizability and wider transparency in visible wavelength region conflict with each other. In order to overcome this drawback, we noticed a 1,3-dithiole ring as the electron donor site[3,4]. It is a 7π electron system, and is easily oxidized to give 6π aromatic system. Thus, a 1,3-dithiole ring can be regarded as an electron donor site. We have investigated SHG of a series of [cyano(alkyloxycarbonyl)methylene]-2-ylidene-1,3-dithiole (1) and their methyl-

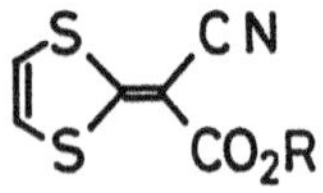
1

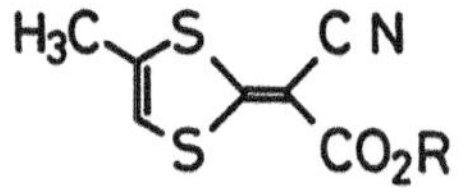
2

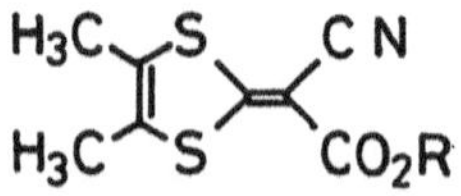
3

Springer Proceedings in Physics, Vol. 36
Nonlinear Optics of Organics and Semiconductors
Editor: T. Kobayashi

(2) and dimethyl-derivatives (3) by a powder technique, and found that they are almost transparent in the visible range, and some of them give large SH intensity comparable to MNA.

2. Experimental

All of the materials were synthesized by the method reported [5]. SH intensities were estimated by powder technique [6] using Nd:YAG laser (1.064 μm) as a fundamental wave.

3. Results and Discussion

All of the molecules in this study are almost colorless. Figure 1 shows an electronic absorption spectra of 1 (R = Et) in methyl alcohol as an example. It has only slight absorptions in the visible range [7]. Table 1 summarizes materials, solvents of recrystallization, SH intensities relative to urea measured by powder method, $\beta(-2\omega;\omega,\omega)$ calculated by Pariser-Parr-Pople molecular orbital method [8], and absorption peak positions and cut-off wavelengths in methyl alcohol. Various aspects of the Table will be illustrated below.

Although most of the SH intensities were obtained by powder method, the β-value of 1 (R = Et) was determined by a dc-SHG method [7]. In this table, nU means that the SH intensity of the material is n times as large as that of urea by powder technique. Large SHG was observed for 2 (10U, R = Et) and 3 (24U, R = Et). The value for the latter material is comparable to MNA. The symbol x mean SHG inactive property. Weak SHG was observed for 1 (R = Me) and 2 (R = n-Pr and i-Pr). However, other materials did not show SHG. It is especially interesting that 3 (R = Et) shows large SHG if it is recrystallized from methyl alcohol. However, it did not show SHG if it is recrystallized from acetone. Recent x-ray crystallographic work revealed that 3 (R = Et) is crystallized out in a triclinic form. Its space group is P1 and $P\bar{1}$ if it is recrystallized from methyl alcohol and from acetone respectively [9]. The lack of the inversion

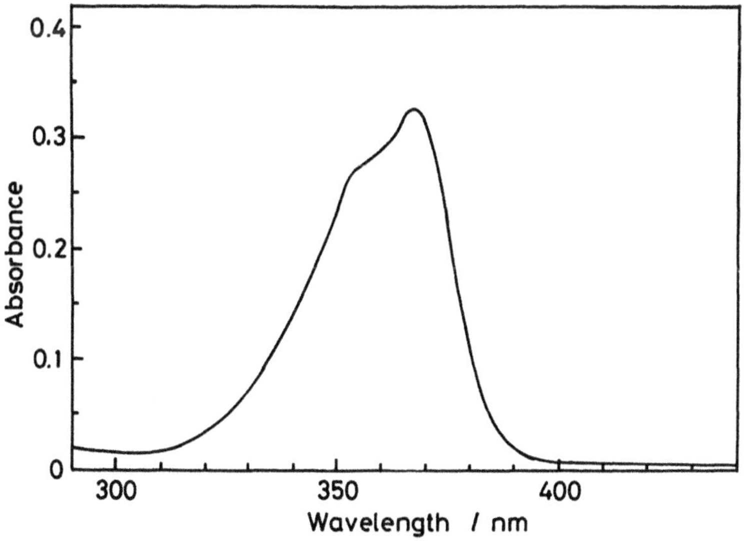

Fig. 1 Electronic absorption spectrum of 1 (R = Et) (1.11×10^{-5} mol dm^{-3}) in methyl alcohol. $\varepsilon_{max} = 29000$ l mol^{-1} cm^{-1}.

Table 1

Materials, solvents of recrystallization, SH intensities relative to urea (powder technique), β-values calculated by PPP method, and absorption peak positions and absorption edge in methyl alcohol.

Materials		Solvents	SHG[a)]	β[b)] (10^{-30} esu)	λ_{max} (nm)	$\lambda_{cut-off}$[c)] (nm)
1	R = Me	MeOH	weak (< 0.1U)	15	368	410
	Et	EtOH	x	13 (dc-SHG)	368	410
	n-Pr	n-PrOH	x	15	368	410
	n-Bu	MeOH	x	15	368	410
	i-Pr	i-PrOH	x	15	368	410
2	Me	MeOH	x	15	369	415
	Et	EtOH	10U	15	369	415
	n-Pr	n-PrOH	weak (< 0.1U)	15	369	415
	n-Bu	MeOH	x	15	369	415
	i-Pr	MeOH	0.5U	15	369	415
3	Me	MeOH	x	15	378	420
	Et	acetone	x	15	378	420
	Et	MeOH	24U	15	378	420
	n-Pr	n-PrOH	x	15	378	420
	n-Bu	MeOH	x	15	378	420
	i-Pr	i-PrOH	x	15	378	420

a) Fundamental wavelength = 1.064 μm. x : no SHG. nU means that the SH intensity is n times as large as that of urea.
b) PPP calculation assuming two-level model.
c) Cut-off wavelength of the methanol solution (5×10^{-5} mol dm^{-3}) with full-scale of the absorbance of 0.05.

symmetry in the crystal for the former case gives rise to SHG. β-Values of 15×10^{-30} esu was calculated for these materials based on the molecular structure in Ref. 9, by taking account of only one excited state [10]. Although the calculation is not so elaborated, the value is relatively close to that of 1 (R = Et) (13×10^{-30} esu) obtained by dc-SHG method [7]. The agreement between the PPP calculation and experiment has also been confirmed recently [11]. Figure 2 shows the calculated π-electron densities in the (a) ground and (b) in the first excited singlet states. This figure (c) also shows the difference in the π-electron densities between the ground and the first excited singlet states. It is evident from the figure that about 0.68 electrons are pulled out from the vinylenedithio portion to the other portion of the molecule in the first excited state. This results in the pronounced increase of the dipole moment in the first excited singlet state, and gives rise to the large SHG of this type of the materials. The absorption peak positions of these materials located in the ultraviolet range around 368 - 378 nm. The cut-off wavelengths were determined by the measurement of the absorbance in methyl alcohol (5×10^{-5} mol dm^{-3}) with the full scale of 0.05. They located around 410 - 420 nm. Summarizing Table 1, we found that these materials possess only slight absorption in the visible region, and some of them showed large SHG comparable to MNA. They are attractive materials for the frequency doubling of infrared semiconductor lasers in the blue wavelength region.

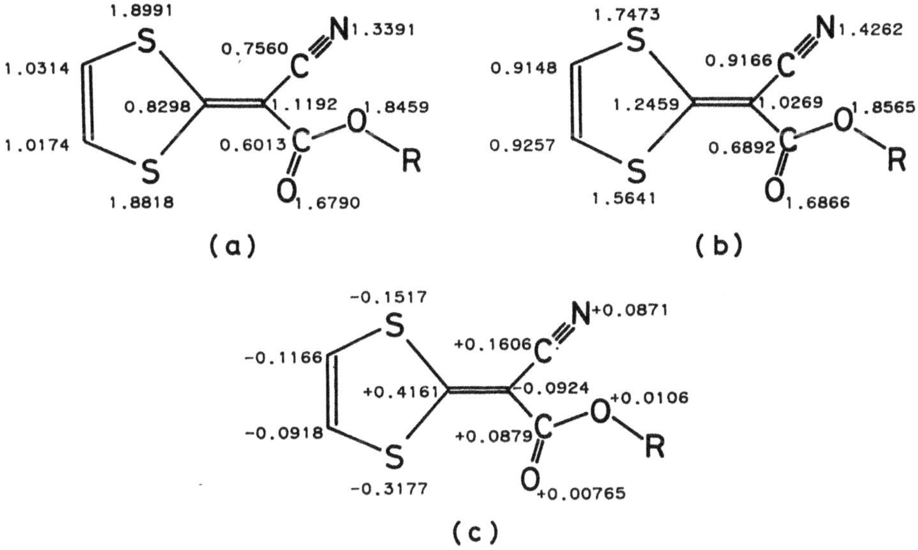

Fig. 2 π-Electron densities of 1 (or 2 or 3) calculated by PPP method in the (a) ground state, and in the (b) first excited singlet state. (c) The difference in the π-electron densities between the first excited singlet state and the ground state. A positive value in (c) means an increase in the π-electron density in the excited state.

References

1. D. S. Chemla and J. Zyss, Nonlinear Optical Properties of Organic Molecules and Crystals, (Academic Press, Inc., Orlando, San Diego, New York, Austin, Boston, London, Sydney, Tokyo and Toronto ,1987).
2. S. J. Lalama and A. F. Garito, Phys. Rev., A20, 1179 (1979).
3. H. E. Katz, K. D. Singer, J. E. Sohn, C. W. Dirk, L. A. King, and H. M. Gordon, J. Am. Chem. Soc., 109, 6561 (1987).
4. M. B-Desce, I. Ledoux, J-M. Lehn, J. Malthete, and J. Zyss, J. Chem. Soc., Chem. Commun., 1988, 737.
5. K. A. Jensen and L. Hendriksen, Acta Chem. Scand., 22, 1107 (1968).
6. S. K. Kurtz and T. T. Perry, J. Appl. Phys., 39, 3798 (1968).
7. S. Umegaki, T. Uemiya, Y. Shimizu, S. Yamanouchi, H. Nakano, T. Nogami, and Y. Shirota, submitted to Jap. J. Appl. Phys.
8. R. Pariser and R. G. Parr, J. Chem. Phys., 21, 466, 767 (1953). J. A. Pople, Trans. Faraday Soc., 46, 1375 (1953).
9. N. Yoshie, H. Yoshioka, K. Nakatsu, T. Nogami, Y. Shirota, Y. Shimizu, T. Uemiya and N. Yasuda, Proceedings of the 56th Annual Meeting of Chemical Society of Japan,1988, 3XIIC11.
10. J. L. Oudar and D. S. Chemla, J. Chem. Phys., 66, 2664 (1977).
11. D. Li, M. A. Ratner and T. J. Marks, J. Am. Chem. Soc., 110, 1707 (1988).

Effect of Esterification on Third Harmonic Generation in Several Dyes

S. Matsumoto, K. Kubodera, T. Kurihara, and T. Kaino

NTT Laboratories, Musashino, Tokyo 180, Japan

1. Introduction

π-electron conjugated organic systems have potential for application to various optical devices because of their large nonlinearity and fast response times. The correlation between optical nonlinearity and molecular structure is fairly well understood in second-order nonlinearity, but it is not yet well understood in third-order nonlinearity. The only guiding principle for enhancing the third-order effect is to extend the conjugated π-bond length [1]. Thus, π-electron conjugated polymers such as polydiacetylene and polyacetylene have mainly been investigated. However, the large $\chi^{(3)}$ of these polymers is due to the resonance effect [2], which may prevent the response time from being so fast. Additionally, the polymers absorb the longer wavelengths in the visible light region. These are the disadvantages in the practical value. We have recently been interested in low molecular weight materials showing a large third-order optical nonlinearity without such longer wavelength absorption. We have already reported an azo dye attached polymer [3], where even if the conjugated length is short, $\chi^{(3)}$ can be large. Low molecular weight compounds can be expected to have larger third-order nonlinearity if they crystalize in preferable crystal structures or preferable molecular alignment that generate third-order nonlinearity.

This study investigated third-order nonlinearity in a series of phenyl azobenzene, stilbene and benzylideneaniline derivatives. We obtained several compounds showing large third harmonic generation (THG). We also found that THG can be enhanced by substituting a group that is not bonded directly to the π-conjugated systems. The reason for this enhancement is also discussed.

2. Experiment

The investigated derivatives are phenyl azobenzene, stilbene, and benzylideneaniline compounds as represented by (1) in Fig. 1. They each have a nitro group as an electron acceptor and an N-ethyl, N-β-hydroxyethylamino group as an electron donor. The hydroxyethyl end group was esterified to several ester compounds such as nitrobenzoate, acrylate and acetate as represented by (2).

The intensity of THG was measured by a powder technique using powdered crystal with a particle size of about 100 μm. THG intensity was also measured for dimethylformamide (DMF) solution of several compounds at a concentration of 2 wt%. A 1 mm thick quartz cell was used for the measurements.

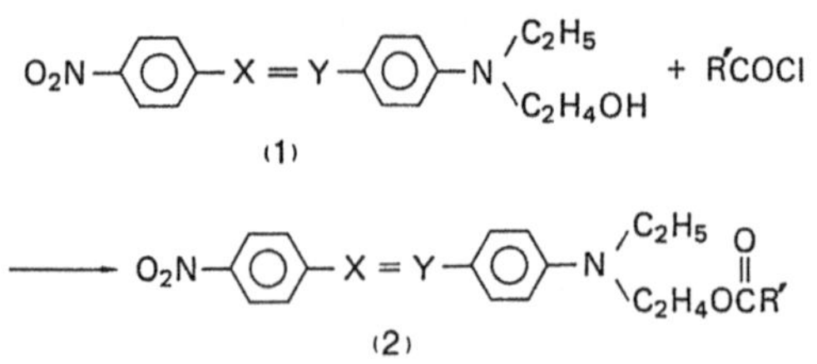

Fig. 1. Molecular structure of investigated compounds.
Phenyl azobenzene : X=Y=N
Stilbene : X=Y=CH
Benzylideneaniline : X=CH, Y=N

Springer Proceedings in Physics, Vol. 36
Nonlinear Optics of Organics and Semiconductors
Editor: T. Kobayashi

3. Results and Discussion

The measurement results are shown in Table 1. An unesterified azo compound (A-1) shows a small THG at 1.06 μm, 1.9 μm, and 2.05 μm. On the other hand, several ester derivatives, especially acetate (A-2), nitrobenzoate (A-3), and acrylate (A-4) show a very large THG. The THG intensity is more than 10 times larger than that for the unesterified compound A-1 at 2.05 μm. Sulfonyl esterification produces little enhancement in THG. For benzylideneaniline derivatives, although a large THG is observed even in unesterified compounds BA-1, a larger THG is generated in ester derivatives, acetate (BA-2) and nitrobenzoate (BA-3). Methylbenzoate (BA-4) has a THG intensity 110 times that of p-nitroaniline (pNA). This is due to the cascade effect (2w + w), because BA-4 generates a second harmonic wave 16 times stronger than urea. THG intensity enhancement is not observed in sulfonyl esters, BA-5 and BA-6. Similar phenomena are found for stilbene derivatives. Ester derivatives, S-2, S-3, and S-4, show larger THGs than unesterified compound S-1. Thus, the enhancement of THG in powdered crystal by esterification is observed in many cases. These results have been unexpected because esterification occurs at the hydroxy group far from the π-electron conjugated system, so it would seem to have little effect on electronic states.

THG intensity in solution is shown in the Table 1. There are large differences between phenyl azobenzene, stilbene and benzylideneaniline compounds. However, esterification gives little changes in all three derivatives. For example, A-2, whose THG intensity is 11 times A-1 in powder, shows a slightly weaker THG than A-1. Thus, enhancement in THG intensity is only observed in powdered crystal, but not in solution. This suggests that the change in crystal structure or molecular alignment due to esterification produces the enhancement of THG intensity.

The esterification of phenyl azobenzene derivatives showed the largest enhancement in THG. The absorption spectra of the compounds in powdered crystal and solution are shown in Fig. 2 and Fig. 3. In solution, the absorption maximum changes slightly to a shorter wavelength by esterification. Molar absorptivity is the same for esterified and unesterified derivatives. On the other hand,

Table 1. THG intensity for phenyl azobenzene, stilbene and benzylideneaniline derivatives.

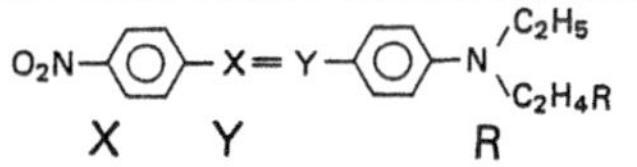

	X	Y	R	THG intensity in powder form (vs pNA=1) 1.06 μm	1.9 μm	2.05 μm	THG in solution $I_{2w}^{1/2}$ (arb. unit) 1.9 μm
A-1	N	N	OH	2	2	4.8	1.0
A-2	N	N	$OCOCH_3$	11	22	-	0.93
A-3	N	N	OCO-Φ-NO_2	21	31	57	-
A-4	N	N	$OCOCH{=}CH_2$	17	-	60	0.93
A-5	N	N	OSO_2-Φ-CH_3	-	4	-	-
A-6	N	N	OSO_2-Φ-NO_2	-	3	-	-
BA-1	CH	N	OH	25	17	-	1.0
BA-2	CH	N	$OCOCH_3$	38	50	-	0.98
BA-3	CH	N	OCO-Φ-NO_2	43	57	44	0.92
BA-4	CH	N	OCO-Φ-CH_3	-	110*	-	-
BA-5	CH	N	OSO_2-Φ-CH_3	-	7	-	-
BA-6	CH	N	OSO_2-Φ-NO_2	-	8	-	-
S-1	CH	CH	OH	14	8	-	2.1
S-2	CH	CH	$OCOCH_3$	33	20	-	1.8
S-3	CH	CH	OCO-Φ-NO_2	18	27	23	2.0
S-4	CH	CH	$OCOCH{=}CH_2$	48	-	-	-

* Second harmonic generation 16 x Urea

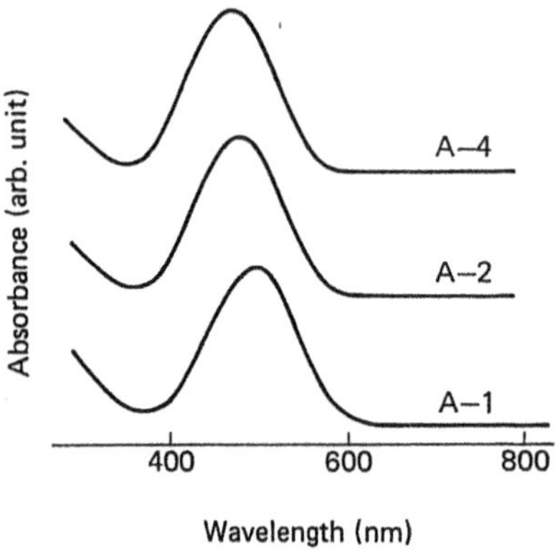

Fig. 2. Absorption spectra for DMF solution of phenyl azobenzene derivatives.

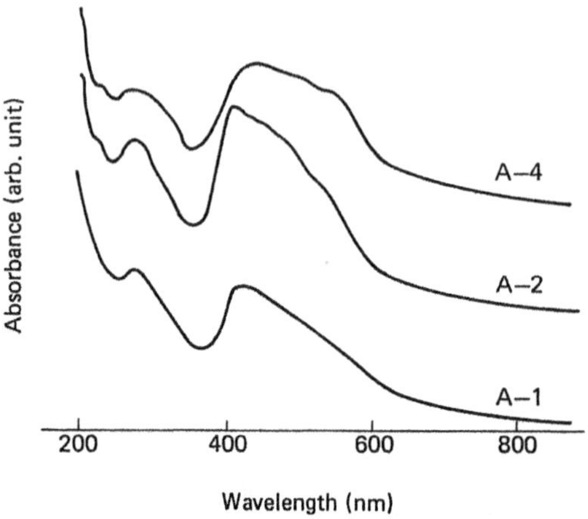

Fig. 3. Absorption spectra for powdered crystal of phenyl azobenzene derivatives in KBr disc. KBr disc thickness: A-1 : 0.31mm, A-2 : 0.21 mm, A-4 : 0.23 mm.

large changes are observed in powdered crystals. Shoulders appear on longer wavelength side of the absorption peak for esterified compounds, A-2 and A-4. The molar absorptivity also seems to be increased by esterification. The difference in absorption spectra between in powdered crystal and in solution is consistent with the results of THG measurements. The changes in absorption spectra mean that the electronic states are very different in esterified and unesterified compound crystals. This is thought to be due to the changes in crystal structure or molecular alignment. There are other examples [4] in similar derivatives, which have clarified that changes in electronic states can be produced by the variations in substituents or crystallization conditions.

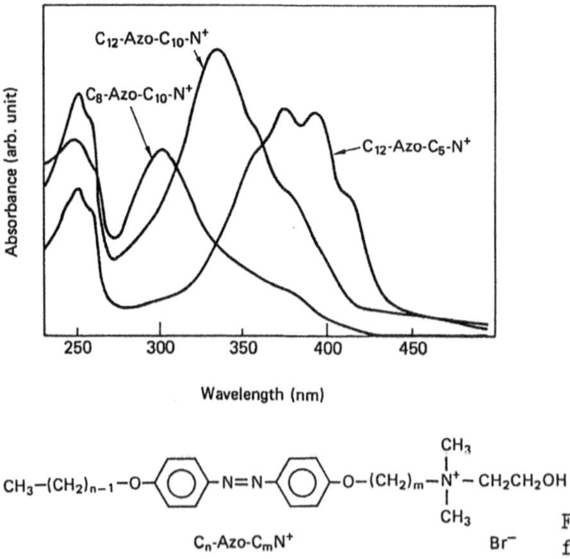

Fig. 4. Absorption spectra for cast films of C_n-Azo-C_mN^+ [5].

Further investigations are necessary to clarify the changes in crystal structure or molecular alignment brought about by esterification. However, there is a report [5] leading to speculation. It is a study on the correlation between the molecular orientation and spectral characteristics in phenyl azobenzene containing compounds shown in Fig 4. It reports that absorption spectra dependent on alkyl chain length m or n reflect the orientation of phenyl azobenzene. The absorption of C_{12}-Azo-C_5N^+ at longer wavelengths is produced by the head-to-tail orientation of the phenyl azobenzene, in other word the J aggregation. The work leads to speculation that the esterification in this study similarly results in the head-to-tail orientation of phenyl azobenzene.

4. Conclusion

The enhancement in THG intensity was shown by the esterification of phenyl azobenzene, stilbene, and benzylideneaniline derivatives. This enhancement is probably due to the changes in crystal structure or molecular alignment by esterification . The largest THG intensity in this work was lower than that in PTS polydiacetylene. However, it should be emphasized that high THG intensity can be achieved even in shorter π-electron conjugated systems by modifying the molecular structure.

Acknowledgments

The authors wish to express their gratitude for the encouragement this work has received from Y. Katayama and H. Hiratsuka. They also wish to express their thanks to S. Tomaru for his valuable discussions.

References

1. J. P. Hermann and J. Ducing, J. Appl. Phys., 45, 5100 (1974)
   K. C. Rustagi and J. Ducing, Opt Commun., 10, 258 (1974)
2. F. Kajzar and J. Messier, Thin Solid Films, 132, 11 (1985)
3. S. Matsumoto, K. Kubodera, T. Kurihara, and T. Kaino, Appl. Phys. Lett., 51, 1 (1987)
4. J. Bernstein, T. E. Anderson, and C. J. Eckharrdt, J. Amer. Chem. Soc., 101, 541 (1979)
5. M. Shimomura, R. Ando, and T. Kunitake, Ber. Bunsenges. Phy. Chem., 87, 1134 (1983)
   H. Watanabe, K. Okuyama, H, Hirabayashi, M. Shimomura, T. Kunitake, S. Kajiyama, and N. Yasuoka, Polymer Preprints, Japan, 34, 652 (1985)

Part VIII

Liquid Crystals

New Theoretical and Experimental Results on Multiwave Mixing and Phase Conjugation in Liquid Crystals and Semiconductors

I.C Khoo

Department of Electrical Engineering, The Pennsylvania State University, University Park, PA 16802, USA

Recently observed nonlinear optical wave mixing effects in liquid crystals and semiconductors are discussed, along with newly developed theoretical frameworks. These theories and experimental results provide new insights into beam/image amplifications, phase conjugations and self-oscillations and are potentially useful in constructing new nonlinear optical devices.

I. Introduction

Current research and development in optical imaging, computing, communication and signal processing and switching have focused considerable efforts on highly efficient nonlinear optical and electro-optical materials. In combination with novel nonlinear optical processes and cw or pulsed lasers covering a wide spectral range, several useful practical devices have been developed. In particular, optical switching elements and logic gates [1,2], infrared-to-visible image converters, real time image correlators, image or beam amplifiers [3], nonlinear guided wave structures, optical memory devices (e.g., associative memory by phase conjugation [4]) and various beam steering and beam combining devices [5] have been proposed and demonstrated.

Concomitant to these device developments, optimization procedures and system requirement considerations have narrowed the "ideal" nonlinear optical materials to a few classes, such as semiconductors, photorefractive crystals and organic materials [6]. In this paper, we will focus on newly observed degenerate four wave mixing processes, and the underlying theories, in materials that possess an intensity-dependent refractive index. To demonstrate the effects, we will use liquid crystals (which possess large orientational and thermal nonlinearities) and the semiconductor silicon (which possesses an electronic nonlinearity) as examples.

II. Nonlinear Optical Properties of Liquid Crystals

Linear optical and electro-optical responses of liquid crystals are the fundamental mechanisms underlying various liquid-crystal-based optical switches, displays and optical modulators. Studies of the nonlinear optical properties of liquid crystals in their liquid crystal phases, and their applications in nonlinear optical devices are relatively new. There are several fundamental mechanisms for optical nonlinearities (a terminology equivalent to intensity-dependent refractive indices) including electronic (one- or two-photon induced) and individual molecular reorientational processes that are associated with picosecond response time and high optical power requirement, and liquid crystal director axis reorientation and collective thermal refractive index changes that are associated with a longer time scale but much lower optical power requirement. All currently developed liquid-crystal-based devices make use of the slower response collective director axis reorientation induced by some dc or low frequency ac field, because the process consumes very little energy and responds on a fairly acceptable time scale (milliseconds to microseconds). Purely optical induced director axis reorientations are

Springer Proceedings in Physics, Vol. 36
Nonlinear Optics of Organics and Semiconductors
Editor: T. Kobayashi

fundamentally similar to these low frequency effects with some important differences that are discussed in detail in reference [7].

Other currently available liquid-crystal-based devices make use of their extremely temperature sensitive refractive indices. Both the ordinary refractive index n_o and the extraordinary refractive index n_e in the nematic phase are critical functions of the temperature near T_c, the nematic isotropic transition temperature. Obviously, if the liquid crystal possesses some finite absorption at a particular laser wavelength, temperature change can be induced through photo-absorption and the resulting refractive index change can be utilized for nonlinear optical processes.

III. Electronic Nonlinearity of the Semiconductor Silicon

An excellent review of optical nonlinearities of semiconductors may be found in reference [8]. In particular, in silicon, one of the mechanisms for nonlinearity is the change in refractive index caused by valence to conduction band transition. In silicon, this transition may be induced by the 1.06μm output of a Nd:YAG laser. Details of the calculation for the third-order nonlinear susceptibility corresponding to this process are discussed in reference [8]. In our recent study [10], this type of nonlinearity, and the associated intensity-dependent absorption loss (due to the generated free carriers) are reexamined in the context of diffracted beam assisted probe beam amplification. Space does not allow us to elaborate on the details, and the interested readers are referred to reference [9] for a discussion of the physical mechanisms accompanying a laser-induced conduction valence band to conduction band transition.

IV. Beam/Image Amplification and Phase Conjugation

Figure 1 depicts schematically the interaction of a reference (or pump beam 1) interacting with an object (or probe beam) on a nonlinear film. The resulting index grating can be read by the pump beam to give a forward propagating image (diffracted beam d_1) or by a beam counter propagating (cf. figure 2) to the pump to give an image beam (phase conjugated) propagating backward along the object beam.

There are several interesting features of this process, which may be generally termed degenerate multiwave mixing. In the phase conjugation

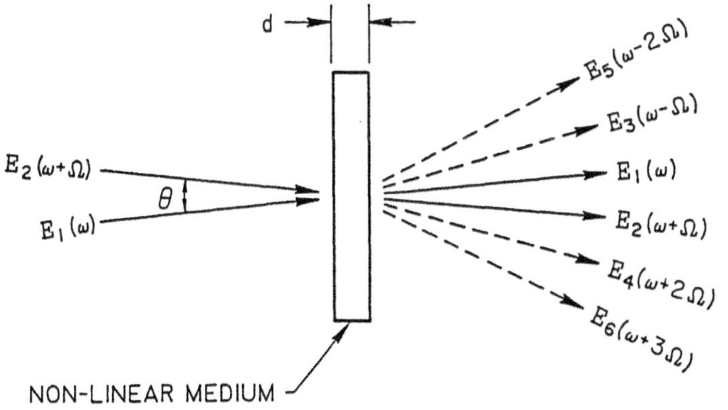

Fig. 1 Schematic of forward wave mixing of a pump and a probe (object) beam. The diffracted beam from the pump (which serves as the reading beam) is the image beam.

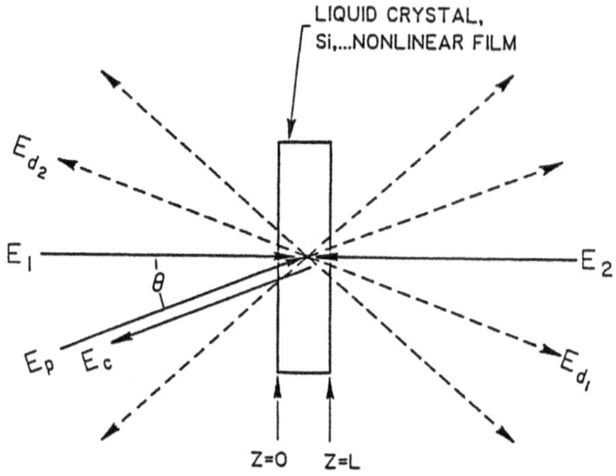

Fig. 2 Schematic of phase conjugation involved a reading beam that is counterpropagating to the pump. If the nonlinear medium is thin, both phase conjugated and diffracted beams are produced.

configuration (figure 2) the forward propagating probe/object beam will experience a gain, whereas the phase conjugated reflection may exceed the incident probe/object beam (i.e., amplified reflection). However, for the forward wave mixing geometry (figure 1) it was generally accepted that the probe beam cannot experience gain unless there is a phase shift between the pump-probe intensity grating and the induced refractive index grating. However, our recent theories and experiments [10] have shown that in a highly nonlinear thin film, the strong pump - weak probe mixing can give rise to substantial probe beam gain if there is a diffracted beam present. Basically, the diffracted beam and the pump beam generate a grating which scatters light from the strong pump beam into the direction of the probe. However, there are several important factors that need to be quantitatively accounted for in order to accurately describe the process. Detailed theories for various types of nonlinearities (diffusive, local, non-local, moving grating, etc.) have been developed [10a-c].

The starting point of these theories is the Maxwell equation, with a nonlinear polarization P_{NL} as the driving source term:

$$\nabla^2 E - \mu_0 \varepsilon \frac{\partial^2 E}{\partial t^2} = \mu_0 \frac{\partial^2 P_{NL}}{\partial t^2} \tag{1}$$

where $\vec{E}$ is the optical electric field and the other symbols have their usual meanings.

The nonlinear polarization P_{NL} describes the material nonlinear response, and thus may assume almost an infinite number of forms. In many cases, it contains the so-called third-order wave mixing terms of the tensor form $\chi^{(3)}$ EEE, where $\chi^{(3)}$ is the nonlinear susceptibility. Among the more frequently dealt with components are terms of the form $\chi^{(3)} |E|^2 E$. Since $|E|^2$ is proportional to the intensity, $\chi^{(3)} |E|^2$ may be identified as an intensity-dependent refractive index change, i.e., the refractive index n of the medium consists of a background n_0 and a term proportional to $|E|^2$:

$$n = n_0 + n^2 |E|^2 \tag{2}$$

and

$$\chi^{(3)} = {}^1/_3\ \varepsilon_0\ n_0 n_2 \quad . \tag{3}$$

Since the spatial dependence of n coincides with the spatial dependence of $|E|^2$ (i.e., the optical field intensity), we term this type of response local.

If the response of the medium to the field is due to heating, or electro-hole generations, etc. which involve diffusion, temporally and spatially dependent recombinations, boundary conditions, etc., then the spatial dependence of n is in general not the same as the field intensity, and the response is termed non-local. In that case, another set of equations describing these nonlocal material responses is needed. In reference [10a,b] the theory for a local type of response is described, where the effects due to a moving grating (frequency shift between the two incident beams) are also included. In reference [10], a nonlocal type of response, corresponding to laser-induced heating and refractive index change, is described in detail and demonstrated with some experimental observations. In reference [9], the nonlinear response is local, but the absorption loss of the medium is extremely intensity dependent, and also gives rise to saturation effects observed in the other studies. In short, for a quantitative description of the effect under study, a correct analysis of the material response must be included.

There are several other important field and material factors that influence these wave mixing processes. Some of these factors are routinely ignored in wave-mixing studies but they are found to have a profound effect on almost all the processes investigated so far. Inclusion of these factors, of course, invariably means more complicacy in the solution of the nonlinear Maxwell equations (e.g., we invariably have to solve the equations by a numerical technique). However, this little extra numerical work pays off very well in terms of quantitative agreement with experimental results; it also leads to new and better fundamental understanding and insights.

To put all of this in a nutshell, we will simply state these factors and their effects. Details may be seen in the references quoted.

From Eq. (1) and the nonlinear polarizations, in general, a set of N complex coupled nonlinear equations for the optical electric field amplitude is obtained. In our studies, the number N is chosen to coincide with the experimentally observed number of exit beams (3, 4, 6, etc.). These equations are of the form

$$i = 1,..,N, \tag{4}$$

$$\frac{\partial E_i}{\partial z} = f_i(E_j;\ E_k E_\ell;\ E_m E_n E_q;\ \omega_i-\omega_j\text{'s};\ \Delta k_i z;\ \phi_i \pm \phi_j \pm \phi_k \ldots,\ n_2, \chi) \tag{5}$$

where E_i, the complex optical field amplitude, is related to the intensity I_i and phase ϕ_i by

$$\left(\frac{1}{2}\sqrt{\frac{\varepsilon}{\mu}}\right)^{\frac{1}{2}} E_j = \sqrt{I_j}\ \exp i\phi_j \quad . \tag{6}$$

The function f_i's on the R.H.S. of these equations in general contain the following terms:

I. Self-phase modulation term of the form $i\chi^{(3)}|E_i|^2 E_i$. By a proper change of variable, these self-phase modulation terms will appear in conjunction with the other phase terms such as II, III.

II. Phase mismatch term $\Delta k_i z$.

III. Phases (ϕ_i) of the optical field and phase shifts $\phi_i \pm \phi_j \pm \phi_k \cdots$

From Eqs. (4) and (5) above, one can rewrite Eq. (4) into two sets of coupled amplitude I_i and phase (ϕi) equations of the form

$$\frac{\partial I_i}{\partial z} = g_i(E_i;\ E_j\ E_k,\ \text{etc.}) \tag{7}$$

$$\frac{\partial \phi_i}{\partial z} = h_i(E_i;\ E_j\ E_k,\ \text{etc.}) \tag{8}$$

from which we can immediately see the importance of accounting properly for all the phase terms involved. Ignoring one or the other always leads to results that differ substantially from the complete treatment. As a function of the interaction length, these phases in a highly nonlinear medium vary substantially and give rise to results not usually predictable by a casual approach.

In combination with these phase terms, another key factor is the medium loss.

IV. Losses in the medium - The loss may be viewed as a complex phase shift. As the beams traverse the medium, these losses may alter their (the phases') magnitudes, and relative magnitudes with respect to one another. The result is a highly interaction-length-dependent gain, self-oscillations, phase conjugation effect, etc.

V. Pump to probe beam intensity ratio - For the beam amplification via multiwave mixing effect, we have shown, in theory and experiments, that the effect is highly dependent on the beam ratio. This dependence is also observed in the phase conjugation reflectivity involving a reading beam from the opposite side of the nonlinear medium (cf. Fig. 5). In phase conjugation, we have also shown for the first time that the presence of the diffracted beam reduces the self-oscillation threshold considerably (Ref. 14).

VI. Frequency shifts, moving grating - If there is a frequency shift between the pump and the probe beam, the intensity grating on the nonlinear medium will move in time, and this produces an index grating that is phase-shifted with respect to the intensity grating. This phase shift, in conjunction with the phase terms I-III above, can augment or diminish the effect under study (e.g., probe gain, diffraction, phase conjugation, etc.). In reference [11a], we provide some preliminary experimental evidence of some of these theoretical predictions.

The factors outlined above are, of course, intertwined with all the other familiar factors like intensity, nonlinear coefficients $\chi^{(3)}$, n_2, and interaction length. We specifically mention these factors I-VI because of their "unexpected" profound effects on the multi-wave mixing mediated probe gain and related processes.

Experiments with infrared lasers (e.g., CO_2) and thermal nonlinearity are particularly interesting [10,11]. We have observed gain exceeding 20 in a 100μm thick nematic (E7) film with low power CO_2 laser pulses. (Pump beam intensity is on the order of 10 Watts/cm^2 and the probe beam is much weaker (by a ratio of about 100).) In the reported experiment [10c], the temperature (T) dependence of the ordinary refractive index $n_o(T)$ is made use of. If one used the extraordinary refractive index $n_e(T)$, a much higher gain can be obtained for the same pump intensity because dn_e/dT is much larger than dn_o/dT. These thermal grating wave mixing effects are basically similar to those reported for visible lasers [7].

Probe amplification in silicon via the multiwave mixing effect described above has also been observed and quantitatively described in reference [9]. The resultant dependence of the probe amplification on the pump-probe beam ratio is in good agreement with theoretical prediction (cf. figure 3, circles and solid line). In figure 3, we have also plotted the phase conjugation reflectivity of the probe beam (which is obtained if the pump beam is retroreflected), as a function of the pump-probe beam ratio. As expected, because of the dependence of the probe beam intensity on the pump-probe beam ratio, the phase conjugation reflectivity also exhibits a similar dependence. As a result of these dependences on the pump-probe ratio (the gain and reflectivity can vary by a factor of five or more, as observed in other experiments), the deduction of the third-order nonlinear susceptibility from the theoretical fit is accurate only if one explicitly accounts for these multiwave mixing effects.

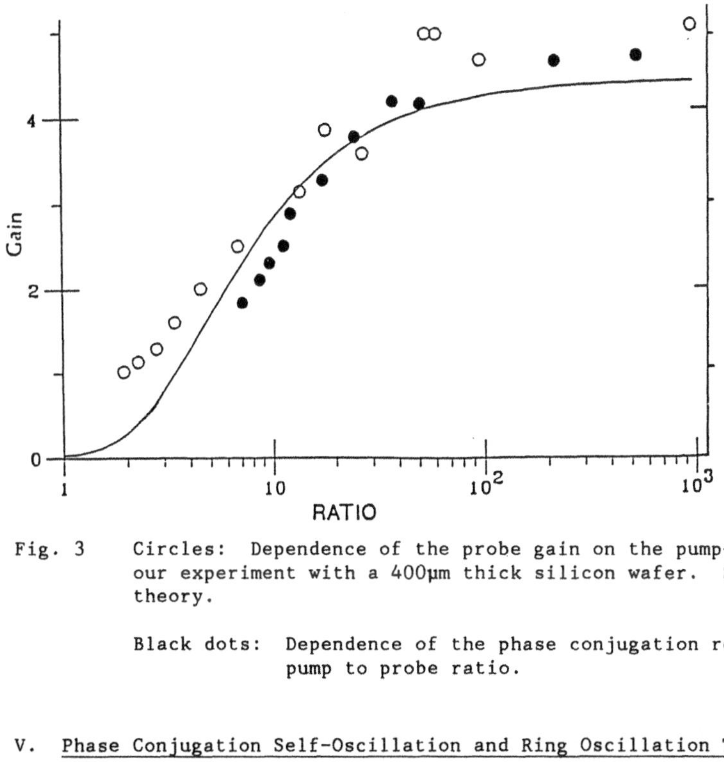

Fig. 3 Circles: Dependence of the probe gain on the pump-probe beam ratio in our experiment with a 400μm thick silicon wafer. Solid line is the theory.

Black dots: Dependence of the phase conjugation reflectivity on the pump to probe ratio.

V. Phase Conjugation Self-Oscillation and Ring Oscillation Thresholds

Because of the large gain experienced by the probe beam (gain of >10 in silicon, >50 in liquid crystals), amplified phase conjugation reflection and self-oscillation can be observed [11]. To quantitatively describe the amplified reflection and self-oscillation, however, one cannot simply employ the usual theory [12] based on degenerate four-wave (pump, probe, retroreflected pump and phase conjugate) mixing for a highly nonlinear thin material. In these thin nonlinear films (e.g., the liquid crystal used is about 100μm thick, and the silicon wafer is on the order of 400μm or less), diffracted beams are inevitably present. Up to second-order diffracted beams are easily observable in both materials. By including the diffracted beam [13] (for the interaction geometry as depicted in figure 2) we found that both the forward probe beam gain and the phase conjugation reflectivity are drastically affected. Among other things, our theory shows that the effect of these couplings with the diffracted beam is to further increase the probe gain and phase conjugation reflectivity, leading to considerably lowered self-oscillation thresholds (figures 4 and 5).

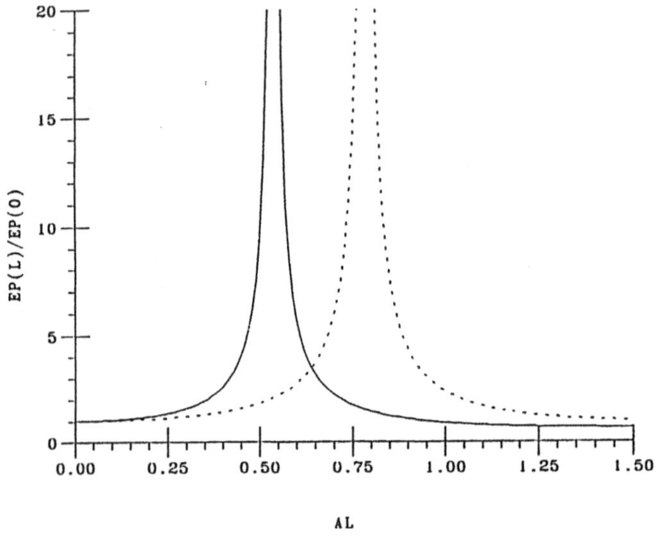

Fig. 4 Plot (full line) of the probe beam gain versus the pump beam intensity parameter $|A|L = \frac{3\pi\omega}{nc} X |E_1E_1| L$, where X is the nonlinear third order susceptibility, L the thickness of the film. The dotted lines are the result of the usual theory that ignores the diffraction coupling.

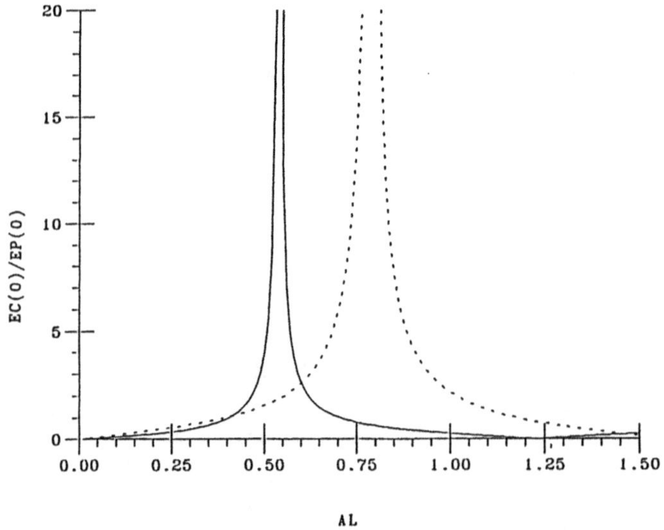

Fig. 5 Plot of the phase conjugated reflectivity of the probe beam versus $|A|L$. Dotted lines are results from the usual theory that ignore the effects due to the side diffraction.

Acknowledgement

This work is supported by a National Science Foundation Grant (ECS871-2078).

References

1. See, for example M. T. Tsao, L. Wang, R. Jin, R. W. Sprague, G. Gigioli, H. M. Kulcke, Y. D. Li, H. M. Chou, H. M. Gibbs and N. Peyghambarian, "Symbolic substitution using ZnS interference filters," Opt. Eng. 26, 41 (1987) and numerous references therein on optical logic gates. See also, K. H. Brenner, A. Huang, and N. Streible, "Digital optical computing with symbolic substitution," Appl. Opt. 25, 3054 (1986).

2. See, for example, S. D. Smith, A. C. Walker, B. S. Wherrett, F.A.P. Tooley, N. Craft, J.G.H. Mathew, M. R. Taghizadeh, I. Redmond and R. J. Campbell, "Restoring optical logic: demonstration of extensible all-optical digital systems," Opt. Eng. 26, 45 (1987) and references therein.

3. See, for example, most of the articles in the special issue on, "Materials and devices for optical information processing," Opt. Eng. 25, 1986 and also in the special issue on "Nonlinear optical materials, devices and applications," Opt. Eng. 24 (1985).

4. Amnon Yariv and Sze-Keung Kwong, Opt. Letts. 11, 186 (1986); B. H. Soffer, G. J. Dunning, Y. Owechko and E. Marom, Opt. Letts. 11, 118 (1986).

5. J. P. Huignard, H. Rajbenbach, Ph. Refregier and L. Solymar, "Wave mixing in photorefractive bismuth silicon oxide crystal and its applications," Opt. A. R. Tanguay, C. Warde, U. Efron, A. M. Glass and others in the special issue on "Materials for optical processing" in JOSA B3 (1986).

6. D.A.B. Miller et al., Opt. Letts. 9, 567 (1984); M. Dagenais and W. F. Sharfin, Appl. Phys. Lett. 46, 230 (1985); T. Venkatesan et al., Opt. Letts. 9, 297 (1984); Williams, D., ed. "Nonlinear optical properties of organic and polymeric materials," (Plenum, NY, 1983).

7. I. C. Khoo, "Nonlinear optics of liquid crystals," in Progress in Optics, Vol. XXVl, E. Wolf, ed. (North-Holland, Amsterdam 1988).

8. R. K. Jain and M. B. Klein, "Degenerate Four Wave Mixings in Semiconductors," in "Optical Phase Conjugation" ed. R. A. Fisher (Academic Press, NY, 1983).

9. I. C. Khoo and R. Normandin, Appl. Phys. Letts. 52, 525 (1988).

10. (a) I. C. Khoo and T. H. Liu, IEEE J. Quant. Electron. QE23, p. 171 (1986).
    (b) T. H. Liu and I. C. Khoo, IEEE J. Quant. Electron. QE23, p. 2020 (1987).
    (c) I. C. Khoo, P. Y. Yan, G. M. Finn, T. H. Liu, and R. R. Michael, J. Opt. Soc. Am. B5, p. 202 (1987).

11. I. C. Khoo, Appl. Phys. Letts, 47, 908, (1985); L. Richard, J. Maurin, and J. P. Huignard, Opt. Comm. 57, 365 (1986).

12. See, for example, D. M. Pepper and A. Yariv, Chapter 2 in "Optical Phase Conjugation," ed. R. A. Fisher (Academic Press, NY, 1983).

13. I. C. Khoo and Yang Zhao, to be published in special issue on Nonlinear Phase Conjugation of IEEE J. Quant. Electron. (Dec. 1988).

Second Harmonic Generation in Ferroelectric Liquid Crystals

A. Taguchi, K. Kajikawa, Y. Ouchi, H. Takezoe, and A. Fukuda

Tokyo Institute of Technology, Department of Organic and Polymeric Materials, O-okayama, Meguro-ku, Tokyo 152, Japan

1. Introduction

Ferroelectric liquid crystals (FLC) have been extensively studied because of their potential application for display devices. For the appearence of ferroelectricity, it is necessary that non-racemic compounds constituted with chiral molecules exhibit a chiral smectic C (Sm C*) phase. In this phase, the directors are uniformly tilted from the layer normal in each layer. The tilted direction, i. e., an azimuth angle, precesses from layer to layer to form a helicoid. Although the overall polarization is zero in an unperturbed state because of the precession, all the dipole moments align under the application of an electric field parallel to the smectic layer. Thus, the directors are in C_2 symmetry, whose C_2 axis is along the direction of the spontaneous polarization.

Because of the symmetry, FLC is subject to even order nonlinear optical effects such as second harmonic generation (SHG). Though cholesteric liquid crystals are also composed of chiral molecules and form a helix, they are locally nematics and do not have spontaneous polarizations. Hence, no SHG is observed in cholesterics. So far, there were a few observations of SHG in nematics, which were attributed to flexoelectric effect/1/ or to quadrupolar mechanism/2/. A preliminary experiment on SHG of FLC has been made by Vtyurin et al. /3/, who confirmed that the SH signal suddenly increased at the phase transition temperature from Sm A to Sm C* by more than one order of magnitude. In this paper, we will present SHG for several FLC's as changing the applied field. It was confirmed that SHG starts to appear upon deforming the helicoid and that the SH light intensity has a correlation with the spontaneous polarization.

2. Experiments

Samples used are several mixtures commercially available and are in Sm C* phase at room temperature. They are listed in Table I together with their spontaneous polarizations, P_s. They were sandwiched between glass plates treated homeotropically. Then,

Table I List of samples and their P_s values.

Mixtures	P_s (nC/cm^2)	Manufacturer
ZLI 3654	29	E. Merck
TKF 8617	23	Teikoku Chemical Industry Co. Ltd.
CS 1013	16	Chisso Corp.
CS 1015	6	Chisso Corp.

Springer Proceedings in Physics, Vol. 36
Nonlinear Optics of Organics and Semiconductors
Editor: T. Kobayashi

smectic layers are parallel to the glass plates. Two sheets of aluminum foil of 15 μm thick were used for electrodes as well as for spacers between the glass plates.

The SHG measurements were performed with a 10 Hz Q-switched Nd:YAG laser (Quanta-Ray, DCR-11) operating 1064 nm. The output laser power was reduced by a couple of ND filters to be 2 mJ. The incident light polarization was chosen using a combination of a λ/4 plate and a Glan-Taylor prism. After blocking the SH light from optical components by a color filter, the incident light was focused on the sample cell. The SH light from the sample cell was collected by a lens from the transmitted direction after passing a $CuSO_4$ solution filter to absorb the incident light. The SH light was detected by a photomultiplier tube through a 532 nm interference filter. The fundamental beam intensity was monitored by a photodiode. These signals were stored and averaged by a digital storage oscilloscope and then were integrated by a microcomputer to evaluate the light intensity.

In order to evaluate the second order nonlinear coefficients, two cell geometries should be adopted as shown in Fig.1: The incident beam is cast upon the cell set vertically from the normal direction (geometry 1) and from the oblique direction (geometry 2). The cell surface normal, z, coincides with the smectic layer normal. An electric field was applied parallel to the cell surface along the y direction.

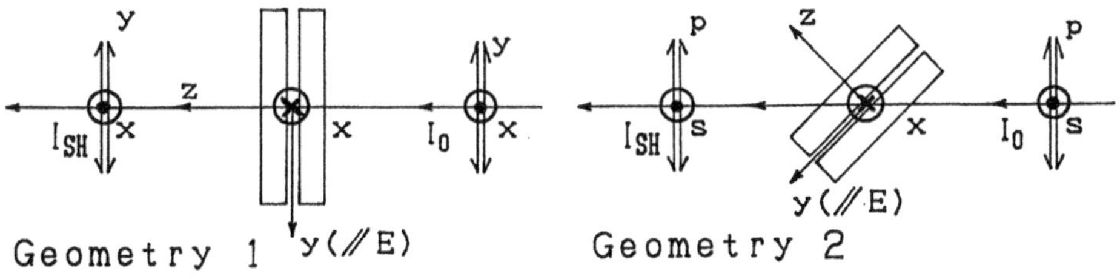

Fig.1 Optical geometries and polarization conditions.

3. Results and Discussion

Figure 2 shows the SH intensity I_{SH} as a function of an applied field. The experiments were made in the geometry 1 using y-polarized light. The ordinates for four mixtures can be compared to each other. The overall profiles in all the mixtures are the same: I_{SH} drastically increased at certain field strengths and gradually decreased after taking maxima.

As mentioned above, the director or the dipole moment forms a helicoid in the unperturbed state. The helicoid is deformed and is unwound completely at a critical field given by /4/,

$$E_c = \pi^4 K_t / 4 L_o^2 P_s, \quad (1)$$

where K_t is a torsional elastic constant for the helicoid, and L_o is the helicoidal pitch in the unperturbed state. Light propagation in such a helicoidal structure is very complex: The eigen polarization states are elliptical polarizations, whose ellipticities and

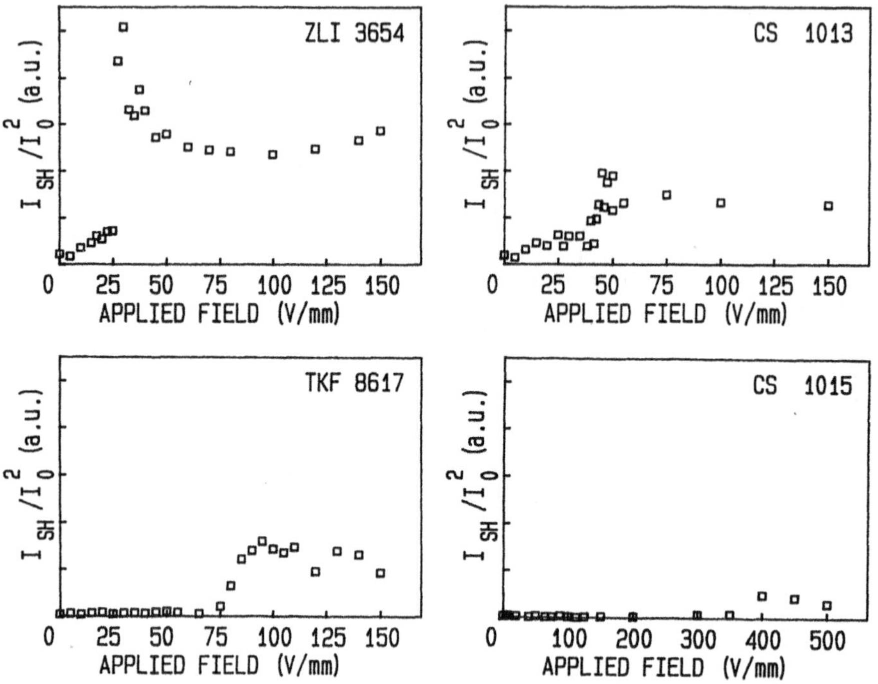

Fig.2 I_{SH} as a function of an applied field in four FLC mixtures.

principal axes change during propagation. It is probably acceptable , however, that the SH light in the unperturbed state is averaged out to be negligible due to the helicoidal structure. It is not easy to determine E_c under a polarizing microscope especially for homeotropic cells. In the present case, it was confirmed that some changes in the cells were observed at field strengths even above ones for maximum I_{SH}. This may give two possible explanations for the appearence of the maxima: 1) Phase matched condition is established in a deformed helicoid. 2) If materials are not phase matched, I_{SH} oscillates as $\sin^2(l\Delta k/2)$, where l is a sample thickness. In the present case, the coherent length, $l_c = \pi/\Delta k$, is comparable with the cell thickness. Therefore, I_{SH} may have maximum, when Δk changes due to the deformation process of the helicoid.

It should be noticed in Fig.2 that the SH signal intensity and the onset voltage for the signal have a correlation with the spontaneous polarization. A discrepancy in the onset voltages between TKF 8617 and CS1013 may be explained by the other material constants such as K_t and L_o.

In the C_2 symmetry as in the present case, the second order nonlinear coefficient d has four independent components:

$$[d] = \begin{bmatrix} 0 & 0 & 0 & d_{14} & 0 & d_{21} \\ d_{21} & d_{22} & d_{23} & 0 & d_{14} & 0 \\ 0 & 0 & 0 & d_{23} & 0 & d_{14} \end{bmatrix}, \qquad (2)$$

Table II Theoretical and observed I_{SH}/I_o^2 in various conditions.

geometry	polarization conditions	theoretical expressions for I_{SH}/I_o^2	observed I_{SH}/I_o^2
	y → y	d_{22}^2	1
1	y → x	0	0.04
	x → y	d_{21}^2	0.32
	x → x	0	0.02
	p → p	$(d_{22}\cos^3\theta + 3d_{23}\sin^2\theta\cos\theta)^2$	
2	p → s	$d_{14}^2\sin^2 2\theta$	
	s → p	$d_{21}^2\cos^2\theta$	
	s → s	0	

provided that the Kleinman's condition is valid. The theoretical expressions for I_{SH}/I_o^2 are summarized in Table II for various combinations of polarizations in the two optical geometries, where θ is the angle between z and the propagation direction. In the geometry 1, the ratio of d_{21}/d_{22} can be determined. Then, in the geometry 2, relative values of the other components to d_{22} can be determined. The observed I_{SH}/I_o^2 in the geometry 1 for ZLI 3654 is shown in Table II. It was confirmed that only negligible signals were observed in the polarization conditions which should not produce I_{SH}. From the measurements, d_{21}/d_{22} was obtained to be 0.57. The measurements in the geometry 2 were also carried out. But, the quantitative analyses are left as future works because of poor reproducibility. As a conclusion, the contribution from the dipole moment along the C_2 axis (y axis) is dominant to SHG as expected. It is also noteworthy that d_{21} shows significant contribution to SHG.

As future studies, 1) determination of the absolute values of d_{ij}, 2) temperature dependence of d_{ij} and 3) establishment of the phase matched condition must be interesting. The second one is important for investigating the correlation between d_{ij} and P_s. The third one should be examined in the helicoidal state by taking account of the k vector of the helicoid. These are our future programs.

Acknowledgment

We acknowledge each company for supplying the commercial mixtures.

References

1. S. Gu, S. K. Saha and G. K. Wong: Mol. Cryst. & Liq. Cryst. 69, 287 (1981)
2. M. I. Barnik, L. M. Blinov, A. M. Dorozhkin and N. M. Shtykov: Mol. Cryst. & Liq. Cryst. 98, 1 (1983)
3. A. N. Vtyurin, B. P. Ermakov, B. I. Ostrovskii and V. F. Shabanov: phys. stat. sol. (b) 107, 397 (1981)
4. R. B. Meyer, L. Liebert, L. Strzelecki and P. Keller: J. Phys. (France) 36, L69 (1975)

Part IX

Device Oriented Research

Measurement of Electrooptic Properties of Organic Thin Films by Attenuated Total Reflection

M. Dumont and Y. Lévy

Institut d'Optique Théorique et Appliquée, Bât. 503, Centre Scientifique, BP 43, F-91406 Orsay Cedex, France

The Attenuated Total Reflection (ATR) method for generating and studying Surface Plasmon Wave (SPW) or guided waves has been used for a long time and has been applied to many experiments on thin dielectric films [1-5]. References [2] and [3] contain a large survey of the subject. The recent increase of interest in nonlinear organic materials, and particularly in organic thin films, has opened a new field of application for this method.

SPW are waves propagating along the interface between a metal and a dielectric, while the guided waves considered in this paper are propagating in a thin dielectric layer. In both cases, the propagation is limited to a 2 D space and SPW appear as a special case of guided waves. In the first part we will briefly recall the structure of these waves and discuss their properties.

The two major features of these surface waves are their extreme sensitivity to the optical properties of the dielectric medium and the large enhancement of the electric field near the boundary or in the guiding film. This second property has been used for producing large nonlinear effects such as second harmonic generation and Raman effect [4, 5]. The high sensitivity to the index of refraction (n) and to the absorption coefficient (κ) of the dielectric material has been used for a long time for measuring the complex index ($\tilde{n} = n + i\kappa$) of thin layers. Recently the method has been extended to the measurement of very small induced variations of n, for instance, variations induced by the light power [6] (see S.K. TRIPATHY'S paper in this issue) or variations induced by a static electric field [7, 8], i.e. the electrooptic effect. In this paper we discuss this last subject and we will give two examples from recent experiments [8, 9] . We will show that the piezoelectric effect can also be studied by this method.

Before starting, let us discuss the order of magnitude of index changes to be measured. If we assume a Pockels effect, the variation of n is given by $2n\Delta n = \chi^{(2)}E$. If $\chi^{(2)} = 10^{-11}$ to 10^{-10} m/V, E = 10^{6} V/m and n = 1.5, one obtains $\Delta n \cong 10^{-5}$. If the layer thickness is t = 1μm, this corresponds to a variation of the optical thickness of $t\Delta n \cong 2\ 10^{-5}\lambda_o$ where λ_o is the wavelength of light. This quantity can be measured interferometrically, e.g. with the Fabry-Pérot method treated by T. KOBAYASHI in this issue. In this case, the guided modes described later on are basically of the same nature as the FP fringes. Now, if the layer thickness is as small as a few tens of Å, $t\Delta n \cong 2 \times 10^{-8}\lambda_o$, which is below the sensitivity of usual interferometric methods. We will show that this effect is easily observable with SPW. The only competitive method is ellipsometry but it is not adapted for electrooptic measurements (due to the need of electrodes).

Springer Proceedings in Physics, Vol. 36
Nonlinear Optics of Organics and Semiconductors
Editor: T. Kobayashi

Surface plasmon waves

Let us consider, first, a plane boundary between two media characterized by the dielectric constants ε_1 and ε_2 (Fig. 1). A surface wave is a wave propagating along the boundary (tangential wave vector k_x) and decreasing exponentially on both sides, perpendicularly to the boundary :

$$E_1 = E_1 \exp(-i\omega t + i\,k_x x + \alpha_1 z) \quad ; \quad E_2 = E_2 \exp(-i\omega t + i\,k_x x - \alpha_2 z) \tag{1}$$

with α_1 and $\alpha_2 > 0$. The Maxwell equations and the continuity conditions at the boundary show that this type of wave exists if light is TM polarized (transverse magnetic : $H_x = H_z = E_y = 0$; $k_x E_{ix} = - k_{iz} E_{iz}$, where $k_{iz} = i\alpha_i$ is the imaginary z component of the wave vector in medium i : $k_i^2 = \varepsilon_i \omega^2/c^2 = k_x^2 - \alpha_i^2$) and if

$$\varepsilon_1 < - \varepsilon_2 < 0. \tag{2}$$

In addition one finds the well known dispersion relation

$$k_x^2 = \varepsilon_1 \varepsilon_2 (\varepsilon_1 + \varepsilon_2)^{-1} k_0^2 \qquad \text{with} \qquad k_o = \omega/c \tag{3}$$

A large negative ε (i.e. an imaginary index $\tilde{n}_1^2 = (i\kappa_1)^2 = \varepsilon_1$) is possible in a perfect metal described by the free electron model. This model gives

$$\varepsilon_1 = 1 - \omega_p^2/\omega^2 \tag{4}$$

where ω_p is the plasma frequency of free electrons. By using this value in (3), one gets the dispersion curve shown in Fig. 2. ω must be smaller than $\omega_p(1 + \varepsilon_2)^{-1/2}$. Real metals are not perfect : their dielectric constant has an imaginary part which produces absorption and, hence, a fast damping of surface waves. For visible light, silver is the best metal with $\varepsilon_{Ag} = \tilde{n}_{Ag}^2 = - 16 + 0.6\, i$ (this value can be affected by metal surface purity). The penetration length of the wave in each medium is given by

$$l_i = \alpha_i^{-1} = \sqrt{-(\varepsilon_1 + \varepsilon_2)/\varepsilon_i^2}\; \lambda_o/(2\pi) \; .$$

In the case of air and silver (for $\lambda_o = 600$ nm) one has $l_{air} \cong 400$ nm and $l_{Ag} \cong 25$ nm.

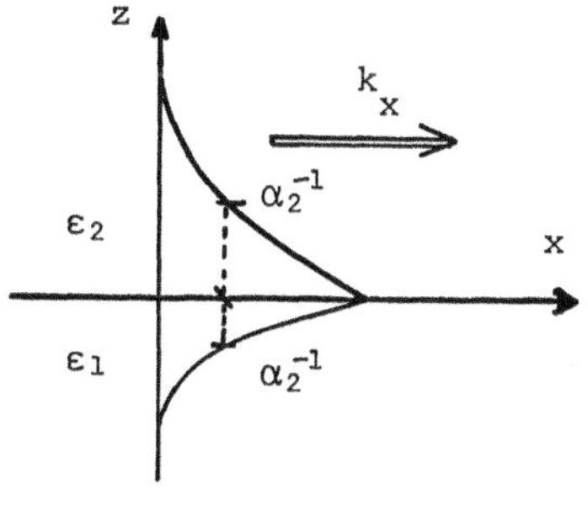

Fig.1 : Structure of surface plasmon waves.

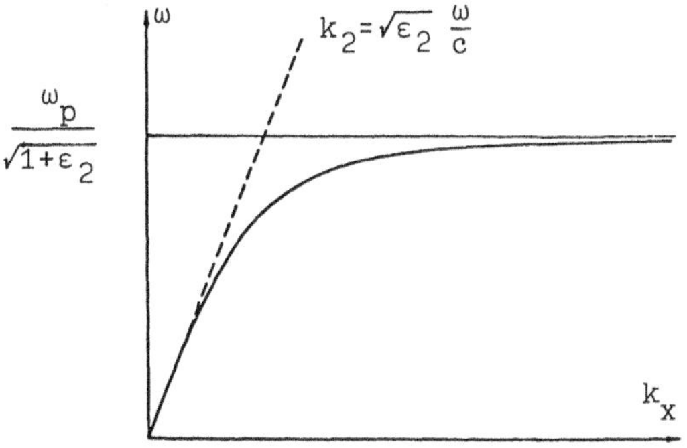

Fig.2 : Dispersion function of SPW.

Guided waves in a thin dielectric film

Let us consider an absorption-free dielectric film of thickness t_2 and index of refraction $n_2=\sqrt{\varepsilon_2}$, between two different media of index of refraction n_1 and n_3 ($< n_2$). It is well known that guided waves may propagate inside this film with the general structure (Figure 3.a)

$$E_1 = \mathcal{E}_1 \exp(ik_x x + \alpha_1 z) \quad ; \quad E_3 = \mathcal{E}_3 \exp(ik_x x - \alpha_3 (z - t_2))$$
$$E_2 = e^{ik_x x}\left(\mathcal{E}_2^+ e^{ik_{2z}z} + \mathcal{E}_2^- e^{-ik_{2z}z}\right) = 2\,|\mathcal{E}_2|\, e^{ik_x x}\cos[k_{2z}(z-z_o)] \,, \tag{5}$$

which corresponds to a plane wave reflected up and down in the film ($k_2 = n_2 k_0$; $k_x = k_2 \sin\theta_2$; $k_{2z} = k_2 \cos\theta_2$) and to two evanescent waves in the surrounding media. This implies total reflection of light on the inner face of the film which forces θ_2 to be larger than a critical value determined by the larger of n_1 and n_3:

$$k_x = k_2 \sin\theta_2 > k_1 = n_1 \omega/c \quad \text{and} \quad k_x > k_3 = n_3 \omega/c \,. \tag{6}$$

Finally, the Fabry-Perot interference condition imposes well-defined values of θ_2 which correspond to the modes of the waveguide :

$$2\pi \lambda_0^{-1}\, 2n_2 t_2 \cos\theta_2 + \Psi_1 + \Psi_3 = 2\, k_{2z} t_2 + \Psi_1 + \Psi_3 = m\, 2\pi \,. \tag{7}$$

m is an integer ($\geqslant 0$) which defines the order of each mode. Ψ_1, and Ψ_3 are the phase shifts due to reflection on medium 1 and 3, respectively. They are given by Fresnel formulae and depend on indices n_i and on θ_2 ; they are different for TM and TE polarizations. The number of modes, limited by (6), is a function of the thickness t_2. As Ψ_1 and Ψ_3 are negative t_2 must be larger than a positive minimum value, even for $m = 0$ (if $n_3 \neq n_1$).

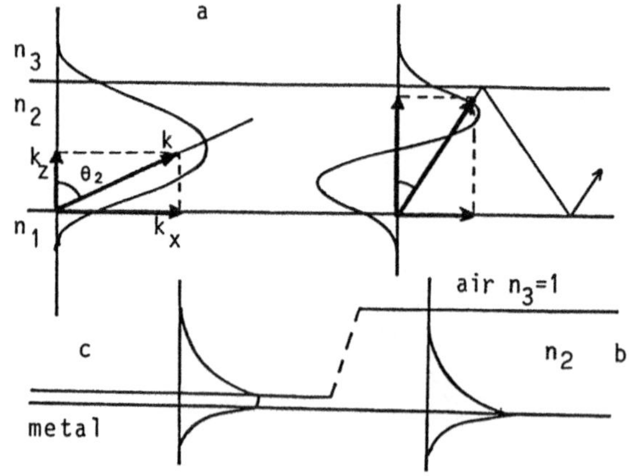

Fig.3-a : Structure of guided modes in a dielectric film (TE_0 and TE_1).
Fig.3-b and 3-c : The two limits of the TM_0 mode when medium 1 is a metal.
b : metal/dielectric SPW ($t \to \infty$).
c : metal/air SPW ($t \to 0$).

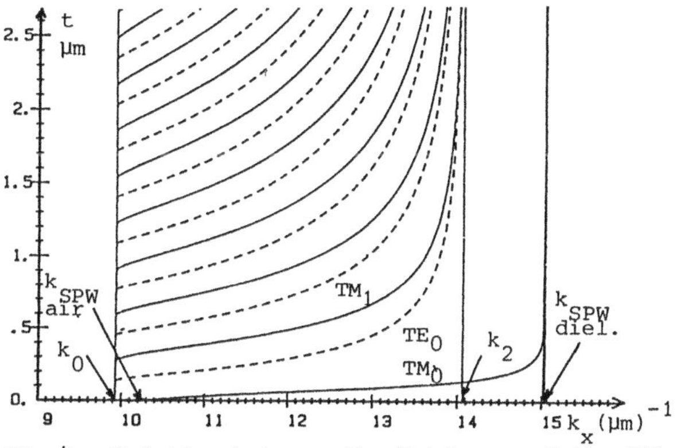

Fig.4 : Relation between the thickness of the film and the tangential wave vector $k_{2x} = k_2 \sin\theta_2$. $\varepsilon_1 = -16$; $n_2 = 1.416$; $n_3 = 1$; $\lambda_0 = 633$ nm.
TM : ——— ; TE : ------

Let us now consider the arrangement of Fig. 3-b where the first material is a perfect metal (ε_1 real < 0 ; $n_1 = i\kappa_1$). Expression (6), for the modes' angular position, is always valid but the Fresnel formulae give a positive value for Ψ_{1TM}. Therefore the TM_0 mode exists even when $t_2 = 0$: it is nothing but the surface plasmon wave at metal/air boundary (Fig.3-c). The TM_0 mode exists for $k_x > k_2$: in that case k_{2z}, Ψ_1 and Ψ_3 are imaginary and the TM_0 mode tends toward the SPW at the metal/dielectric boundary when $t_2 \to \infty$. Thus the TM_0 mode will be called the SPW mode. Fig. 4 gives an example of the relation between k_x and t_2 (n_2 fixed).

If medium 1 is not a perfect metal, waves are damped as they propagate along the wave guide (k_x has an imaginary part). If medium 3 is also a metal, Ψ_3 has also to be modified. In this case, there is no lower limit for θ_2 since transmission in medium 3 is not possible. Furthermore, some complications arise with the TM_0 mode (the SPW mode still exists but it is now the TM_1 mode).

Attenuated Total Reflection (ATR) method

We have now to choose a method for producing SPW and guided waves, and for measuring k_x (i.e. θ_2 and the damping coefficient), in order to take advantage of the very large sensitivity of these waves for measuring organic films' optical properties. No direct coupling is possible with free running waves in medium 3 since $k_{3x} \leqslant k_3 < k_x$. One method is to use a grating structure which increases the k vector of the incident light (as explained by S.K. TRIPATHY in this issue) but the most common method is the Attenuated Total Reflection (ATR) technique.

The KRETSCHMANN [1] ATR experimental method is shown in Fig. 5. A thin metallic layer (most often a silver layer) is deposited on the base of a glass prism (index n_p), and overlaid with the dielectric film to be studied. A light beam is reflected on the metallic layer, through the prism, with an incidence angle θ_p. This light beam produces an evanescent wave in the metal, at the boundary with the prism. Due to the low thickness of the metallic layer, this evanescent wave can induce waves in other media. All waves must have the same k_x value which is defined by the incidence angle in the prism : $k_x = k_p \sin\theta_p$ (with $k_p = n_p k_0$) [11].

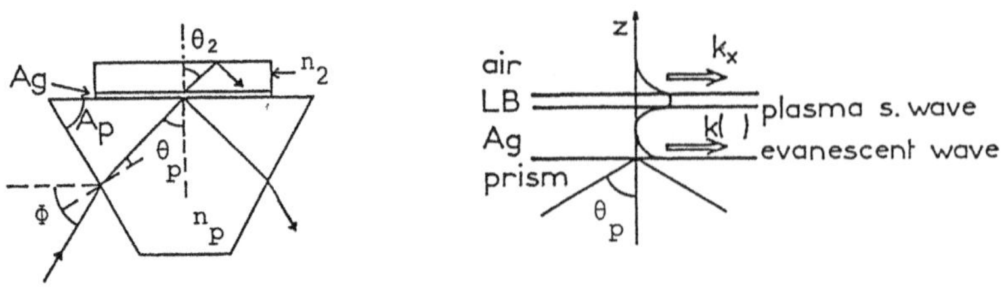

Fig.5 : KRETSCHMANN configuration for producing guided waves and SPW.

If $k_x < k_3$, a real transmitted wave is produced in medium 3 : the metallic layer is a partially reflecting mirror. If $k_x > k_3$ the light is totally reflected (R=1 for ideal lossless materials) except if k_x is nearly equal to the wave vector k_{2x} of one guided mode. In that case a resonance occurs and a large amount of energy is transferred to the guided wave or to the SPW. As the metal and, possibly, the dielectric layer absorb light, the guided wave energy is dissipated and is not restored to the reflected beam. Thus the SPW and the guided modes appear as dips in the reflectivity curve of the layers, as a function of the incidence angle θ_p. For a bare silver layer of optimum thickness (~ 500 Å), the SPW resonance produces a total extinction of the reflected beam and the width of the dip is approximately 0.3 degree. Gold and aluminium can also be used but the resonances are much broader.

As in all resonant phenomena, the width of resonances increases and their depth decreases when damping (absorption) increases. The exact shape of the reflectivity curve is a function of the thickness of both layers and of the complex index of refraction of the four media. It can be calculated by a computer program using the usual methods dealing with stratified media (Fresnel formulae at each interface and propagation equations in each layer). Nevertheless, when the dips are sharp enough (low absorption), the measurement of their positions allows a good determination of n_2 and t_2 from (7) (if n_1 and t_1 are known). The presence of the prism modifies very slighty Ψ_1 in (7), and hence the position of resonances, but in first approximation one can neglect this effect (if silver thickness >> $1/\alpha_{Ag}$). The independent determination of n_2 and t_2 is easy when the thickness allows the existence of several modes but more difficult for thin layers with only the SPW mode (a careful analysis of its shape is necessary). When t_2 is as small as a few tens of Å, this independent determination is impossible.

Experimental method for electrooptic measurements

For convenience, the silver layer and the organic film are deposited on a glass plate (index n_p) which is put in optical contact with the prism through an index matching oil. The silver layer is made by vacuum evaporation and the organic film can be obtained by any way giving a good homogeneity of the material and a good flatness of the surface (LB layers, casted, evaporated or spin deposited films). Fig. 6 shows the experimental arrangement. The prism is mounted on a computer driven goniometer which allows a large scanning of the incidence angle of a fixed collimated light beam (a stabilized He-Ne laser in our experiments). A photodiode detector is mounted on a rotating arm following the reflected beam. Reflectivity

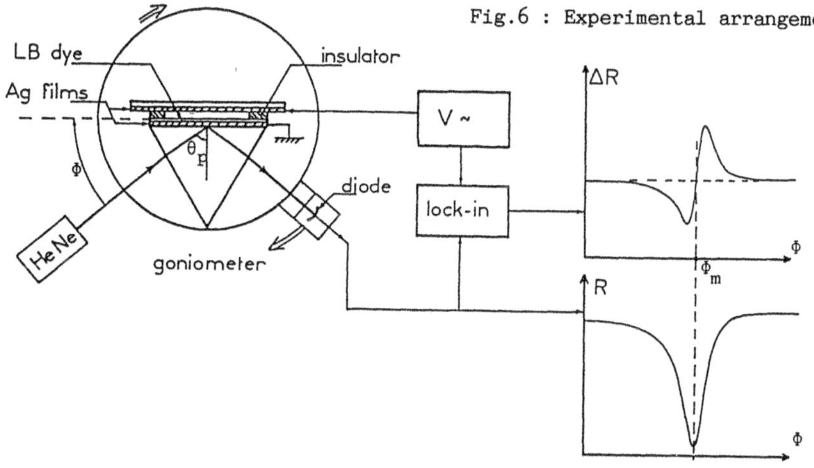

Fig.6 : Experimental arrangement.

is directly recorded by the microcomputer as a function of the external angle, Φ, which can be easily related to the internal angle, θ_2, used in Eq. (7) (Fig. 5).

For studying electrooptic coefficients (Pockels or Kerr effect), one has to measure the variations of n induced by an external electric field. The KRETSCHMANN geometry is the only one which is convenient for these measurements : indeed, the silver layer can be used as a first electrode and the second electrode can be located in free space above the stack of layers. For studying the SPW dip with very thin films, the second electrode must be far enough away to avoid perturbations of the surface wave ($d \gg 1/\alpha_{air}$) . Practically the 2nd electrode is a glass plate coated with a thin silver layer, pressed against the sample and maintained at a distance $d \sim 10$ µm by spacers (thin plastic film). d is measured by observing the Fabry-Pérot fringes produced by scattered light, between the two silver layers. For guided modes the electrode may be evaporated directly on the organic film provided that one takes into account the modification of Ψ_3 in (7).

The Pockels ($\Delta\varepsilon = \chi^{(2)}E$) or Kerr effect ($\Delta\varepsilon = \chi^{(3)}E^2$) is observed by measuring the differential reflectivity : $\Delta R = R(E)-R(E=0)$. Since this variation is very small ($\sim 10^{-5}$) and much smaller than laser noise, the electric field is modulated ($\Omega \sim 1$kHz) and the resulting modulation of R is detected with a lock-in amplifier (fig. 6). ΔR is modulated at the fundamental frequency Ω in the case of the Pockels effect and at the second harmonic frequency 2Ω for the Kerr effect. In the following, we will discuss only the Pockels effect. In the most general case $\chi^{(2)}$ is complex ($= \chi'^{(2)} + i\chi''^{(2)}$) and the shape of ΔR is produced by a variation of the complex index ($\tilde{n} = n + i\kappa$) :

$$\Delta R = (\partial R/\partial n)\, \Delta n + (\partial R/\partial \kappa)\, \Delta\kappa \qquad (8)$$

with $\chi'^{(2)}E = 2n\Delta n - 2\kappa\Delta\kappa$ and $\chi''^{(2)}E = 2n\,\Delta\kappa + 2\kappa\,\Delta n$. In this general case the computer fit of R and ΔR allows a measurement of Δn and $\Delta\kappa$, and hence of $\chi'^{(2)}$ and $\chi''^{(2)}$. Nevertheless, when absorption is negligible (κ and $\chi''^{(2)} = 0$), a very useful approximation can be used. Indeed, one can show that a small variation of n produces a shift of resonances according to Eq. (7), but no change in the shape of the dips. Thus each resonance m can be written as $R_m(\phi - \phi_m)$ in which only ϕ_m (the dip angular position) is a function of E_o. For each mode m, one has

$$\Delta R_m = (\partial R_m/\partial \phi_m)\ (\partial \phi_m/\partial n)\ \Delta n = -\ (\partial R_m/\partial \phi)\ (\partial \phi_m/\partial n)\ \chi^{(2)} E_o/(2n) \tag{9}$$

ΔR_m appears to be proportional to $\chi^{(2)}$ and to the derivative of $R_m(\phi)$. $\partial\phi_m/\partial n = (\partial\phi_m/\partial\theta_{2m})\ (\partial\theta_{2m}/\partial n)$ is easily calculated from (7) and from the geometric relations between ϕ and θ_{2m}: it is a function of n_2, t and θ_{2m}. One can obtain a fast estimation of the order of magnitude of $\chi^{(2)}$ by a simple examination of R and ΔR.

Study of LANGMUIR-BLODGETT films of an azo dye [8]

The DPNA molecule (see ref. 10 for a more precise description) is synthesized by ICI. It consists of a donor group and an acceptor group providing a permanent dipole and a large β factor. One end of the molecule is a hydrophilic radical and the other is a hydrophobic chain. It is therefore possible to deposit these molecules by the LANGMUIR-BLODGETT technique. Our samples were prepared at Thomson-CSF (LCR).

We first tested a monolayer : Fig. 7 shows the SPW dip and the associated differential reflectivity. As expected, the electrooptic effect is large as all molecules are oriented in the same way. Then we tried multilayers with the most common Y structure, i.e. with molecules orientation reversed from one layer to the next one. Fig. 8 shows the position of the SPW dip as a function of the number of layers. The best fit is obtained for $\tilde{n}_2 = 1.70 + 0.02i$ and an average thickness of 19.2 Å for each layer. As expected, the overall electrooptic effect (measured by the surface second-order susceptibility $\chi_s^{(2)} = t_2\chi^{(2)}$) is constant for all odd numbers of layers (Fig. 9).

In order to get a large $\chi_s^{(2)}$ we tried the Z structure for which all molecules are parallel and sum up their electrooptic properties. Fig. 8 shows that the structure is different (assuming n=1.7 one finds 22 Å for each layer), but contrary to the predictions, the electrooptic effect does not increase with the number of layers. The same result has been obtained for the second harmonic generation [10]. The most probable interpretation is that Z layers are not stable and are completely disordered.

For obtaining stable highly non-centrosymmetric films, Y structures have been built with two types of molecules altlernatively (from ICI). The first type has its hydrophobic tail on the donor group and the second type has it on the acceptor group : in a Y alternated structure all dipoles are parallel. A few samples have been tested and showed an increase of $\chi_s^{(2)}$ with the number of layers.

The anisotropy of the electrooptic effect has been evidenced by using a sample with 121 layers, exhibiting one TM guided mode, and a sample with 75 layers, exhibiting one TE mode. The TM mode gives the same value of $\chi_s^{(2)}$ as the SPW mode did in thin layers : indeed, in both cases, the main component of the optical field is E_z (parallel to E_o and to the molecular axis) and we mainly measure $\chi_{zzz}^{(2)}$. On the

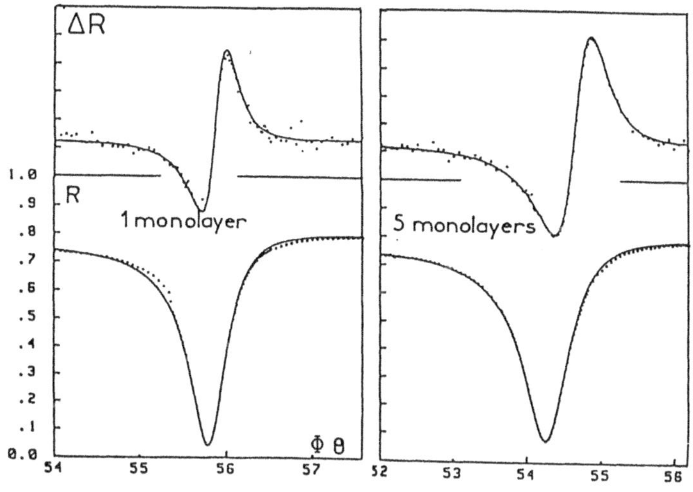

Fig.7 : Experimental points with computer fitted curves.

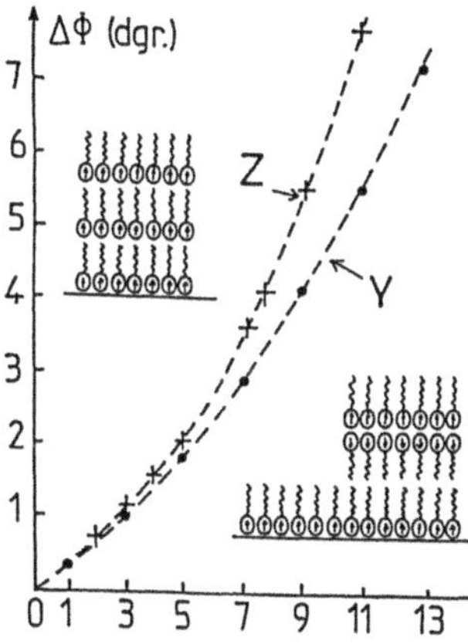

Fig.8

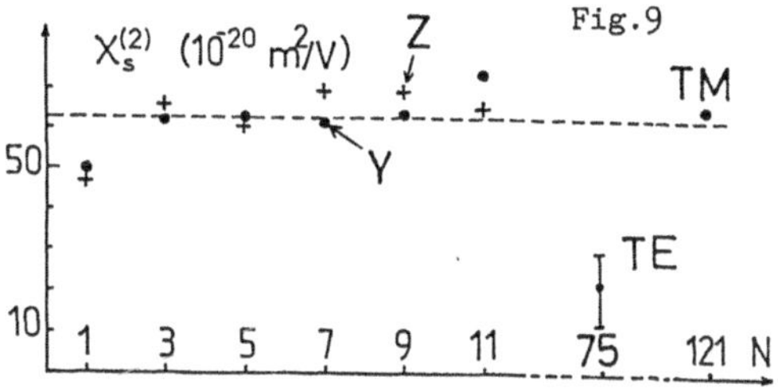

Fig.9

Fig.8 : Angular shift of the SPW dip versus the number of LB layers.
Fig.9 : Surface second-order susceptibility versus the number of LB layers.

other hand, the TE modes gives a much smaller value of $\chi_s^{(2)}$, since the optical field is E_x (perpendicular to E_o and to molecules) : we measure $\chi_{xzx}^{(2)}$.

Study of Ferroelectric material [9]

The second material we have studied in collaboration with Ph. ROBIN (Thomson-CSF) is a copolymer of vinylidene fluoride and trifluoroethylene, P(VDF-TrFe), from ATOCHEM, which is well known for its ferroelectric and pyroelectric applications. This material is also known to be efficient for SHG. The film is prepared by spin coating the silver layer with a solution of polymer in methylethylcetone. The film is then dried (100°C, 1 hour) and crystallized (130°C, 1/2h) and the second electrode (aluminium) is directly evaporated on it. Thicknesses from 1.5 to 2.5 µm have been obtained and exhibit several TM and TE modes.

The non-centrosymmetry of the film is induced by poling it with a large electric field (80 MV/m at room temperature). We first verified that the sign of the ΔR signal changes when the poling sign is changed and that it is possible to observe the ferroelectric hysteresis cycle on electrooptic properties as well as it has been observed with SHG (cf. paper by D. BROUSSOUX in this issue).

Fig. 10 shows a record of 7 successive TM modes fitted with the computer program (n=1.416, t=2.237 µm). The associated ΔR signal exhibits a surprising property : the sign of the resonances is different for large and small angle modes. This property cannot be explained by a simple electrooptic effect, but theoretical curves (Fig. 11) and the computer fit in Fig. 10 show that the piezoelectric behaviour of the material (variation of thickness with E_o) explains it simply. Fig. 11 shows that the turning point Φ_o where the sign of ΔR is reversed, is very sensitive to $\Delta n/\Delta t$. The physical meaning of this can be understood with the previous low losses approximation. Then (9) must be written

Fig.11

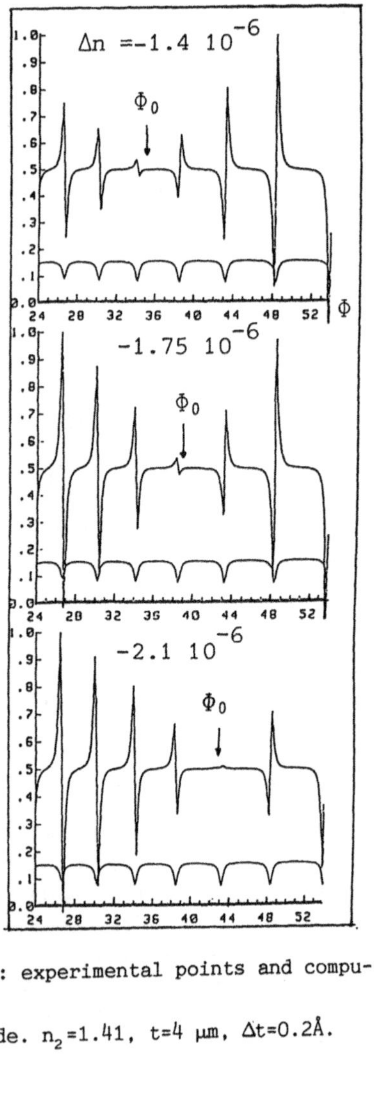

Fig.10

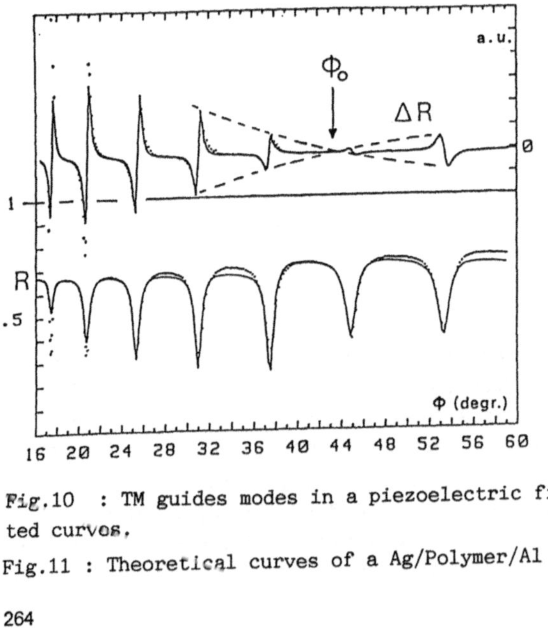

Fig.10 : TM guides modes in a piezoelectric film : experimental points and computed curves.

Fig.11 : Theoretical curves of a Ag/Polymer/Al guide. n_2=1.41, t=4 µm, Δt=0.2Å.

$$\Delta R_m(\phi) = -(\partial R/\partial\phi)\,\Delta\phi_m = -(\partial R_m/\partial\phi)\,[(\partial\phi_m/\partial n)\,\Delta n + (\partial\phi_m/\partial t)\,\Delta t]$$

$$= -(\partial R_m/\partial\phi)\,[K_n(\phi_m)\,\Delta n/n + K_t(\phi_m)\,\Delta t/t] \quad . \tag{10}$$

K_n and K_t are calculated from (7) ; they are shown in Fig. 12. The ratio $\Delta n/\Delta t$ which fits the experimental value of the turning point ϕ_o is easy to obtain. One can go further in the interpretation of experimental results by using the Lorentz-Lorenz formula for the index of refraction :

$$n^2 - 1 = \chi = N\alpha(1 - N\alpha/3)^{-1} \tag{11}$$

where α is the polarizability of molecules and N their density. The static electric field acts on n by two means : by modifying α ($\Delta\alpha = \beta E$) and by changing the density ($N \propto 1/t$ since the material is clamped on the substrate). Then Δn, in (10), is the sum of two contributions obtained by differentiating (11) :

$$\Delta n_1 = -\,n\,b\,\Delta t/t \quad ; \quad \Delta n_2 = n\,b\,\Delta\alpha/\alpha \quad \text{with} \quad b = (n^2+2)\,(n^2-1)\,(6\,n^2)^{-1} . \tag{12}$$

Then according to (10), the amplitude of the differential signal for mode m (or, in other words, the translation of mode induced by a given electric field) is

$$\Delta\phi_m = [K_n(\Delta n_1 + \Delta n_2)/n + K_t\Delta t/t] = n^{-1}\left[\left(K_n - b^{-1}K_t\right)\Delta n_1 + K_n\Delta n_2\right]. \tag{13}$$

The contribution of the piezoelectric effect is determined by $K_p = K_n - b^{-1}K_t$, which is drawn in Fig.12. The zero of K_p is different from the experimental value : this difference is the contribution of the pure electrooptic effect (Δn_2). By writing $\Delta\phi_m(\phi_o) = 0$, one obtains the ratio $\Delta n_2/\Delta n_1$ and the relative amplitude of ΔR resonances defined by $K = [K_p\Delta n_1 + K_n\Delta n_2]\,[\Delta n_1 + \Delta n_2]^{-1}$, as shown, in Fig. 12. From the computer fit we obtain

$$E_0^{-1}\,\Delta t/t = -\,13 \text{ pm/V} \qquad \text{and} \qquad \chi^{(2)}_{exp} = 2n\,E^{-1}\Delta n = 16 \text{ pm/V} \quad .$$

From the piezoelectric effect, we calculate

$$\chi^{(2)}_{pz} = 2n\,E_0^{-1}\Delta n = -\,2n^2 b\,E_0^{-1}\Delta t/t = 17.3 \text{ pm/V} \quad ; \quad \chi^{(2)}_{eo} = \chi^{(2)}_{exp} - \chi^{(2)}_{pz} = -\,1.3 \text{ pm/V}.$$

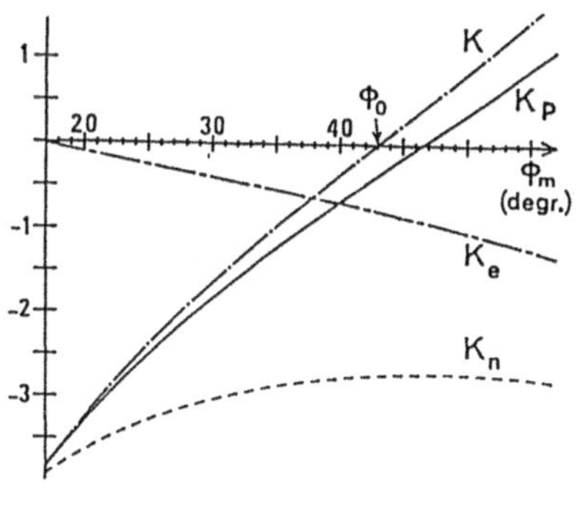

Fig.12 : Angular shift of modes (or amplitude of ΔR_m) induced by $\Delta n[K_n]$, $\Delta t[K_t]$ and piezoelectric effect $[K_p]$ and for experimental values [K].

We conclude that the piezoelectric effect is dominant and explains the reversing of the sign of ΔR resonances. The value of $\chi^{(2)}_{eo}$ is too small to be reliable in view of the two approximations made for its estimation : first, we have neglected the anisotropy of the material and, second, the Lorentz-Lorenz formula is doubtful due to the anisotropy and to the microcrystallite structure of the material. The simultaneous careful analysis of TM and TE modes could give the anisotropy of n and of $\chi^{(2)}$: this study is in progress. The fact that $\chi^{(2)}$ appears as comparable for TM and TE modes is in agreement with its piezoelectric origin.

Conclusion

The ATR method is a very sensitive method for measuring the complex index of refraction and the complex electrooptic coefficients of thin organic films. For thick layers (few μm), which present several TM and TE modes, it is easy to determine also the thickness of the layer and its variations with the applied field (piezoelectricity, electrostriction). For thin layers, which exhibit only the SPW resonance, the separate determination of n and t and of Δn and Δt is more difficult and becomes impossible when the thickness is of the order of a few tens of Å.

Acknowledgments : All samples used in both experiments reported in this paper have been prepared by E. CHASTAING. We are very indebted to her and to the Chemistry Department of Thomson-CSF (LCR) for this fruitful collaboration.

References

1. E. KRETSCHAMNN and H. RAETHER, Z. Naturf., 23 a, 2135 (1968).
   E. KRETSCHMANN, Z. Phys. 241, 313 (1971).
2. Electromagnetic Surface Modes, edited by A.D. BOARDMAN, John Wiley and Sons (1982).
3. Surface Polaritons - Electromagnetic Waves at Surfaces and Interfaces, edited by W.M. AGRANOVICH and D.L. MILLS, Noth-Holland (1982).
4. Y. LEVY, Can. J. Phys., 58, 1525 (1980).
5. D. RICARD, Ann. Phys. Fr. 8, 273 (1983).
6. G.M. CARTER, Y.J. CHEN and S.K. TRIPATHY, Appl. Phys. Lett. 43, 891, (1983).
7. G.H. CROSS, I.R. GIRLING, I.R. PETERSON, N.A. CADE, Electron. Lett. 22, 1111 (1986).
8. J.C. LOULERGUE, M. DUMONT, Y. LEVY, P. ROBIN, J.P. POCHOLLE and M. PAPUCHON, Thin Solid Films, 160, 399 (1988).
9. V. DENTAN, Y. LEVY, M. DUMONT, P. ROBIN, and E. CHASTAING, Optics Comm. (to be published).
10. I. LEDOUX, D. JOSSE, P. VIDAKOVIC, J. ZYSS, R.A. HANN, P.F. GORDON, B.D. BOTHWELL, S.K. GUPTA, S. ALLEN, P. ROBIN, E. CHASTAING, J.C. DUBOIS, Europhys. Lett. 3, 803 (1987).
11. In a third method (A. OTTO and W. SOHLER, Optics Comm., 3, 254 (1971)), the dielectric film is located between the prism and the silver. The coupling is through evanescent waves in the dielectric film. This method is not convenient in our case.

Phase-Matched Second-Harmonic Generation in Fibre Waveguides by Electric-Field Modulation of the Third-Order Susceptibility

R. Kashyap

British Telecom Research Laboratories,
Martlesham Heath, Ipswich, Suffolk IP5 7RE, UK

1 Introduction

Frequency doubling has recently been observed in phosphorous doped single-mode fibres by several workers [1] [2] [3]. Since silica exhibits a centre of inversion and thus lacks an electric dipole allowed second order nonlinearity, $\chi^{(2)}$, it has been postulated that the non-linear interaction is due to the electric quadrupole and magnetic dipole moments [4]. This process is phase matched by a self-organised grating of colour-centres [3], and as such there is little choice in the selection of particular interacting modes. Phase matching has been discussed [5] and dc-field-induced second-harmonic generation also considered [6], where phase matching is achieved by proper *design* of the fibre. It has also been suggested that efficient nonlinear devices may be fabricated by incorporating high $\chi^{(3)}$ media as the core of a fibre and using a periodic electrode grating structure for phase matching [7]. The asymmetry is induced by the electric field and phase-matching is achieved by the spatial periodicity of the field.

This paper presents results on phase-matched electric-field-induced second-harmonic generation in ordinary telecommunications type silica fibres, initial results of which were reported recently [8]. A *single* spatially-periodic electrode-structure is used and rotated to achieve phase matching conditions. The most efficient mode combination is observed experimentally, i.e from the fundamental LP_{01}^{ω} mode to the second-harmonic $LP_{01}^{2\omega}$ mode. This is generally the most difficult when the fibre is single-mode at the fundamental-wavelength, since the phase mismatch is the largest of all the mode interactions with the LP_{01}^{ω} mode. However, the benefit of the largest field-overlap integral makes this interaction highly attractive. Phase-matching is possible over the entire length of the grating region by automatic static field reversal after each coherence length. Additionally, phase matching to all the propagating higher order second-harmonic $LP_{11}^{2\omega}$, $LP_{21}^{2\omega}$ and the $LP_{02}^{2\omega}$ modes is also shown. Since one fibre was slightly bi-moded, interactions between the second mode, LP_{11}^{ω}, to other higher order second-harmonic modes is also demonstrated. Although the conversion efficiency at present is small, the principle of phase matching is demonstrated and the possibilities of improvements discussed. This technique can also be applied to waveguides made with oriented polymers and other non-linear materials.

2 Theory

The second-harmonic power from a fibre of length L, with a third order nonlinearity of $\chi^{(3)}$ and for an applied spatially-periodic static electric field amplitude of E_0, is given by [7]

$$P_{2\omega} = [\frac{3}{4}\epsilon_0 \omega \chi^{(3)} B_N E_0 P_\omega L \ \mathrm{sinc}(\Delta\beta l/2) \ \ I\,]^2 \qquad (1)$$

Springer Proceedings in Physics, Vol. 36
Nonlinear Optics of Organics and Semiconductors
Editor: T. Kobayashi

where, B_N is the Fourier coefficient of the spatial frequency of the periodic static field, P_ω is the power in the fundamental mode, ω is the fundamental angular frequency, $\Delta\beta$ is the phase mismatch and I is the overlap integral between the static field, the fundamental and second-harmonic transverse mode fields.

The effective second order nonlinearity is then given by $\chi^{(3)}B_N E_0$. For phase matching, the spatially periodic static field pitch is $\Lambda = 2l_c/N$, where l_c is the effective coherence length as a result of the mismatch between the phase velocities of the fundamental and second harmonic guided modes, and N is the number of the spatial harmonic of Λ. In general the static field is not uniform. Additionally, the selection rules for the mode interactions are governed by symmetry of the modes and sign of the fields. For non-uniform static fields, the forbidden interactions in the linearly polarised modes' (LP) approximation [9] are no longer valid and virtually all mode interactions thus have a finite overlap integral. The strength of the interaction will also depend on the square of the Fourier coefficient of the spatial harmonic of the applied field (eqn (1)) which will generally be less than 1. In this study the fundamental spatial harmonic, N =1 was used.

A schematic of the device is show in Figure 1. A *single* inter-digitated electrode structure is used so that the application of a voltage ensures that the field between the electrodes reverses sign every half period. This field reversal allows the growth of the second-harmonic over the entire length of the device.

The potential distribution across the electrodes in relation to the core of the fibre was calculated numerically and is shown in Figure 2. The overlap integral in the core for this device can be shown to be

$$I_{core} \propto \int_0^r \int_0^{2\pi} (J_\mu^\omega)^2 \cdot J_\nu^{2\omega} \cos^2(\mu\phi) \cdot \cos(\nu\phi) \times \exp\left(-[(\cos \phi) r/\rho + 1] \Gamma \right) \cdot r \, dr \, d\phi \qquad (2)$$

where, the J-Bessel functions of order μ, ν, describe the optical fields in the core at the appropriate frequencies, ω and 2ω. The azimuthal angle is ϕ, and r is the radial position. $\Gamma = 2\pi\rho/\Lambda$, where $\rho = R + d$ is the distance of the core axis from the

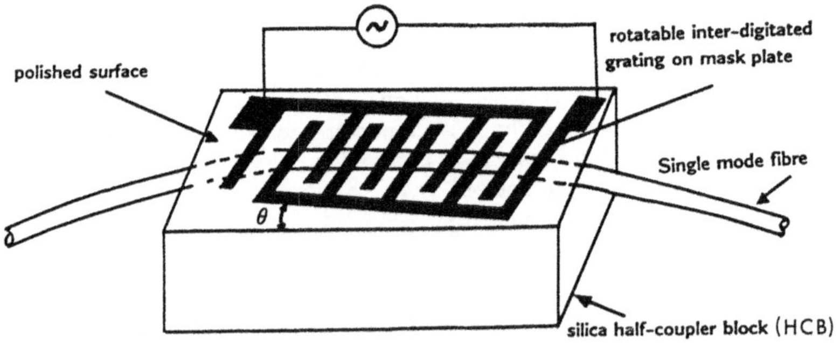

Figure 1 Schematic of phase-matchable electric-field induced second harmonic generator. The electrode grating is simply placed on the surface of the silica block containing the polished fibre, and rotated to achieve phase matching. The spatial static field penetrates into the fibre core which is close to the surface.

electrode surface. A similar function as equation (2) describes the overlap integral in the cladding, but with J, replaced with K-Bessel functions. The field decays exponentially with distance from the electrode, and therefore the field across the core is not uniform. The cross-section of the device is shown in Figure 3.

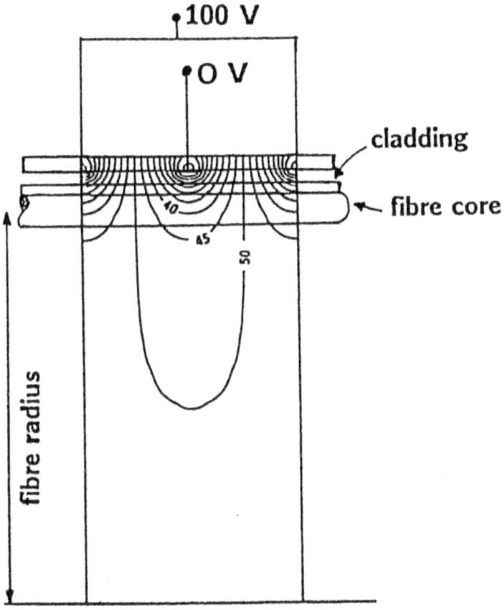

Figure 2 Computed potential distribution across fibre core showing 5V equi-potential lines. The direction of E_x changes sign from electrode to electrode.

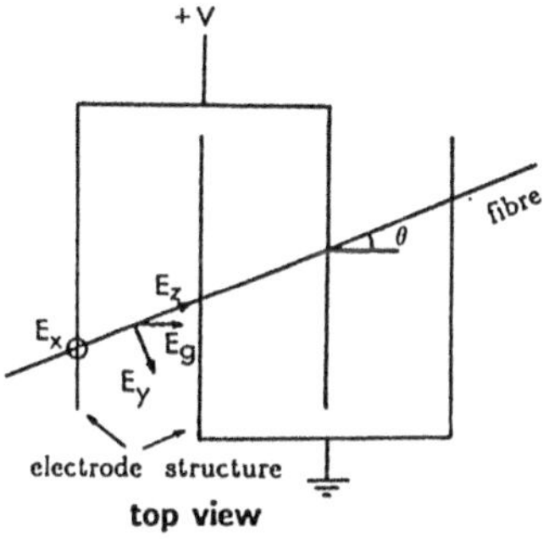

Figure 2a Top view of device showing the grating field, E_g, which can be resolved into two fields, E_y and E_z shown in the figure.

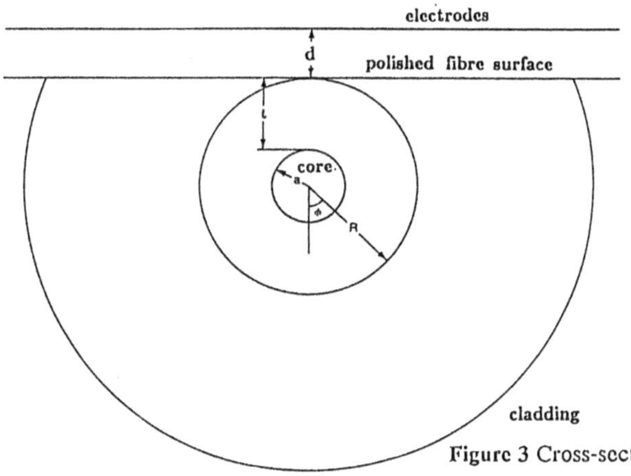

Figure 3 Cross-section of electric-field induced SHG device.

3 Experimental Details

The cladding of a single-mode silica fibre is removed by polishing to within a few microns of the edge of the core, and the electrode structure is lowered on to its surface. A Mylar spacer $\sim 2\ \mu m$ in thickness is sandwiched in between and a drop of low index

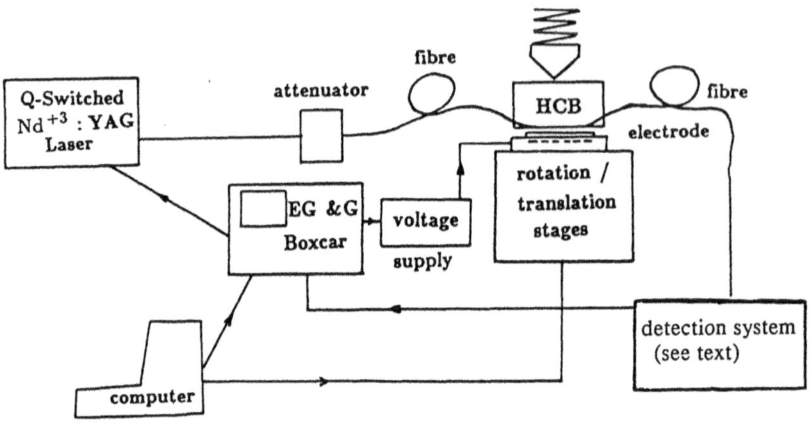

Figure 4 Schematic of experimental layout. A pressure spring holds the electrode structure and the fibre-device together.

dielectric liquid is allowed to fill the gap. In order to avoid catastrophic damage [10], care is taken so that the evanescent field does not encounter the metal electrode grating. Experimental layout is shown in Figure 4.

The output from a Q-switched Nd:YAG laser operating at 1.064 μm was launched into one end of the single-mode silica fibre. The output was focused on to a photo-multiplier tube, after fundamental wavelength-blocking and second-harmonic band-pass filters. The data was logged using a digital Box-car averager in the static-gate configuration. The fibre end could be viewed through a tetravar telescope arrangement to observe the second-harmonic modes. The Q-switch driver with a repetition rate of 1 kHz was synchronised to a high voltage sinusoidal-source.

The voltage applied to the electrodes was arranged to be a maximum at the arrival time of the Q-switched optical pulse. A voltage of up to 350V peak-peak was used. The input optical-power was varied to a maximum of about 1.8 kW.

The grating was selected from one of three masks detailed in Table 1. The choice of a small pitch grating enabled phase-matching in many different types of fibre for all propagating second-harmonic modes. Several fibre devices were measured. The radius of curvature of the fibre in the polished region was 250 mm for device 'D1'. All other devices ('D2' - 'D4') had a flat region where the fibre core was parallel to the top surface of the silica block.

4 Measurements

Figure 5 shows the second-harmonic data for device 'D1' as a function of the grating rotation angle using electrode-mask A (Table 1) with a pitch of 40 μm. The continuous curve shows the second-harmonic generated with phase-matching of $LP_{01}^{\omega} \rightarrow LP_{01}^{2\omega}$ at an angle of 21.34°. Comparison with other second-harmonic modes designated on the figure highlights the relative efficiency of this interaction. The coherence length of $LP_{01}^{\omega} \rightarrow LP_{01}^{2\omega}$ was measured to be 21.47 $\pm$0.17 μm. The coherence lengths for the interaction of the LP_{01}^{ω} mode to the $LP_{11}^{2\omega}$, $LP_{21}^{2\omega}$ and $LP_{02}^{2\omega}$ modes were measured to be 22.98, 25.86 and 27.57 μm, respectively.

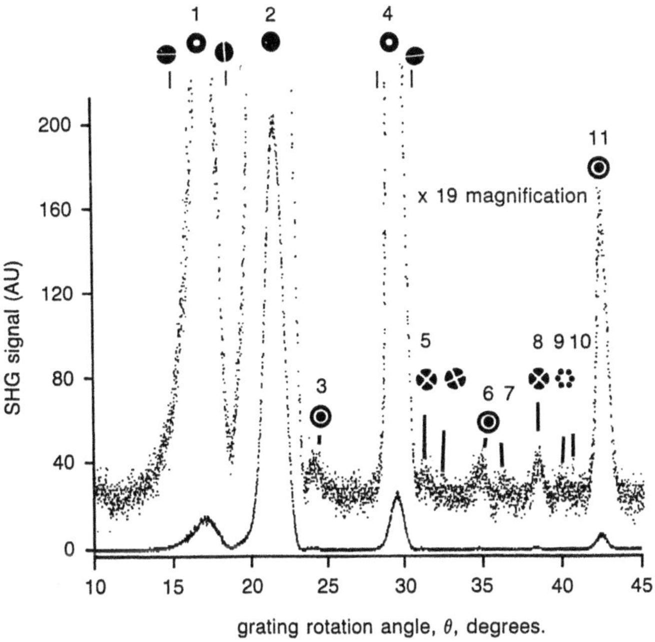

Figure 5 Second harmonic signal for device 'D1' as a function of electrode angle, θ. The expanded vertical scale highlights all the phase-matched interactions, as listed in Table 2. $LP_{01}^{\omega} \rightarrow LP_{01}^{2\omega}$ is the most favourable interaction. $LP_{11}^{\omega} \rightarrow LP_{02}^{2\omega}$ and $LP_{01}^{\omega} \rightarrow LP_{31}^{2\omega}$ were not observed. The former requires the smallest pitch of all the interactions and was not possible with the grating used for the measurement, while the overlap integral for latter interaction was very small. The numbering for the interactions is as follows: 1 = $LP_{11}^{\omega} \rightarrow LP_{11}^{2\omega}$, 2 = $LP_{01}^{\omega} \rightarrow LP_{01}^{2\omega}$ and $LP_{11}^{\omega} \rightarrow LP_{21}^{2\omega}$, 3 = $LP_{11}^{\omega} \rightarrow LP_{02}^{2\omega}$, 4 = $LP_{01}^{\omega} \rightarrow LP_{11}^{2\omega}$, 5 = ? $\rightarrow LP_{21}^{2\omega}$, 6 = ? $\rightarrow LP_{02}^{2\omega}$, 7 = ? $\rightarrow LP_{02}^{2\omega}$, 8 = $LP_{01}^{\omega} \rightarrow LP_{21}^{2\omega}$, 9 = $LP_{11}^{\omega} \rightarrow LP_{31}^{2\omega}$, 10 = ? $\rightarrow LP_{31}^{2\omega}$ and 11 = $LP_{01}^{\omega} \rightarrow LP_{02}^{2\omega}$.

TABLE 1: Data on Periodic Electrode Structures

Mask	Pitch, Λ (μm)	Electrode Width, b (μm)	Electrode space, c (μm)
A	40	15.6	4.4
B	40	4.4	15.6
C	32	4.4	11.6

Measurements and calculation on device 'D1' are detailed in Table 2. A sinc^2 fuction was fitted to the main $LP_{01}^{2\omega}$ peak. The FWHM angular tuning bandwidth, $\Delta\theta = 2.8\Lambda \;/\; (\pi LN \tan\theta)$.

An effective interaction length of 3.6mm was calculated from the width at half maximum. A maximum conversion efficiency of 4×10^{-6} % was measured for a peak input power of 1.67 kW. For a second device, 'D4' with a flat 4mm fibre section and using grating 'C', a conversion efficiency of 4×10^{-4}% was measured for 1.37 kW and 215 V pk-pk applied. Device details and measurements on all devices with interaction lengths up to 9.8mm are summarised in Table 3.

The computed overlap integral ratios and the measured values are expected to be in good agreement only for the strong mode interactions with the lowest order fundamental wavelength mode (LP_{01}^{ω}). The ratios are approximately within a factor of

TABLE 2: Comparison of measured and computed coherence-lengths and overlap integrals for seven mode interactions. $LP_{11}^{\omega} \rightarrow LP_{21}^{2\omega}$ coincides with $LP_{01}^{\omega} \rightarrow LP_{01}^{2\omega}$ and was therefore not resolved. $LP_{11}^{\omega} \rightarrow LP_{01}^{2\omega}$ could not be achieved since the grating pitch was too large. $LP_{01}^{\omega} \rightarrow LP_{31}^{2\omega}$ was not measured. Measured data is from 'D1'.

Interaction (Figure 5)	Modes	PITCH (μm) Meas ($2l_c$)	Calc	Meas. Overlap Ratios	Calculated Overlap Ratios *
2	$LP_{01}^{\omega} \rightarrow LP_{01}^{2\omega}$	42.98	43.39	1	1
4	$LP_{01}^{\omega} \rightarrow LP_{11}^{2\omega}$	45.96	47.00	0.117	4.79×10^{-2}
8	$LP_{01}^{\omega} \rightarrow LP_{21}^{2\omega}$	51.71	52.59	0.005	3.49×10^{-4}
11	$LP_{01}^{\omega} \rightarrow LP_{02}^{2\omega}$	55.14	54.56	0.0311	1.25×10^{-2}
-	$LP_{01}^{\omega} \rightarrow LP_{31}^{2\omega}$	-	61.01	-	8.78×10^{-7}
-	$LP_{11}^{\omega} \rightarrow LP_{01}^{2\omega}$	-	36.49	-	2.48
1	$LP_{11}^{\omega} \rightarrow LP_{11}^{2\omega}$	41.83	39.00	1	1
2	$LP_{11}^{\omega} \rightarrow LP_{21}^{2\omega}$	-	42.72	-	3.27
3	$LP_{11}^{\omega} \rightarrow LP_{02}^{2\omega}$	43.90	44.04	0.061	2.46
9	$LP_{11}^{\omega} \rightarrow LP_{31}^{2\omega}$	53.00	48.18	0.0131	0.189

* $\delta n = 0.0045$, $a = 4$ μm

TABLE 3: Summary of Measurements on Devices

DEVICE →	D1	D2	D3	D4
Meas. η% ($LP_{01}^{\omega} \rightarrow LP_{01}^{2\omega}$)	4×10^{-6}	1.18×10^{-4}	3.57×10^{-4}	4×10^{-4}
Computed η%	1.4×10^{-5}	6.77×10^{-5}	1.58×10^{-4}	8.64×10^{-5}
P_{in}, kW	1.67	1.80	1.40	1.40
E_0, Vm^{-1}	1.75×10^{7}	1.75×10^{7}	1.75×10^{7}	1.34×10^{7}
(Grating) θ_{pm}°	(B)21.34	(A)22.23	(A)8	(C)43.1
Coherence length, l_c, μm ($LP_{01}^{\omega} \rightarrow LP_{01}^{2\omega}$)	21.47	21.61	20.20	21.91
FWHM, $\Delta\theta_{pm}$°	1.45	0.63	1.48	0.44
Phase-match length, mm	3.6(250mm rad)	8.3	9.8	4.0
Fourier coef., B_1	0.776	0.420	0.420	0.811
Distance, d, μm (see Fig 3)	5	2	4	2
Measured fibre Δn	4.5×10^{-3}	4.5×10^{-3}	5.5×10^{-3}	4.5×10^{-3}
Fibre core-radius, μm	4	4	2.25	4

three for interactions 4 and 11. For interactions with modes near to cut-off, the agreement is not good since the overlap integral is very sensitive to the exact value of the propagation constant.

A choice of the correct pitch, Λ, and therefore a small phase matching angle, θ, ensures a large tuning bandwidth for a given device length. This can be seen by comparing devices 'D2' and 'D3' in Table 3. The Fourier coefficient, B_1, is largest for electrode 'C' used with device 'D4'. The computed coefficients for devices 'D2' and 'D3' are identical since the same electrode, 'A', was used for both.

Although 'D3' was double the length of 'D4', the conversion efficiency achieved in both the devices was similar. For device 'D4', the fundamental Fourier coefficient, B_1 was twice that for device 'D3', and the overlap integral of all the fields for the most efficient interaction in 'D4' were calculated to be similar. Also, 'D3' was twice as long as 'D4'. Since the conversion efficiency is proportional to the square of these parameters [see eqn (1)], the efficiencies are nearly the same. The details are shown in Table 3.

Measurements of phase-matched second-harmonic power as a function of applied static fields showed excellent square law dependence, up to the maximum voltage without saturation effects, as shown in Figure 6. Power supply restrictions prevented the application of higher fields. Second-harmonic power also showed good quadratic dependence with respect to input fundamental power. Owing to the large angle of the grating (near 45 °),two components of the static field contribute significantly to the generation of the second-harmonic (see figure 2, 2a). The harmonic will have an orthogonal polarisation and be $\pi/2$ out of phase with that generated by $\chi^{(3)}_{xxxx}$ or $\chi^{(3)}_{xyyx}$. It can be shown that [11] if the two static field components are equal, the second-harmonic power generated is independent of the polarisation of the input fundamental

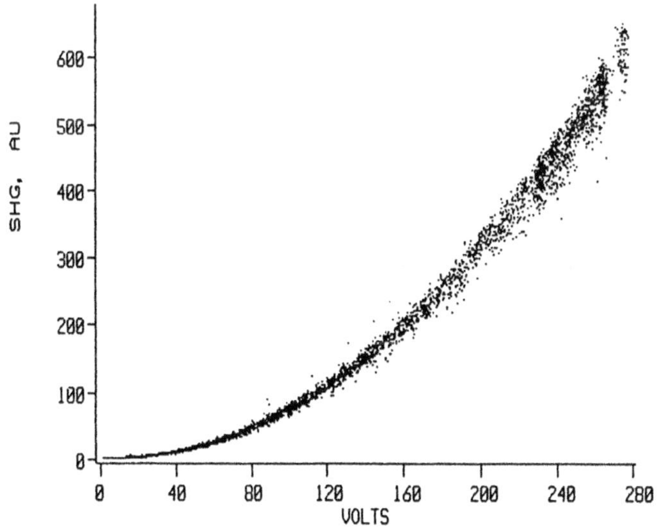

Figure 6 Second harmonic power as a function of applied volts. Comparison with a quadratic curve shows excellent agreement. The maximum voltage was limited by the power supply used, and was equivalent to a field of 1.75×10^7 V m^{-1}. The signal shown is for the most efficient phase-matched mode-combination, $LP^{\omega}_{01} \rightarrow LP^{2\omega}_{01}$. The grating used for this experiment was 'C'.

wavelength. This phenomenon can result in second-harmonic generators with reduced polarisation sensitivity when the two static-field components are nearly equal.

Table 4 lists the active $\chi^{(3)}$ coefficients operating in the device and summarises the salient features. There is scope for improvement of the overlap integrals and device optimisation which would allow phase-matching over fibre lengths of several centimetres. The conversion efficiencies expected for a 10 cm device using glass with two orders of magnitude larger nonlinearity [12], is calculated to be about 10% for an input power of 1 watt average, allowing compact all-fibre frequency translators to be fabricated.

TABLE 4: Relative magnitudes and phases of the active $\chi^{(3)}_{ijkl}E^0_l$ tensors as a result of the components of the applied DC electric-field. * dependent on electrode parameters. Values given assume equivalence of the static field components with E^0_x.

$\chi^{(3)}_{ijkl}E_0$	Magnitude Relative to $\chi^{(3)}_{xxxx}E^0_x$	Relative Phase
$\chi^{(3)}_{xxxx}E^0_x$	1	0
$\chi^{(3)}_{xxyy}E^0_y$	~ 2/3*	$\pi/2$
$\chi^{(3)}_{xyyx}E^0_x$	1/3	0
$\chi^{(3)}_{xxzz}E^0_z$	~ 2/3*	$\pi/2$
$\chi^{(3)}_{xzzx}E^0_x$	~ 1/3*	0
$\chi^{(3)}_{yyyy}E^0_y$	~ 1*	$\pi/2$
$\chi^{(3)}_{yyxx}E^0_x$	2/3	0
$\chi^{(3)}_{yxxy}E^0_y$	~ 1/3*	$\pi/2$
$\chi^{(3)}_{yyzz}E^0_z$	~ 2/3*	$\pi/2$
$\chi^{(3)}_{yzzy}E^0_y$	~ 1/3*	$\pi/2$
$\chi^{(3)}_{zzzz}E^0_z$	~ 1*	$\pi/2$
$\chi^{(3)}_{zzxx}E^0_x$	2/3	0
$\chi^{(3)}_{zxxz}E^0_z$	~ 1/3*	$\pi/2$
$\chi^{(3)}_{zzyy}E^0_y$	~ 2/3*	$\pi/2$
$\chi^{(3)}_{zyyz}E^0_z$	~ 1/3*	$\pi/2$

5 Conclusions

Phase-matched electric-field-induced second-harmonic generation has been demonstrated in optical fibres for the first time. A *single* inter-digitated electrode is used to induce a spatially-varying effective second-order non-linear susceptibility. Phase-matching of all propagating fundamental and second-harmonic modes has been achieved, including the most efficient $LP_{01}^{\omega} \rightarrow LP_{01}^{2\omega}$ interaction, with a maximum conversion efficiency of 4.0×10^{-4}% for a device 4 mm long. Device lengths up to 9.8 mm have also been phase matched. This technique could be applied to generate different sources, intra-cavity frequency doubling in fibre-lasers and efficient frequency mixing in optical fibres at relatively low-powers. It can also be used with other non-linear materials incorporated in waveguides, such as doped and poled-polymers.

The author would like to thank C A Millar for numerous discussions and suggestions on the manuscript. This publication is by permission of the Director of Research & Technology (British Telecommunications plc).

REFERENCES

1. Osterberg U & Margulis W, *Opts Letts 11,* pp516-518, 1986.
2. Farries M C et al, *Electronics Letts, 23,* pp322-324, 1987.
3. Stolen R H & Tom H W K, *Opts Letts 12,* 585, 1987.
4. Payne F P, *Electronics Letts, 23* (23), pp1215-1216, 5th November 1987.
5. Terhune R W & Weinberger D A, *JOSA-B, 4* (5), May 1987.
6. Weinberger D A & Terhune R W, *CLEO'87,* Session TUHH3, pp78-79, Baltimore, April 27-May 1, 1987.
7. Kashyap R, in Proc of *SPIE* Symposium on 'Molecular and Polymeric Optoelectronic Materials: Fundamentals and Applications', San Diego, California, Volume 682, pp170-178, 21-22 August 1986.
8. Kashyap R, paper MP-25, In Proc of *The XVI International Conf on Quantum Electronics,* Tokyo, Japan, July 18-21, 1988.
9. Gloge D, *Appl Opts 10,* pp 2252-2258, Oct. 1971.
10. Kashyap R & Blow K J, *Electron Letts, 24* (1), 7th January 1988.
11. Kashyap R, 'Phase Matched Periodic-Electric-Field Induced Second-Harmonic Generation in Optical Fibres', to be published in *J Opt Soc Am B, 6* (3), March 1989.
12. Hall D W, Borrelli N F & Dumbaugh W H, paper PD27-1, *CLEO'88,* 25-29, April 1988, Anaheim, California, USA.

Second Harmonic and Parametric Difference Frequency Generation in Organic Nonlinear Waveguides

K. Sasaki[1], *T. Kinoshita*[1], *H. Sasabe*[2], *A. Yamada*[2], *and A.F. Garito*[2]

[1]Faculty of Science and Technology, Keio University, 3-14-1 Hiyoshi, Yokohama 223, Japan

[2]International Frontier Research System, Institute of Physical and Chemical Research, 2-1 Hirosawa, Wako, Saitama Pref. 350-01, Japan

1. Introduction

In recent years, many charming organic materials with large optical nonlinearities have been found by theoretical molecular design and experimental synthesis[1][2]. Furthermore, these materials promise to realize practical device applications in the field of optoelectronics. Especially, low threshold intensity of excitation of the materials for nonlinearities is one of the most important in connection with semiconductor lasers. But there are many difficulties to overcome before practical applications are realized. For example, large molecular nonlinear succeptibilities are often cancelled or suppressed during the growth of bulk crystals via positioning and orientation of molecules. Also, most organic crystals are mechanically weak in comparison with inorganic crystals, and need new processing technologies. The concept of phase matching in optical wave mixing is vitally important. Many organic crystals are phase matchable but cannot efficiently use the large diagonal second harmonic generation (SHG) coefficients, for example, d_{11} of 2-methyl-4-nitroaniline(MNA), in the bulk phase matching.

Phase matching in a waveguide is an important method to phase match organic crystals with large diagonal coefficients with an additional virtue of energy concentration of guided waves. To date, we have investigated second and third order nonlinear optica in slab-type waveguide structures[3]-[7]. In this paper, we report MNA thin single crystal films which we used prepared in narrow gaps of faced substrates by the zone-melt recrystallization are used as slab-type waveguides. The MNA plate like crystal belongs to the monoclinic class and has a parallelogram shape. The largest element of SHG tensor d_{11} can be applied to the TE wave in the waveguide as previously reported.

Phase matching of SHG can be realized by adjusting waveguide thickness with lateral movement of tapered MNA crystal film waveguides. In this case a hybrid waveguide of MNA crystal film and rf sputtered Corning 7059 film on a substrate is useful for the phase matching of the SHG.

Parametric difference frequency generation is observed in the same structure by mixing the tunable Rh 6G dye laser(0.58-0.60 μm) and He-Ne laser(1.15 μm).

The efficient generation of a difference frequency wave(the idler wave) needs also the phase match condition. This is approximately realized by adjusting the

Springer Proceedings in Physics, Vol. 36
Nonlinear Optics of Organics and Semiconductors
Editor: T. Kobayashi

tapered waveguide or wavelength of Rh 6G dye laser. Also degenerate optical parametric amplification is observed by mixing of Nd;YAG laser(1.064 μm) and the SHG(0.543 μm) in the MNA waveguide.

2. Preparation of Tapered Thin Film MNA Crystals.

The preparation of tapered thin MNA single crystal film is carried out by a two step process. First, two fused quartz substrates, held face to face with a tapered gap polycarbonate spacer film(0.6 μm), is inserted into molten MNA. For the SHG experiment, one substrate is covered with an rf sputtered Corning 7059 glass film. The gap fills with molten MNA by capillary action. After this stage the MNA crystal is mosaic-like with randomly oriented microcrystals. Then the faced substrates are passed through a precisely controlled temperature gradient with a glass vessel at the rate of 40 mm/day corresponding to the zone-melt recrystallization as shown in Fig 1.

The quality of the prepared MNA thin single crystal depends on several factors; temperature distribution in the glass vessel, down speed of the faced substrates, and purity of the raw MNA powder. Unwanted thermal decomposition is mainly attributed to inherent impurities. The x-ray diffraction of the crystal film shows that the crystal has a monoclinic structure whose (0,1,0) plane is parallel to the substrate surface.Observation with a crossed-Nicols system gives the optical axis[8]. Also, the optical axis(roughly corresponding to d_{11} direction) of the crystal film grows parallel to equi-temperature line in the glass vessel at a certain probability.

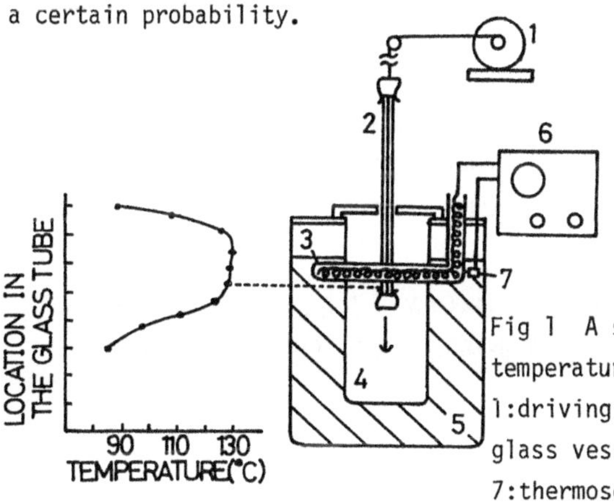

Fig 1 A schematic recrystallization system with temperature distribution in the glass vessel; 1:driving motor,2:faced substrates,3:heater,4: glass vessel,5:oil bath,6:temperature controller, 7:thermosensor

3. SHG in the Hybrid Waveguide

It is well-known that guided waves have a proper effective index of refraction in connection with each cross-sectional field distribution at given waveguide parameters. The dispersion relation between effective index and thickness of a slab-type waveguide is determined by wavelength of the guided wave, the polarization modes and guided wave modes. Phase matching in SHG is realized at the degenerate points of the dispersion curves for two wavelengths, the fundamental and the second harmonic wave. The quantitative expression for the SHG power under the above condition is given as follows:

$$P(2\omega) = (d_{11}^2 L^2 \omega^2 / 16w) \cdot [P(\omega)]^2 S^2 [\sin(L\Delta\beta/2)/(L\Delta\beta/2)]^2 \quad .....(1)$$

where $P(2\omega)$ is the SHG output, ω is the angular frequency of the fundamental wave, w is the beam width of the guided wave, L is the waveguide length, $P(\omega)$ is the input power of the fundamental wave, S is the spatial coupling of the field distributions of two waves, and $\Delta\beta = 2\beta(\omega) - \beta(2\omega)$ is the difference of two propagation constants(β/k_o is the effective indices,and k_o is the wave number in vacuum). The spatial coupling factor is given as follows:

$$S = \int_{-\infty}^{\infty} [\, \varepsilon_y^n(\omega)^2 \;\; \varepsilon_y^m(2\omega)\,] dy \quad ....(2)$$

where $\varepsilon_y^n(\omega)$ is the field distribution of the nth order fundamental wave and $\varepsilon_y^n(2\omega)$ is the mth order doubled frequency wave. In the equation (1) two factors S and $\Delta\beta$ are very important. Namely $\Delta\beta = 0$ means the phase matching condition for the 4 layer waveguide as shown in Fig 2.

In this section, our discussion is focused mainly on the phase match condition for the structure. In the structure, guided waves are classified into two cases:

(i) effective index of refraction n_{eff} is larger than the index of the Corning 7059 film but smaller than the index of MNA ($n_x > n_{eff} > n_c$).

(ii) n_{eff} is larger than the index of fused quartz substrate, n_s, but smaller than n_c ($n_c > n_{eff} > n_s$).

For the case (i) the waveguide is approximately a 3 layer waveguide where only the MNA layer plays a role as the waveguide. Combinations of modes between

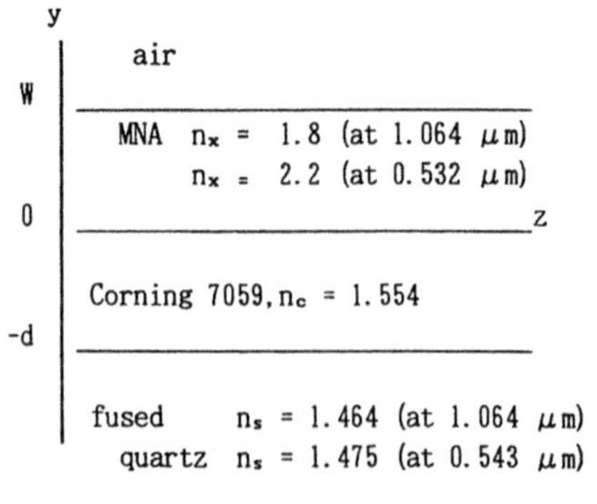

Fig 2. Cross-sectional structure of the 4 layer waveguide.

the even and the odd modes of the field distributions of the fundamental and the second harmonic wave give small values of the spatial coupling factor as the result of the integrations in Eq. (2). But case (ii) is different from case (i)[9].

In case(ii) MNA layer and Corning glass make a hybrid waveguide where the field distributions of the guided waves spread over both layers. This situation is complicated as the structure works as a hybrid waveguide to the fundamental and a 3 layer waveguide to the second-harmonic wave simultaneously. This property can be applied to advantageously adjust the spatial coupling factor around the cutoff thickness of the main waveguide, MNA layer. Additionally new phase matchable points arise around the cutoff of effective indices of refraction.

The experiment is carried out by the setup shown in Fig 3. The fundamental wave from Nd;YAG laser is coupled into the hybrid waveguide by an FD 11 prism.

Phase match condition of the SHG means a coincidence of effective indices at a certain waveguide thickness in the dispersion curves. Actually this can be given by lateral movement of the tapered waveguide with adjusting incident angle θ which corresponds to the effective index. In our experiment the maximum SHG is observed at $\beta/k_0 = 1.462$ with MNA waveguide thickness of 2.52 μm corresponding to the circled point (degeneration between 7th fundamental and 20th doubled) in Fig 4. The estimated conversion efficiency is a few percent at a one hundred watt fundamental wave after input coupling with 0.1mm of guide length and beam size.

Summary of this section is as follows:

1) the 4 layer hybrid waveguide structure is available for phase matching of SHG around the cutoff of effective index of refraction.
2) a hybrid waveguide with a combination of high index layer can give a freedom of designing high conversion efficiency device.
3) cheaper FD 11 prism can be used as coupler for the hybrid waveguide instead of expensive rutile prism.

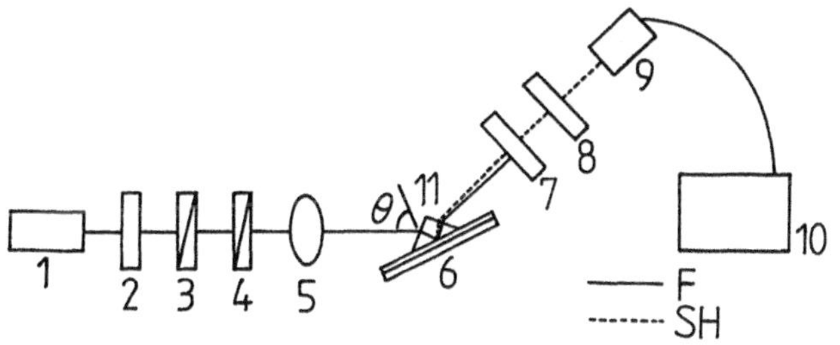

Fig 3. Experimental setup for SHG. 1:Nd;YAG laser,2:IR filter, 3,4:polarizer,5:lens,6:waveguide,7:beam stopper,8:interference filter,9:detector,10:power meter,11:FD 11 prism.

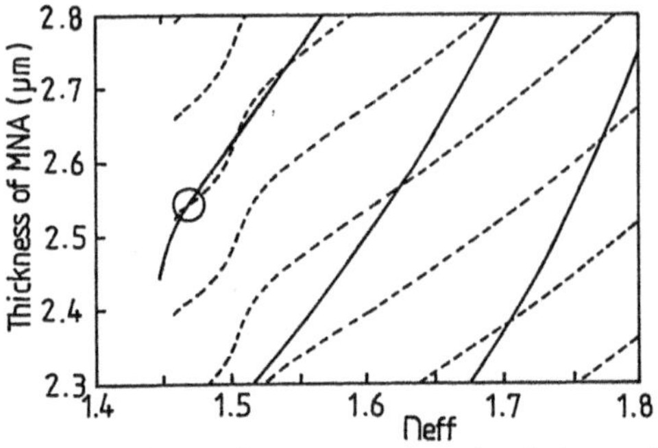

Fig 4. Dispersion curve for the 4 layer hybrid waveguide. d = 0.379 μm, solid lines:fundamental, dashed lines: doubled wave.

4. Parametric Difference Frequency Generation and Degenerate Parametric Amplification in an MNA Waveguide

Parametric difference frequency generation and parametric amplification in nonlinear wave mixing are also promising for many important practical applications in optoelectronics and long distance optical communication. In parametric sum and difference frequency generation the high second order nonlinearity of MNA crystals is useful. When a signal wave with an angular frequency ω_s inpinges into an excited highly nonlinear material pumped by strong pumping wave ω_p, a new wave with an angular frequency $\omega_i = \omega_p - \omega_s$ is generated with accompanying amplification of the signal wave as shown in Fig 5 where ω_i is called the idler wave.

Parametric amplification and parametric difference frequency generation with interaction length L are given by the coupled wave theory[10] as follows:

$$\varepsilon_s(L) = \varepsilon_s(0)\exp(-\alpha L - i\Delta\beta/2)[\cosh(gL)+(i\Delta\beta/2g)\sinh(gL)] \quad \text{.....(3)}$$

$$\varepsilon^*_i(L) = i(\Gamma\gamma/g)\,\varepsilon_s(0)\exp(-\alpha L - i\Delta\beta/2)\sinh(gL) \quad \text{............(4)}$$

where α is the attenuation constant, * means the expression of complex conjugate,

$$g^2 = \Gamma^2 - (\Delta\beta/2)^2 \quad \text{....(5)}, \qquad \Delta\beta = \beta_p - \beta_s - \beta_i \quad \text{....(6)}.$$

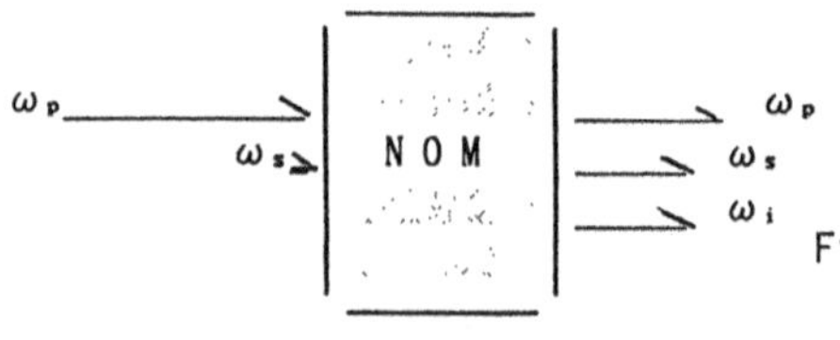

Fig 5. A schematic diagram of the wave mixing.

$\beta_R = n_R\omega_R/c$, (R = p,s and i) ...(7), n_R is effective index of refraction,

c is velosity of light in vacuum, $\gamma^2 = [\omega_i z_i W_s \varepsilon_p^*(z)]/[\omega_s z_s W_i \varepsilon(z)]$...(8),

$z_R = \omega_R\mu_o/\beta_R$ ...(9), $W_R = \int_{-\infty}^{\infty}\varepsilon_R^n(y)\,\varepsilon_R^m(y)dy$.....(10)

$\Gamma^2 = [\,\omega_p^2(1-\delta^2)\,d_{11}\varepsilon_o z_p z_s z_i I_{pis}^2\,P_p]/(8WW_pW_sW_i)$.......(11)

$\delta^2 = 1 - (4\omega_i\omega_s/\omega_p^2)$...(12), $P_p = |\varepsilon_p|^2\,\omega_p\,W/2z_p$.....(13)

$I_{psi} = \int_0^W \varepsilon_p(y)\varepsilon_s(y)\varepsilon_i(y)dy$........(14),

W is the waveguide thickness. By (6), $\Delta\beta = 0$ gives the phase match condition of the system as

$$n_p\omega_p = n_s\omega_s + n_i\omega_i \quad ....(15),$$

and (3),(4) are transformed into

$$\varepsilon_s(L) = \varepsilon_s(0)\exp(-\alpha L)\cosh(-\alpha L).......(16)$$

$$\varepsilon_i^*(L) = i\,\gamma\varepsilon_s(0)\exp(-\alpha L)\sinh(-\alpha L).....(17).$$

The equations of power for each wave respectively are

$$P_s(L) = P_s(0)\exp(-2\alpha L)\cosh^2(\alpha L)......(18),$$

$$P_i(L) = \gamma^2\,P_s(0)\exp(-2\alpha L)\sinh^2(\alpha L).....(19).$$

The power gain of the signal wave is

$$G_s(L) = P_s(L)/P_s(0)-1 = \sinh^2(\Gamma L)\;.....(20).$$

The equation (19) corresponds to the output of the idler wave and (20) means the gain of the parametric amplification. At the degenerate state($\omega_i=\omega_s=\omega_p/2$) the idler wave is included in the amplified signal wave which means the most efficient optical parametric amplification(OPA). In this case the total wave is

$$\varepsilon_t(L) = \varepsilon_s(L) + \varepsilon_i^*(L) = \varepsilon_s(0)\exp(-\alpha L)[\cosh(\Gamma L) + \sinh(\Gamma L)].....(21).$$

Then the total power of OPA is

$$P_t(L) = P_t(0)\exp(-\alpha L)[\cosh^2(\Gamma L) + \sinh^2(\Gamma L)] = P_t(0)\exp(-\alpha L)\cosh(2\Gamma L)...(22).$$

The experimental setup for the optical parametric difference frequency generation is shown in Fig 6. A tunable Rh 6G dye laser feeds the pumping wave to the MNA waveguide together with the signal wave of a He-Ne laser(1.15 μm).

The phase matching at λ_p = 0.579 μm, λ_s = 1.15 μm and λ_i = 1.173 μm is realized by the same method as in the SHG experiment. λ_i is determined by a monochrometer. The phase matching peak is located at the waveguide thickness of W = 0.497 μm in Fig 7.

This phase matching corresponds to the order of the propagation modes (p,s,i) = (2,0,0) in the dispersion relation. The degenerate OPA is carried out by the experimental setup as shown in Fig 8.

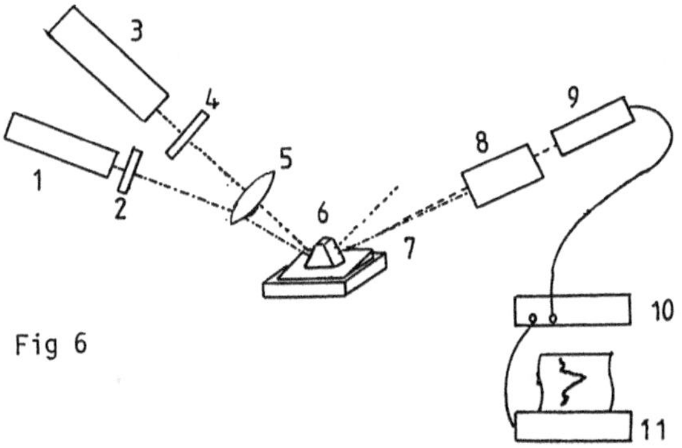

Fig 6

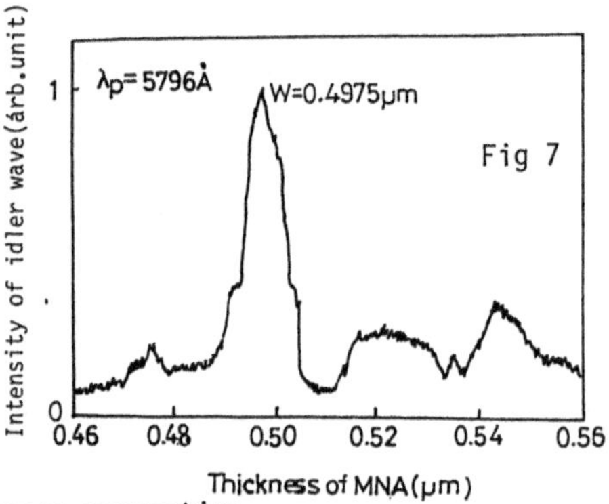

Fig 7

Fig 6. Experimental setup for difference frequency generation. 1:He-Ne laser,2:IR pass filter,3:dye laser,4:aperture,5:lens 6:rutile prism,7:MNA waveguide,8:monochrometer,9:photomultiplier,10:boxcar integrator,11:pen recorder.

Fig 7. Phase matching of the idler wave.

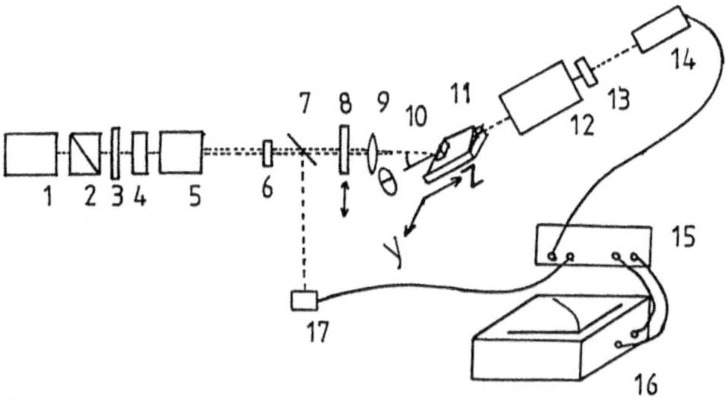

Fig 8. Schematic explanation of the OPA. 1:Nd;YAG laser, 2: polarizer,3:IR pass filter,4:$CuSO_4$ cell,5:KDP SHG crystal, 6:IR attenuator,7:splitter,8:IR pass filter,9:lens,10: prism,11:MNA waveguide,12:monochrometer,13:IR pass filter, 14:photomultiplier,15:boxcar integrator,16:X-Y plotter, 17:photo-diode.

From (22) the power amplification of the degenerate wave is

$$A \equiv P_s(L)/P_s(0) = \cosh(2\Gamma L) \text{ ....(23)},$$

and the gain factor is

$$\Gamma = (1/2L)\ln[A + (A^2-1)^{1/2}] \text{ ....(24)}.$$

In our experiment, L= 1mm, the maximum amplification is A= 1.23 by measurement of the output in Fig 8, then $\Gamma = 3.33\ cm^{-1}$. This value is rather small in comparison with 14.0 for the bulk POM by Zyss[11] but in Sohler's experiment in Ti difused $LiNbO_3$ waveguide $\Gamma = 0.308\ cm^{-1}$[12][13]. Our gain factor has an ambiguity associated with estimation of pumping power and interaction length. From the view point of the nonlinear optical constant, actual value of Γ for the MNA waveguide is larger than the above values.

Acknowledgments

The authors wish to express their thanks to Dr.C.Grossman for his suggestion for English expressions used in this paper and for valuable discussion. We would also like to thank the graduate students, K.Takeshige,H.Takara,M.Ishida,S.Kanba, T.Toda and M.Hiwatashi for their experimental support.

References

[1] D.J.Williams, ed., Nonlinear Optical Properties of Organic and Polymeric Materials, ACS Symp. Ser., 233 (1983).

[2] D.S.Chemla and J.Zyss, Nonlinear Optical Properties of Organic Molecules and Crystals. vol. 1 & 2, Academic Press Inc.,(1987).

[3] K.Sasaki,T.Kinoshita and N.Karasawa, Appl. Phys. Lett.,45, 333 (1984).

[4] K.Sasaki,H.Kawagishi and Y.Ishijima, Appl. Phys. Lett.,47,783 (1985).

[5] H.Itoh,K.Hotta,H.Takara and K.Sasaki, Appl. Opt.,25,1491 (1986).

[6] H.Itoh,K.Hotta,H.Takara and K.Sasaki, Opt. Commun., 59,299 (1986).

[7] K.Sasaki,K.Fujii,T.Tomioka and T.Kinoshita, J. Opt. Soc. Am. B, 5,457 (1988).

[8] H.Itoh,Y.Ishijima,K.Takeshige,S.Kanba and K.Sasaki, Appl. Opt., 26,17 (1987)

[9] C.Liao and G.I.Stegeman, Appl. Phys. Lett.,44,164 (1984).

[10] A.Yariv,IEEE J. Quantum Electron.QE 9(9),919(1973).

[11] J.Zyss, IEEE, J. Quantum Electron. QE 21(8), 1286 (1986).

[12] W.Sohler and H.Suche, Appl. Phys. Lett., 37,255 (1980).

[13] W.Sohler and H.Suche, Proc. SPIE, 408, 163 (1983).

Second Order Optical Nonlinearities in Organic Polymer Films and Crystal Cored Fibres

R. Heckingbottom, J.R. Hill, G.E. Holdcroft, P.L. Dunn, P. Pantelis, and J.D. Rush

British Telecom Research Laboratories,
Martlesham Heath, Ipswich, Suffolk IP5 7RE, UK

Recent progress on developing optically nonlinear polymers for device applications is reported. A simplified theoretical model is shown to provide a sufficient guide in selecting chromophores. Internal fields of $\geq 10^8$ V/m, reported previously for a piezoelectric polymer host used in guest/host systems, are shown to be stable over eighteen months. A single component acrylic ester polymer has been used to fabricate a demonstration modulator. Coefficients comparable with those of lithium niobate have been obtained and the device has remained constant in operation over a six months period.

Crystal cored fibres of DAN in silica now show SHG at >1% efficiency - an increase of 10^3x in the past year. Measurement of sideways scatter at both the fundamental and the second harmonic is increasing the understanding of the remaining problems.

1 INTRODUCTION

The potential of organic materials for applications in optical processing, through their high molecular coefficients of optical nonlinearity, has been well recognised for several years now/1/. Although there is still significant scope for improved coefficients through optimised molecular design, attention is increasingly being focussed on the effective incorporation of known molecules into useable films and waveguides. The three main options - single crystals, polymer films and Langmuir-Blodgett films - all offer both promise and problems.

Single crystals offer the prospect of perfect molecular packing and extremely low loss. Unfortunately, as is well known, the molecules with the highest coefficients often form crystals that are centrosymmetric so that the second order susceptibility - the property where organics currently show the greatest improvement over inorganics - falls to zero. Even in noncentrosymmetric crystals, the packing is not always optimised so that the full potential of the high molecular coefficients is not realised in the crystal properties. In practice the growth of highly perfect crystals presents additional challenges.

Polymer films offer the most tractable technology and here the current challenges are (a) to minimise the dilution inherent in synthesising a suitable polymer incorporating the optically nonlinear molecules and (b) to maximise the degree of ordering of these molecules by application of external fields. The solutions to both problems need to be compatible both with the necessary device processing and a competitive device lifetime.

Langmuir-Blodgett films also suffer from the dilution inherent in synthesising a film-forming molecule containing the optically nonlinear moiety, but in principle they offer highly ordered molecular packing with little self-cancellation of the molecular coefficients. Current research is concerned with the technology of suitably stable films, consistent with high coefficients and low optical scattering.

This paper reports recent results on both optically nonlinear polymer films and single crystal growth in fibre capilliaries. Also, at the molecular level

Springer Proceedings in Physics, Vol. 36
Nonlinear Optics of Organics and Semiconductors
Editor: T. Kobayashi

underpinning both these approaches, it is shown that a relatively simple theoretical model is sufficient to give accurate calculation of the second order hyperpolarisability - the first nonlinear coefficient.

2 THEORY

The relevance of the two level approximation involving only the ground and first excited states, together with the full tensorial expression for β, has already been reported/2/. The present work identifies a quick and effective method to obtain the necessary input parameters for the above approach. The starting point is the Pariser-Parr-Pople (PPP) self-consistent field method which is known to be reliable in predicting absorption spectra in large unsaturated molecules. The method assumes that σ & π electrons can be treated separately and that electronic excitation processes depend largely on the latter. The calculations are confined to planar molecules and it is assumed that solvent interactions are small and can be absorbed into semi-empirical parameters. Initial data for the required values of electron densities and bond orders are calculated using Huckel molecular orbitals and are then improved by iterative procedures to give the required convergence. This approach gives a sufficiently accurate description of the ground state but an additional configuration interaction treatment is required to obtain a comparable representation of the excited state.

The method was used to calculate the ground state dipole (μ_g), the first excited state dipole, the energy and intensity of the first electronic transition and the β values for all the molecules considered by DOCHERTY et al./3/ in their full CNDO calculations. Good agreement with experimental values was found in all cases, as shown for the β values in figure 1. The main advantage of this approach is that calculations only take a short time on a 32 bit microcomputer compared with an extended period on a mainframe for a complete CNDO treatment.

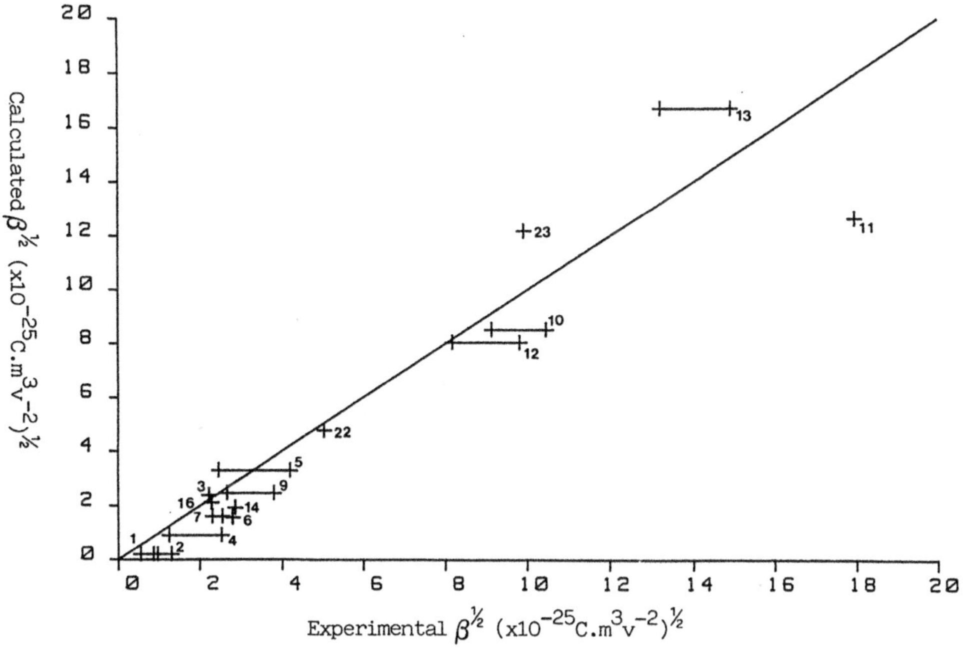

Figure 1. Comparison of calculated and experimental values for the second order hyperpolarisability, β. Compounds numbered as in DOCHERTY et al./3/.

3 POLYMER FILMS

3.1 Guest Host Polymer Systems

Much progress has recently been reported on a range of optically nonlinear guest molecules in a poly(vinylidene fluoride/trifluoro-ethylene)copolymer host/4/. The host has the advantages that it can be poled at room temperature and that the internal field should be particularly stable as it is provided by oriented micro crystallites. The key features are the β coefficient of the guest molecule, its solubility limit in the host, its degree of orientation in the applied field and the long term stability of the guest host system. Several of these factors are now well understood. Good correlation is shown between the nonlinear properties calculated by the method outlined earlier and measured values as illustrated in table 1. Internal field strengths greater than 10^8V/m were generated by the room temperature poling process/5/ resulting in a degree of molecular orientation. Second harmonic coefficients of 2.5 pm/V have been reported/6/, which is about 10% of the largest coefficient of lithium niobate. The main limitation therefore is in the solubility limit of the high β coefficient molecules, which is currently reported to be just in excess of 10%/7/.

Table 1

The relationship between molecular properties of guest molecules and their second harmonic coefficient in nonlinear optical films.

Guest molecule (5% by weight)	Calculated $\mu_g.\beta$ (10^{-80} $C^2m^4V^{-2}$)	Molecular weight ($g\ mol^{-1}$)	d_{33} due to guest (10^{-12} mV^{-1})
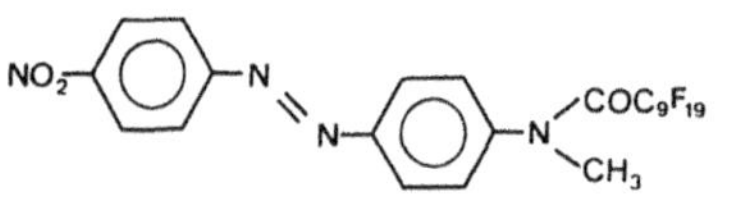	2,800	752	0.24
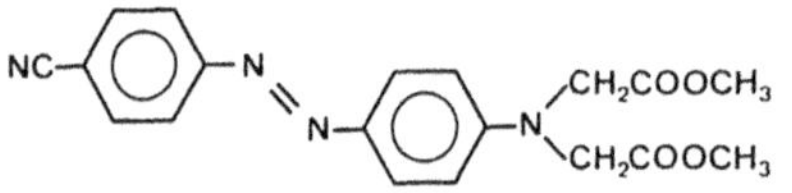	4,200	366	1.22
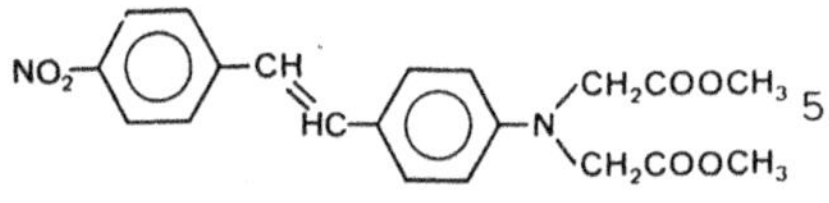	5,200	386	1.67

Stability of the system depends on both the retention of the internal field to maintain molecular ordering and the prevention of guest precipitation. The effects have been assessed by monitoring the internal field in a separate series of experiments. The internal field was determined directly by measuring the resulting change of absorption of the guest at a much lower concentration (0.4% by weight)/5/. The value of the internal field (Table (2)) was constant over a period of one and a half years, within the estimated 10% error of the experiment. Thermal annealing of the films at 80°C increased the internal field, which resulted in both an improved orientation of the guest and a larger second harmonic coefficient.

Further improvements in this approach will therefore require enhanced stability of solutions of guest molecules at even higher concentrations.

Table 2

Long term stability of internal fields in copoly(vinylidene fluoride/trifluorethylene)		
Sample	Internal Field after 24 hrs (V/m)	Internal Field after 18 months (V/m)
Poled as cast	1.13×10^{8}	1.15×10^{8}
Poled after casting and anneal at 80°C	1.55×10^{8}	1.48×10^{8}

3.2 Single Component Polymers

Single component polymers offer a practical method of achieving high concentrations of the optically nonlinear molecular moieties without the risk of precipitation. Although in principle the nonlinear moiety can be incorporated in either the main chain or the side chain, the latter is more common in practice. Variation in the side chain length can be used to form liquid crystal polymers with an increased coefficient through increased ordering in response to a given applied field/8/. However light scattering can be a problem in polymers of this kind and the work reported here involved an isotropic side chain polymer with only a short spacer separating the nonlinear moiety from the polymer backbone.

Poly 4-(4'-nitrophenylazo)-NN-methyl-ethyl-acrylylester, in which the nonlinear moiety represents over 75% of the polymer by weight, was prepared for study. Its structure is shown in figure 2.

The synthesis has been described recently/9/. The polymer was obtained in 40% yield and shown, after purification to have a low molecular weight approaching 5,000. Its glass transition temperature was measured as 82°C giving a reasonable margin for use in devices for room temperature operation.

As stated earlier, it is an advantage of polymers that they are in principle suitable for processing for device fabrication. In practice however, much work is still needed to develop materials that are optimised for the full range of necessary processing steps including multi-layer deposition, waveguide definition and electrode formation. The acrylate polymer has been used to form a demonstrator modulator and although one of the advantages of this structure was that it allowed some of the processing steps to be bypassed, it also showed the polymer's suitability for several key fabrication steps. The system and the device are shown in figure 3. The interdigital electrodes were photolithographically defined and chemically etched in an indium tin oxide coating on a glass substrate. The interdigital lines were 20μm wide with 20μm spacing and covered a total of $1cm^2$. The supply lines were broader and served to introduce both the poling and signal voltages. The polymer was melted at 170°C and spread under a cover slip to form a clear film a few microns thick. The film was then reheated to 110°C and a potential

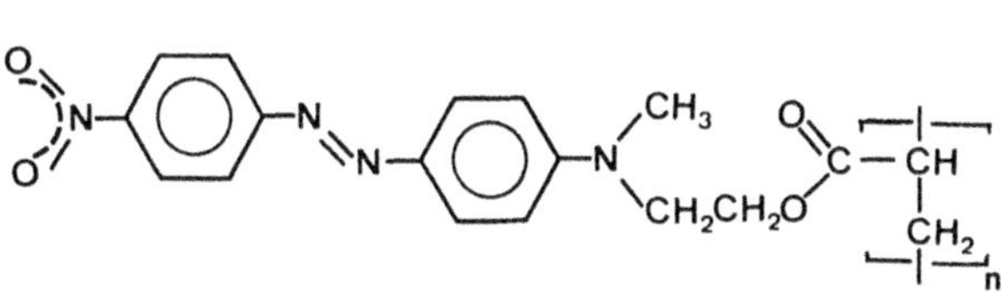

Figure 2. The structural formula for poly 4-(4'-nitrophenylazo)-NN-methyl-ethyl-acrylylester

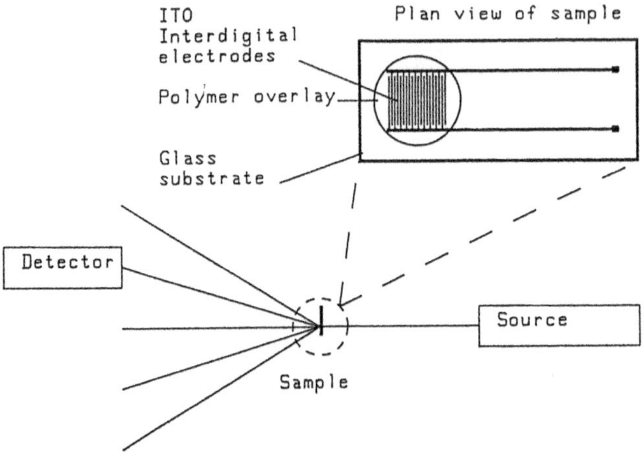

Figure 3. Schematic of the optically nonlinear polymer demonstrator modulator

of 200V applied across the electrodes while the sample was cooled through the glass temperature to room temperature. The polymer between the electrodes is poled during this stage and the ordering that creates the noncentrosymmetric structure also changes the refractive index in this region, forming an optical diffraction grating.

In the demonstration system, light from a 1mw HeNe 633nm laser source with electric vector horizontally polarised strikes the poled polymer sample normal to its surface, positioned with the interdigital electrodes in the vertical plane. The resulting Fraunhofer diffraction pattern was viewed on a screen and the light from one diffracted beam measured with a photodiode. Application of a varying ac electric field to the electrodes caused further small matching changes in the refractive index of the oriented polymer. The resulting change in the diffraction efficiency caused a corresponding variation in the intensity of the light incident on the photodiode. The depth of modulation of the output was linearly proportional to the input voltage up to 170 volts rms, the maximum value of the signal generator. The response of the polymer is expected to be sub-picosecond as the linear electro-optic effect arises from the polarization of molecular orbitals but the demonstration was limited to 20kHz owing to the low bandwidth of the photodiode integral amplifier. In this demonstrator system the fields are applied non-uniformly through the polymer. Despite this however it was possible to estimate the nonlinear coefficient of the polymer per unit poling field (in V/m) as follows. A liquid cell, having identical interdigital electrodes was constructed and filled with nitrobenzene. The electro-optic modulation resulting from the Kerr effect was then determined/10/. The unknown geometrical detail thus cancelled in the two measurements and the known value of the Kerr constant for nitrobenzene could be related directly to the nonlinear coefficient of the polymer. It was estimated that the poling field of 1×10^7 V/m applied through the near surface regions of the polymer in the demonstrator structure would have given an electro-optic coefficient (r_{33}) of $\simeq$15pm/V at 633nm. A more detailed account of this work is published elsewhere/9/.

The polymer system thus demonstrates several attractive features. Its glass temperature, Tg, at 82°C is well above room temperature and the device has remained constant in operation for over six months. It is also stable to high electric fields for poling well above Tg and the resulting refractive index change on poling gives a simple direct process for waveguide definition. Finally, its stability above its melting point allows thin film formation by the convenient melt spreading technique. Further work is now being directed towards the development of compatible cladding layers, solvent spreading of multilayer structures, plasma etching of waveguides and electrode deposition/11/.

4 CRYSTAL CORED FIBRES

Significant progress in crystal cored fibres implies the fulfilment of at least three basic requirements - single crystals with high nonlinear coefficients, suitable alignment of a major coefficient with respect to the crystal axis and high quality crystal growth within the fibre. A successful technology will also require the control of suitable core/cladding combinations and geometry to give single mode fibre with good phase matching . Earlier work/12/ using benzil has shown promising control of most of the necessary aspects but the basic electro-optic coefficients of the material are not high enough to warrant more than scientific study. More recently DAN 2-(N,N-dimethylamino)-5-nitroacetanilide has been reported to have much superior nonlinear coefficients with a second harmonic coversion efficiency as high as 20% for a Q switched laser operating at λ=1.06μm /13/. Over the past few years we have studied the growth of a range of organic crystal cores in silica fibre capilliaries and here too DAN has given the most promising results so far. As a result, most work has concentrated on this material.

Single crystal growth in silica capilliaries with diameters in the range 2-20μm have been grown using a modified Bridgman technique/14/. X-ray analysis has shown that the a-axis of the crystal lies along the fibre axis with the net d_{eff} favourably aligned almost perpendicular to the fibre core. Crystal quality has improved significantly in the past year and optical performance has advanced even more dramatically as indicated in table 3 which also shows early results for a wide range of other organics.

Table 3

Measured nonlinear properties of crystal cored fibres

Material	Average Power mW	Throughput %	SHG %
Acetamide	10	10	$<10^{-6}$
Benzil	10	20	$<10^{-6}$
ABP	10	10	$<10^{-6}$
MNA	10	0.5	$5x10^{-5}$
CMONS	10	1.0	$5x10^{-5}$
NPAN	10	0.2	$7x10^{-5}$
NPP	10	0.3	$7x10^{-5}$
DAN (initial)	10	0.2	$8x10^{-3}$
DAN (Spring 87)	10	1.5	$1x10^{-2}$
DAN (Summer 88)	1	15	1.5

All measurements were made using a Q switched Nd/YAG laser at λ=1.06μm with 200ns pulses at 0.5kHz. The definitions used are:-
throughput = $P_w$out(core)/$P_w$in(total) and SHG = P_{2w}out(total)/$P_w$in(total), where P_w is the power at frequency w. The latest results (at 1 mW), on 6μm diameter core fibres, are two orders of magnitude better than the previous best reported values/14/.

Recent progress has therefore been most encouraging. Additionally the crystal core geometry lends itself to convenient incorporation in all-fibre systems and the silica or higher refractive index glass cladding provides the basis of a robust device. None-the-less significant further progress will be required before practical systems, particularly those using convenient semiconductor lasers, could be contemplated. Current studies are accordingly concentrating on identifying the remaining materials problems and their removal.

Initial growth in capilliaries with new organic materials often results in crystal cores with readily observable defects such as cracks, striations and both transverse and longitudinal voids. Careful optimisation of growth conditions can eliminate such problems however and some of the best DAN crystals show no obvious defects under the microscope. They also show good extinction contrast when viewed through crossed polarisers in the microscope, indicating an absence of variable strain within the crystals. Equally clearly, from the results in table 3, most of the incident light is being either scattered or absorbed. We have recently established a system for measuring the sideways scatter from crystal cored fibres in order to investigate more closely the variation in scatter with state and composition of the core/16/. In poorer quality DAN crystals, scattering of both the fundamental and the second harmonic shows a series of peaks with distance along the fibre corresponding to the random occurrence of large scattering centres. The behaviour in better DAN crystals is more complex as illustrated in figure 4. There is still some evidence of occasional major scatter centres but there is also a steady increase of scattered second harmonic with distance along the fibre. This suggests a persistent more homogeneous source of scattering which none-the-less allows finite build-up of second harmonic intensity along the fibre. A core of supercooled liquid DAN can also be produced inside a narrow capilliary and such a fibre is a particularly useful diagnostic variation. The almost negligible sideways scatter for this sample, also shown in fig 4, confirms that the scatter mechanisms are not intrinsic to the system - from roughness of the inner walls of the capilliary for instance. Also the measured throughput was as high as 40% indicating that absorption is not a major problem. Measurements on benzil, the crystalline material in table 3 with the highest throughput, then showed a similar low level of sideways scatter - a particularly encouraging result, indicating that these problems are not intrinsic to the solid organic/silica system. It should be stressed that these sideways scatter studies are under development and are only semiquantitative at this stage but already they are providing valuable insights into the problems that are yet to be solved.

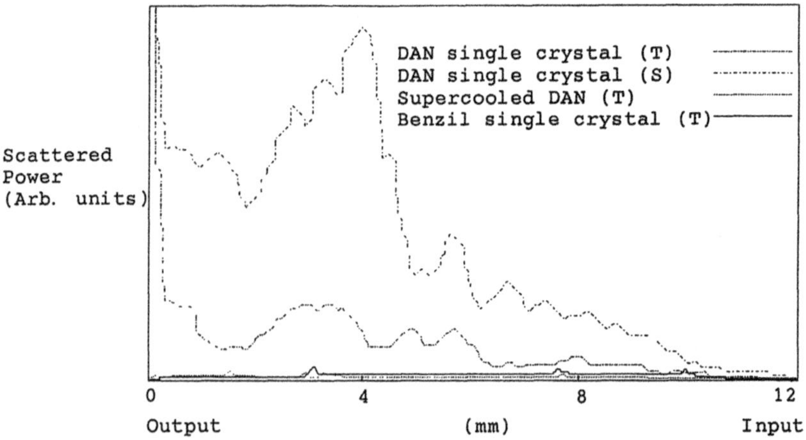

Figure 4. Sideways scatter from crystal cored fibres. Throughput (T) at 1064 nm, SHG at 532 nm.

Conclusions

Recent progress with single component acrylic ester polymers indicates that a range of devices with performance comparable with lithium niobate is now possible using these or similar materials as the active layer. Emergence of these initial devices will require the successful application of conventional aspects of device processing to these new materials. The performance of crystal cored fibres based on DAN now show SHG efficiency$\geq$1%, an improvement by two orders of magnitude in the past year. Detailed assessment of sideways scatter from the fibres is providing new evidence on the nature of current limits to performance.

Acknowledgements

Acknowledgement is made to the Director of Research and Technology of British Telecom for permission to publish this paper.

References

1. D. J. Williams ed: Nonlinear Optical Properties of Organic and Polymeric Materials, ACS Symp. Ser. 233 (American Chemical Society, Washington 1983).

2. J. L. Oudar, J. Zyss: Phys. Rev. A 26, 2016 (1982).

3. V. L. Docherty, D. Pugh, J. O. Morley: J. Chem. Soc., Faraday Trans. 2 81, 1179 (1985).

4. P. Pantelis, J. R. Hill, G. J. Davies: In Nonlinear Optical and Electroactive Polymers, ed. by P N Prasad, D. R. Ulrich (Plenum Publishing Corp, New York 1987) pp.229-241.

5. J. R. Hill, P. L. Dunn, G. J. Davies, S. N. Oliver, P. Pantelis, J. D. Rush: Electron. Lett. 23, 700 (1987).

6. J. R. Hill, P. Pantelis, G. J. Davies: Ferroelectrics 76, 435 (1987).

7. J. R. Hill, P. Pantelis, P. L. Dunn, S. N. Oliver, G. J. Davies: Proceedings of 1st International Conference on Electrical, Optical and Acoustic Properties of Polymers, Canterbury UK, in press (Plastics and Rubber Institute, London 1988).

8. P. Le Barny: Thin Solid Films 152, 99 (1987).

9. J. R. Hill, P. Pantelis, F. Abbasi, P. Hodge: J. Appl. Phys. 64, in press (1988).

10. J. N. Polky, J. H. Harris: Appl. Phys. Lett. 21, 307 (1972).

11. P. Pantelis, J. R. Hill, S. N. Oliver, G. J. Davies: Br. Telecom. Technol. J. 6, No 3, 5 (1988).

12. B. K. Nayar: In Nonlinear Optical Properties of Organic and Polymeric Materials, ed. by D. J. Williams, ACS Symp. Ser. 233 (American Chemical Society, Washington 1983) pp153-166.

13. P. A. Norman, D. Bloor, J. S. Obhi, S. A. Karaulov, M. B. Hursthouse, P. V. Kolinsky, R. J. Jones, S. R. Hall: J. Opt. Soc. Am. B, 4, 1013, (1987).

14. G. E. Holdcroft, B. K. Nayar, D. O'Brien, J. D. Rush, M. A. G. Halliwell, R. Kashyap, K. I. White: Proceedings Conference Lasers and Electro-Optics, Paper THE4, (Optical Society of America, Washington DC, 1987).

15. P. Kerkoc, Ch. Bossard, H. Arend, P. Gunter: Proceedings Conference Lasers and Electro-Optics, Paper WM39, (Optical Society of America, Anaheim, 1988)

16. G. E. Holdcroft, P. L. Dunn, J. D. Rush, M. A. G. Halliwell: Proceedings Conference Laser and Electro-Optics, Paper WM55, (Optical Society of America, Anaheim, 1988)

Frontier Materials for Organic Nonlinear Optics

T. Wada, S. Yamada, Y. Matsuoka, C.H. Grossman, K. Shigehara, H. Sasabe, A. Yamada, and A.F. Garito

Frontier Research Program, RIKEN (The Institute of Physical and Chemical Research), Hirosawa 2-1, Wako, Saitama 351-01, Japan

1. Introduction

Organic molecules which have conjugated π-electron systems exhibit unique optoelectronic responses such as photoconductive, photovoltaic and photocatalytic behavior. These photoactive materials have attracted not only pure but also industrial research. For example, the following practical applications have been made: an organic photoconductor as a key device in a photocopy machine, device fabrication such as photo-, EB, or X-ray resists and the encapsulation of devices (electro-magnetic insulation) and so on. During a photochemical process via excited states, unfavorable side reactions of organic molecules such as photoinduced isomerization, optical breaching and/or degradation may be sometimes induced and then the chemical structures of initial photoresponsive materials may change irreversibly. Unless this irreversible change can be prevented, organic photoactive materials must be regarded as unreliable for practical applications. On the other hand, the optical nonlinear effects, especially in nonresonant processes, appear in the optical transparent region. Under irradiation by a high power laser, nonlinear polarization can be induced in the optical nonlinear media. The second- and third-order optical nonlinear effects are characterized by corresponding macroscopic second- and third-order susceptibilities, e.g., $\chi^{(2)}$ and $\chi^{(3)}$, respectively. Dielectric crystals such as potassium titanyl phosphate (KTP) and $LiNbO_3$ are typical optical nonlinear media and are practically used for a frequency doubler and substances of an optical waveguide. Recently, it has been established experimentally and theoretically that organic intramolecular charge transfer compounds have anomalously large optical nonlinearity [1] and show ultrafast response [2].

A lot of effort has been put into finding new organic materials which show large second-order optical nonlinear responses. From the chemistry viewpoint, systematic ideas of molecular design have succeeded in the enhancement of microscopic second-order optical susceptibility to some degree. On the other hand, from the physics viewpoint, molecular orientation in the crystalline state significantly influences the macroscopic second-order susceptibility. Several approaches have been taken to control the molecular orientation, such as introducing steric hindrance [3], hydrogen bonding [4], a chiral center [5] and a combination [6], and reducing the electrostatic dipole-dipole interaction [7].

Polydiacetylenes (PDA) obtained by solid state polymerization were firstly reported to have a large $\chi^{(3)}$ comparable to that of germanium, especially parallel to the main chain [8]. Optical nonlinear response based on third-order optical nonlinearity has a

Springer Proceedings in Physics, Vol. 36
Nonlinear Optics of Organics and Semiconductors
Editor: T. Kobayashi

lot of interest because of its potential possibility for fast optical switches, transient nonlinear memories, optical transistors and so on. For these applications, it is important to establish thin film fabrication. Several approaches such as epitaxial growth [9], liquid crystal polymerization [10], the solution-shear technique [11] and the Langmuir-Blodgett (LB) method [12] have been used to get a thin film of polydiacetylene.

In the field of optical telecommunication and storage applications, several polymeric systems have already been developed, that is, plastic optical fibers, optical discs and aspherical lenses. However, these are not active but passive media themselves. Optical active media which show optical nonlinear behavior will play an important role in the next generation of computing systems and image processing. Development of promising optical nonlinear active materials is a key issue in the field of optoelectronics applications.

In this paper we will describe novel materials which have large optical nonlinearity and good processability.

2. Organic Single Crystal for Second Harmonic Generation

After the demonstration of the powder test by Kurtz et al. [13], a large number of organic molecules have been found to possess second-order optical nonlinearity. However, only a few of them could be crystallized in high quality and investigated for phase matched second harmonic generation (PMSHG) properties. Among these materials are the well-known crystals of 2-methyl-4-nitroaniline (MNA) [3], urea [4], methyl-(2,4-dinitrophenyl)-aminopropanate (MAP) [5], N-(4-nitorophenyl)-(L)-prolinol (NPP) [6], 3-methyl-4-nitropyridine-1-oxide (POM) [7] and so on. With regard to the applications as a frequency doubler of a diode laser, crystals should meet several requirements, such as a large phase-matchable optical nonlinear coefficient, a wide optical window around the visible region or shorter cut-off wavelength (less than 400 nm), mechanical and chemical stability, and a high damage threshold. Some of these requirements based on molecular properties can be satisfied with the assistance of molecular design. For example, absorption edges of intramolecular charge transfer compounds can be controlled by selecting the combination of electron donative and acceptive substituents. To enhance the microscopic optical susceptibility, an extended π-electron system (e.g., styrene or stilbene skeleton) is better than a benzene ring. On the other hand, crystal properties such as bulk phase matching are difficult to control. On searching for new organic structures that possess nonlinear optical properties, cyanoacetate compounds were found to show PMSHG and have a large SHG coefficient and a short cut-off wavelength comparable with that of POM [14]. The crystal structure of piperoniridene cyanoacetate ethylester was determined to have a triclinic unit cell and space group P_1. Strong interaction between polar functional groups of adjacent molecules in this crystal causes almost antiparallel orientation in the crystal unit cell. From the viewpoint of crystal engineering ortho-substituted structures were selected to retain a non-vanishing molecular level second-order susceptibility. Following these guiding principles, we found ortho-methoxy benzalmalono-nitrile, here called DIVA, which is an abbreviation for <u>di</u>cyano-<u>v</u>inyl <u>a</u>nisole, to possess large second-order nonlinearities and good crystal qualities.

3. Crystallographic Structure of DIVA

New organic optical nonlinear compounds which possess a methoxy group as an electron donative and a dicyanovinyl group as an electron acceptive substituent were investigated and the crystallographic structure of ortho- and para-methoxy-benzalmalononitrile was investigated by an X-ray diffraction method. Although both of them were determined to have a monoclinic unit cell, the para-substituted compound is centrosymmetric with space group $P2_1/a$. As shown in Fig. 1, a "network" interaction in the crystalline state between polar functional groups of adjacent molecules is observed and causes antiparallel molecular orientation in the crystal unit cell. Along the c-axis this planar molecule is close packed with a lattice constant of 3.922 Å. On the other hand, the ortho-substituted compound is a non-centrosymmetric structure with space group $P2_1$. Polar substituents were introduced to one side of a benzene ring in order to generate a non-centrosymmetric structure. This highly anisotropic structure resulted in non-centrosymmetric stacking in the crystalline state as shown in Fig. 2. The molecular dipole moment roughly points from the dicyanovinyl to the methoxy group. Non-vanishing components of molecular polarizability exist in the b-axis direction.

This crystal has a unique advantage; both type I and type II PMSHG were observed on naturally grown faces [15]. It is difficult to cut and polish the organic crystals which exhibit poor mechanical strength compared with inorganic ones. In this particular case, this procedure is not needed in order to achieve bulk PMSHG. Solvent evaporation from a saturated solution and vapor sublimation growth techniques are available to grow single crystals of DIVA. From Maker fringe measurement, the SHG

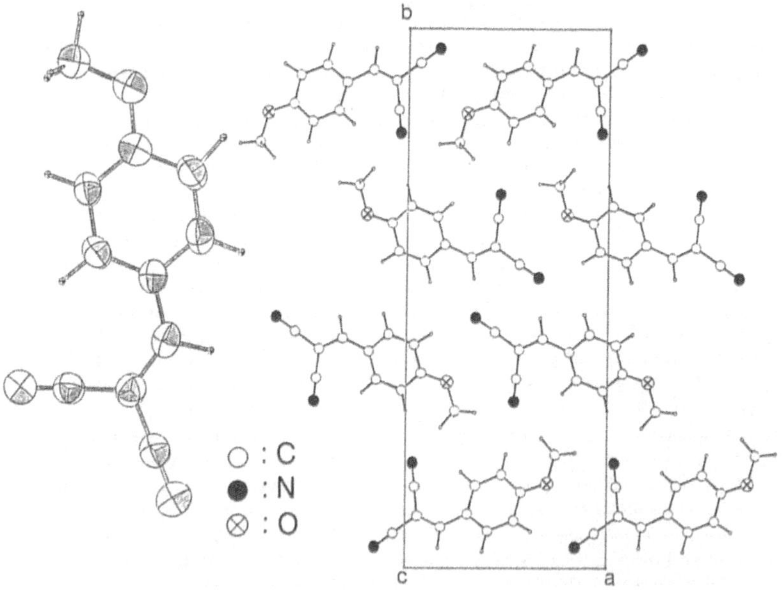

Fig. 1 The molecular structure of para-methoxy-benzalmalononitrile determined by X-ray crystallography and the projection along the c-axis showing a monoclinic unit cell with the centrosymmetric $P2_1/a$ space group.

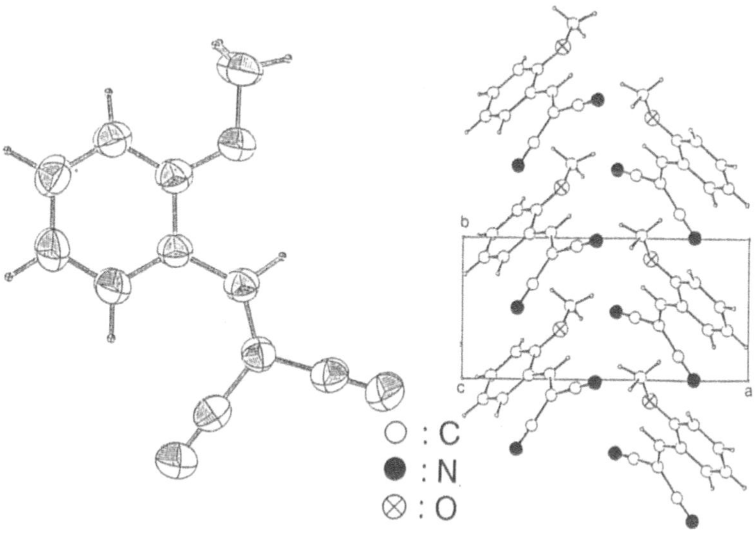

Fig. 2 The molecular structure of ortho-methoxy-benzalmalononitrile (DIVA) and the projection along the c-axis showing a net dipole alignment along the crystallographyic b-axis.

coefficient d_{22} was obtained as 2.6×10^{-8} esu and the effective one as 4.9×10^{-8} esu at a wavelength of 1.064 µm [15]. These values are comparable to those of POM and MAP. Details of PMSHG properties and Maker fringes were presented in another contribution to this volume.

4. Macrocyclic Compounds for Third-Order Optical Nonlinear Response

Reducing the dimensionality was suggested to enhance the third-order susceptibility. For example polydiacetylene is one of the good candidates for one-dimensional conjugated π-electron systems [16]. Based on this idea, other conjugated polymers such as polyacetylene [17], ladder polymer [18], poly-p-phenylene vinylene synthesized via a precursor polymer route [19] and polythiophene made by electrochemical polymerization [20] were established as third-order optical nonlinear active materials. However, these polymers are not stable in air or do not have processability, that is, they are insoluble in conventional organic solvents or it is difficult to obtain a thickness-controlled thin film.

Current theoretical and experimental studies show that it is not necessary to extend the conjugation length infinitely in order to enhance the third-order optical susceptibility [21]. Therefore the macrocyclic compound phthalocyanines (Pc) which have 2-dimensional π-electron systems have been studied in our laboratory. These macrocyclic compounds show exceptional thermal and chemical stabilities and are easily crystallized and sublimed. Thin films of Pc can be obtained by physical vapor deposition (PVD) and the film thickness is easily controlled by the deposition rate and time in the range of 10 nm to several micrometers.

A huge number of different metallophthalocyanines (PcMs) and their derivatives have been synthesized and their optoelectronic properties widely studied. Unsubstituted phthalocyanines are

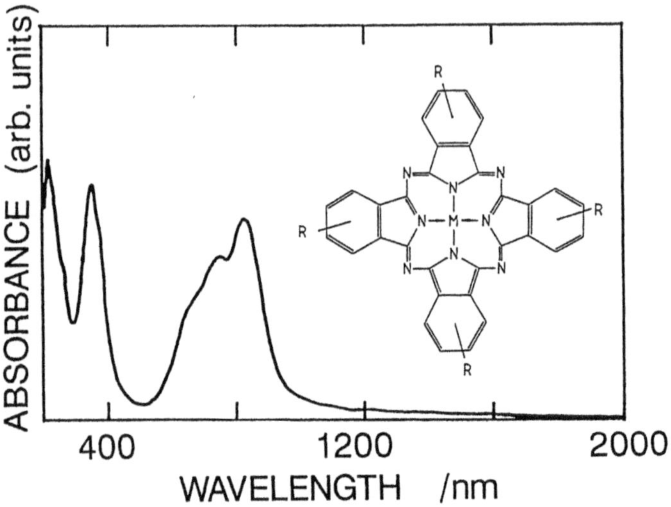

Fig. 3 Structure of metallophthalocyanines (PcMs) and typical electronic spectrum for a PcVO thin film fabricated by PVD.

usually insoluble in common organic solvents. However, chemical modification is available to control their physical properties. For example, alkoxyphthalocyanine derivatives ($R{=}OC_nH_{2n+1}$ in Fig. 3) are soluble in chloroform and hence thin films of molecularly doped polymer can be easily made by conventional polymer processing.

In the electronic spectra of phthalocyanines, two characteristic absorption bands are well established, e.g., Soret band (300-400 nm) and Q band (600-800nm). The latter is sensitive to the environment of the Pc rings. This inhomogeneous broadening has been well studied in terms of photochemical hole burning. Figure 3 shows a typical absorption spectrum for a vanadyl phthalocyanine (PcVO) thin film prepared by PVD. In the condensed state, the broadening of the Q band was observed, which indicated strong coupling to lattice vibrations. Moreover, depending upon the central metal, the features of the Q band in the condensed state changed remarkably [22].

5. Third Harmonic Generation on Phthalocyanine Thin Films

Transmitted third harmonic generation (THG) from the thin film deposited on a quartz substrate was observed at a wavelength of 1.907 µm obtained by Raman shifting through a hydrogen gas cell. Figure 4 shows the typical Maker fringes of TH intensity in a quartz substrate (ca. 1mm thick) (a) and a PcVO thin film (ca. 51.4 nm thick) (b). In the case of a quartz substrate, the coherence length obtained from the interference pattern is in good agreement with the value in the literature [23]. However, in the case of PcVO, a monotonic decrease in the intensity of TH was observed by rotation of the sample. Therefore, the change of the optical path length is much shorter than the coherence length. Under this condition, equation (1) can be used for the calculation of $\chi^{(3)}$ following the discussion by Kajzar et al. [24],[25].

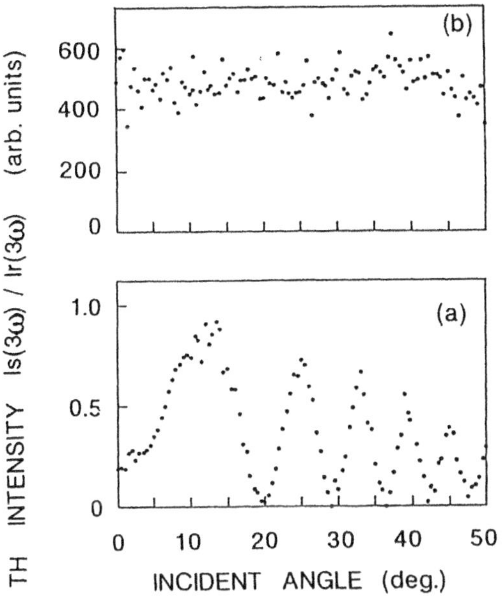

Fig. 4 Maker fringe THG for a quartz substrate (1mm thick) (a) and a PcVO thin film deposited on a quartz substrate (51.4 nm thick) (b) as a function of incident angle at a wavelength of 1.907 µm.

$$J_{3\omega} = \frac{2304 \pi^6 J_\omega^3}{C^2} \left| \frac{A \chi^{(3)}}{N_\omega + N_{3\omega}} \right|^2 \left(\frac{1}{\lambda_\omega} \right)^2 \quad (1)$$

where J_ω is the fundamental beam intensity, N_ω and $N_{3\omega}$ are refractive indices at fundamental and harmonic frequencies, and A is a factor arising from transmission and boundary conditions. We applied the values of refractive indices from the literature [26]. By analyzing Maker fringes, we obtained the data of $\chi^{(3)}$ in the order M= VO, Sn >> H_2, Ni, Co. The value of $\chi^{(3)}$ for a PcVO thin film was estimated to be 1.85×10^{-10} esu, which is two orders of magnitude larger than that of a metal-free Pc (PcH_2) film [22].

In the case of PcH_2, reabsorption of TH light cannot be neglected because the peak position of the Q band is matched to the wavelength of TH at the condition of 1.907 µm fundamental. The effect of the central metal on third-order susceptibilities is not clear yet. In general, central metals hardly affect the electronic state of Pc but the molecular stacking of Pc rings.

Recently, Ho et al. observed THG from the reflection of chloro-gallium (PcGa(Cl)) and fluoro-aluminum phthalocyanine (PcAl(F)) thin films at a wavelength of 1.064 µm. They discussed the scattering loss of the polycrystalline surface compared with the (100) face of a polished silicon crystal for the reference and estimated the value of $\chi^{(3)}$ to be 2.5×10^{-11} and 5×10^{-11} esu for PcAl(F) and PcGa(F), respectively. [27]. Under the condition of 1.064 µm fundamental, the TH light was strongly absorbed at the Soret band (molecular absorption coefficient is about 10^4), while under the condition of 1.907 µm fundamental, transmitted TH light could be observed; however, the absorption of TH light cannot be neglected. Detailed studies of the dispersion of $\chi^{(3)}$ in resonant and nonresonant conditions and time-dependence experiments are required. A large value of $\chi^{(3)}$ may be possible through resonant enhancement.

As shown in Fig. 3, there is an optical window around 500 nm and DC electric field induced second harmonic generation (DC-SHG) can be observed at a wavelength of 1.064 μm. The second harmonic intensity (at 532 nm) from tetrakis(n-pentoxy) phthalocyanine ($Pc(R_4)H_2$: $M=H_2$, $R= OC_5H_{11}$ in Fig. 3) dispersed in PMMA was proportional to the square of the applied electric field. Concentration dependence of THG was investigated at a wavelength of 1.907 μm. The TH intensity was in proportion to the square of the Pc concentration at the low concentration range.

6. Implementation of Optical Waveguide Structures

Two types of optical waveguides were examined, that is, a molecularly doped polymer as the guiding layer and a slab waveguide containing a Pc thin film as a top layer with refractive index higher than the surrounding media. In the case of the first type, molecularly doped polymer films are tough enough to apply prism coupling techniques for coupling the external radiation field into a polymer film. At a proper angle of incidence, light can propagate in a guiding polymer layer at a wavelength of 1.064 μm. In the case of molecularly doped polymers, the value of $\chi^{(3)}$ can be controlled by the concentration of Pc. This value increased with increasing Pc concentration and was 10^{-13} esu at the weight ratio of $Pc(R_4)H_2$ to PMMA of 20:80 under the condition of 1.907 μm fundamental. However, aggregation of Pc was observed at a high Pc concentration range. The propagation losses are believed high in such a phase separated media. There is a limit to incorporation of optical nonlinear active molecules into a polymer matrix without phase separation.

For the second type, we used a slab waveguide [28] consisting of quartz substrate as cladding, Corning-7059 glass thin layer as a guiding layer, and a PcVO thin layer as a top layer which has a gradient of thickness along the y-axis as shown in Fig. 5. Laser light (1.064 μm) was coupled and decoupled to a glass layer by a prism coupling technique and the output intensities were normalized by the intensity of reflection from the surface of the coupling prism. Figure 5 shows normalized output intensities of the TE_0 mode at a wavelength of 1.064 μm as a function of PcVO thickness. The output intensities were almost constant when the thickness of top layer was zero. With an increase in thickness of the top layer, the output intensity decreased remarkably and then increased gradually at more than the critical thickness (about 100 nm thick in Fig. 5 (b)) [22]. The simulation of light propagation shows that the output intensity changes periodically as a function of the top layer thickness and goes to zero at a critical thickness. This phenomenon can be interpreted as follows; when the field distribution of the propagation mode in region I is similar to that of the guiding layer in region II, the output intensity becomes maximum, otherwise light cannot propagate well from region I to region II and/or from region II to region III along the z-axis in Fig. 5. The details of this simulation will be published in elsewhere.

When the incident intensity (Ii) was increased by one order of magnitude, the minimum intensity point shifted slightly toward a thinner top layer. At this minimum point (about 80 nm thick in Fig. 5 (c)), the ratio of output to input power decreased on increasing the input power around a high power range. The glass

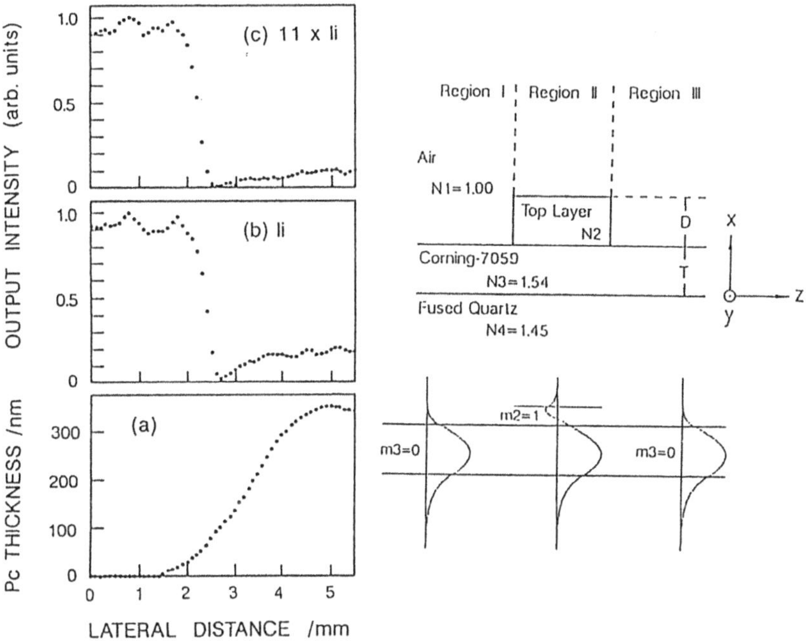

Fig. 5 Output intensities of propagating light at a wavelength of 1.064 µm as a function of PcVO thickness and the geometry of a slab waveguide. The PcVO top layer has a gradient of thickness along the y-axis (a). The field distribution of the propagation mode was calculated with the parameters used in this figure and PcVO thickness of 400 nm.

optical waveguide itself, i.e., without a PcVO top layer, did not exhibit this nonlinear behavior. The propagation constant changed under high power conditions. In this case the nonlinear refractive index can be made to appear in a PcVO film which has a large $\chi^{(3)}$. We also observed a propagation loss due not to absorption but to imperfect energy transport between different regions with a step-index change, and imperfect structure causing a rough surface and boundary. Further investigation of the transient response is needed to clarify this behavior.

Compared with other conjugated polymers such as polyacetylene and polydiacetylene, this system has a lot of advantages, that is, processability, thermal stability and easy fabrication as a film by spin coating and/or casting, and conventional vacuum deposition techniques.

Recently Si semiconductor processing techniques have been applied to functionalization of organic materials. The epitaxial growth of PcCu or tetracyanoquinodimethane (TCNQ) on the cleavage surface of mica or KCl single crystals has been tried under ultra-high vacuum. By the vacuum deposition technique at cryogenic temperatures, Seki and Itoh [29] succeeded in the control of molecular orientation of anthracene perpendicular or parallel to the substrate depending upon the surface state and annealing conditions. This molecular beam epitaxy technique can be extended to phthalocyanine systems, which have thin film processability under high vacuum. It is necessary to couple thin film fabrication to microlithography techniques for applications as optical waveguides. These novel optical nonlinear materials have compatibility to the required processing in terms of device fabrication.

7. References

1. A. F. Garito, K. D. Singer, C. C. Teng: In Nonlinear Optical Properties of Organic and Polymeric Materials, ed. by D. J. Williams, ACS Symp. Ser. #233, (ACS, Washington DC, 1983) p.1
2. G. M. Carter, J. V. Hryniewicz, M. K. Thakur, Y. J. Chen, S. E. Meyler: Appl. Phys. Lett. 49, 998 (1986)
3. B. F. Levine, C. G. Bethea, C. D. Thurmond, R. T. Lynch, J. L. Bernstein: J. Appl. Phys. 50, 2523.
4. J. Zyss, G. Berthier: J. Chem. Phys. 77, 3635 (1982)
5. J. L. Oudar, R. Hierle: J. Appl. Phys. 48, 2699 (1977)
6. J. Zyss, J. F. Nicoud, M. Coquillay: J. Chem. Phys. 81, 4160 (1984)
7. J. Zyss, D. S. Chemla, J. F. Nicoud: J. Chem. Phys. 74, 4800 (1981)
8. C. Sauteret, J. -P. Hermann, R. Frey, F. Pradere, J. Ducuing, R. H. Baughman, R. R. Chance: Phy. Rev. Lett. 36, 956 (1976)
9. S. E. Rickert, J. B. Lando, S. Ching: In Nonlinear Optical Properties of Organic and Polymeric Materials, ed. by D. J. Williams, ACS Symp.Ser. #233, (ACS, Washington DC, 1983), p. 229
10. A. F. Garito, C. C. Teng, K. Y. Wong, O. Zammani'Khamiri: Mol. Cryst. Liq. Cryst. 106, 219 (1984)
11. M. Thakur, S. Meyler: Macromolecules 18, 2341 (1985).
12. F. Kajzar, J. Messier, J. Zyss, I. Ledoux: Opt. Comm., 45, 133 (1983)
13. S. K. Kurtz, T. T. Perry: J. Appl. Phys., 39, 3798 (1968)
14. Y. Mori, M. Okamoto, T. Wada, H. Sasabe: In Mat. Res. Soc. Symp. Proc., 109 345 (1988)
15. C. H. Grossman, T. Wada, S. Yamada, A. Yamada, A. F. Garito, H. Sasabe: In IQEC'88 Technical Digest, (1988) p.134
16. G. P. Agrawal, C. Cojan, C. Flytzanis: Phys. Rev. B17, 776 (1978)
17. A. J. Heeger, D. Moses, M. Sinclair: Synth. Metals, 17, 343 (1987)
18. D. N. Rao, J. Swiatiewicz, P. Chopra, S. K. Ghoshal, P. N. Prasad: Appl. Phys. Lett., 48, 1187 (1986)
19. T. Kaino, K.-I. Kubodera, S. Tomaru, T. Kurihara, S. Saito, T. Tsutsui, S. Tokito: Electron. Lett., 23, 1095 (1987)
20. T. Sugiyama, T. Wada, H. Sasabe: Synthetic Metals in press
21. A. F. Garito, J. R. Heflin, K. Y. Wong, O. Zammani-Khamiri: In Organic Materials for Nonliner Optics ed. by D. J. Ando, D. Bloor in press
22. T. Wada, Y. Matsuoka, K. Shigehara, A. Yamada, A. F. Garito, H. Sasabe: In Mat. Res. Soc. Symp. Proc. in press
23. G. M. Meredith, B. Buchalter, C. Hanzlik: J. Chem. Phys. 78 1533 (1978)
24. F. Kajzar, J. Messier: Phys. Rev. A 32, 2352 (1985)
25. F. Kajzar, J. Messier, C. Rosilio: J. Appl. Phys. 60, 3040 (1986)
26. B. H. Schechtman, W. E. Spicer: J. Mol. Spectros. 33, 28 (1970)
27. Z. Z. Ho, C. Y. Ju, W. M. Hetherington III: J. Appl. Phys. 62, 716 (1987)
28. K. Sasaki, Y. Kudo, H. Watanabe, O. Hamano: Thin Solid Films 89, 297 (1982)
29. H. Seki, U. Itoh: J. Chem. Phys., 72, 2166 (1980)

Time-of-Flight Measurement of Excitonic Polaritons in a GaAs Quantum-Well Waveguide

K. Ogawa, T. Katsuyama, and H. Nakamura

Central Research Laboratory, Hitachi Ltd., Kokubunji, Tokyo 185, Japan

1. Introduction

Excitonic polaritons are formed when electromagnetic waves incident into semiconductor crystals are coupled to excitons at low temperatures [1]. Experimental evidence for excitonic polaritons has been presented in bulk CuCl and GaAs crystals [2, 3]. However, no experimental evidence for excitonic polaritons in quantum wells (QWs) has been obtained yet, although a number of studies have investigated optical properties of excitons in QWs. This is probably because the optical properties have been studied with incident light being transmitting perpendicular to the QWs. Translational motion of excitons in QWs is possible in the direction parallel to QW layers, and quantum-well excitonic polaritons (QWEPs) will propagate only in this direction.

If QWEPs are formed, a large delay of the light transmitted through QW layers is expected, which reflects the decrease in group velocity of the light. The present paper reports a time-of-flight method with light propagation parallel to a GaAs QW for observation of QWEPs. A waveguide-type structure of a QW sample was employed to facilitate time-of-flight measurements with the parallel propagation.

2. Sample Preparation

A waveguide sample containing a QW was grown by molecular-beam epitaxy on a semi-insulating GaAs substrate. Its structure is illustrated in Fig. 1. It formed a leaky waveguide, in which the refractive index of the core was smaller than that of the cladding [4]. A 50-Å GaAs QW was sandwiched between 1.8-μm $Al_{0.3}Ga_{0.7}As$ barrier layers. These layers constituted the core of the waveguide. The cladding was GaAs layers. The waveguide was designed so that only the fundamental mode could transmit [4]. Superlattice layers

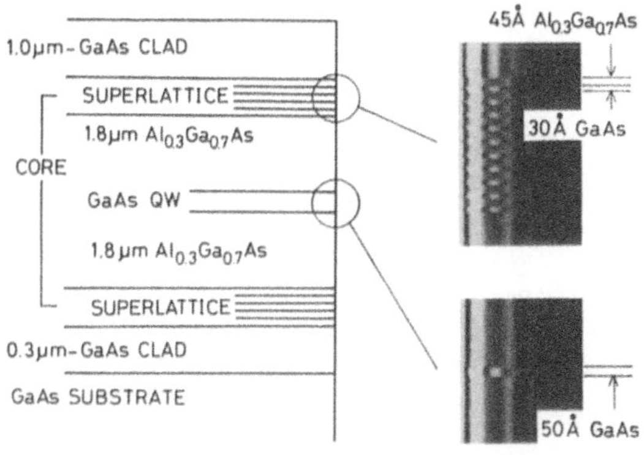

Fig. 1. Cross section of GaAs quantum-well waveguide sample and transmission micrographs of the sample. Micrographs were taken by composition analysis by thickness-fringe method.

Springer Proceedings in Physics, Vol. 36
Nonlinear Optics of Organics and Semiconductors
Editor: T. Kobayashi © Springer-Verlag Berlin, Heidelberg 1989

were introduced between the AlGaAs core and the GaAs cladding to trap crystal defects.

An excellent quantum-well waveguide was fabricated as shown by the transmission micrographs in Fig. 1. These micrographs were taken using the composition analysis by thickness-fringe method [5]. The sample was mounted on a Si substrate.

3. Experimental Apparatus

A picosecond time-of-flight measurement system was built. The light source was a Pyridine-1 dye laser (Spectra-Physics 375B) having 10-ps pulse width. The laser beam was split into the probe and reference beams. The probe beam was guided to the QW waveguide sample which was cooled to 6.0 K in an optical cryostat (Air Products LT-3-110). The propagation delay time of the probe pulse transmitted through the waveguide sample was provided by the time difference between the probe and reference pulses using a streak camera (Hamamatsu C1587). The delay time was measured as a function of the photon energy of the laser light.

4. Experimental Data and Discussion

The optical-absorption spectrum of the QW waveguide has been taken at 6.0 K, as shown in Fig. 2. The absorption lines at the lower photon energies, 1.602 eV, 1.625 eV and 1.642 eV are considered to be associated with two-dimensional excitons in the GaAs QW. These lines are also observed in a GaAs QW waveguide sample containing no superlattice layers. The rest of the absorption lines, at photon energies higher than 1.65 eV, are caused by excitons in the superlattice layers.

From the polarization-sensitive measurements, the line at 1.602 eV is identified as a heavy-hole (hh) exciton line, and that at 1.642 eV as a light-hole (lh) exciton line. The broad line at 1.625 eV between the above two lines seems to be composed of hh- and lh-exciton lines.

In Fig. 3, the delay time of the probe beam transmitted out of the waveguide is plotted against the photon energy of the laser light. The delay time of 0 ps corresponds to ordinary light propagation without interaction with QW excitons. Delay time increases drastically at almost the same photon energies as the excitonic absorption lines in Fig. 2. This shows that the group

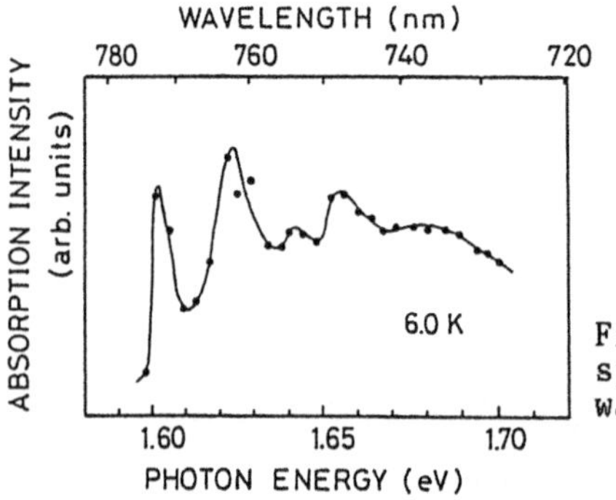

Fig. 2 Optical-absorption spectrum of the quantum-well waveguide measured at 6.0 K.

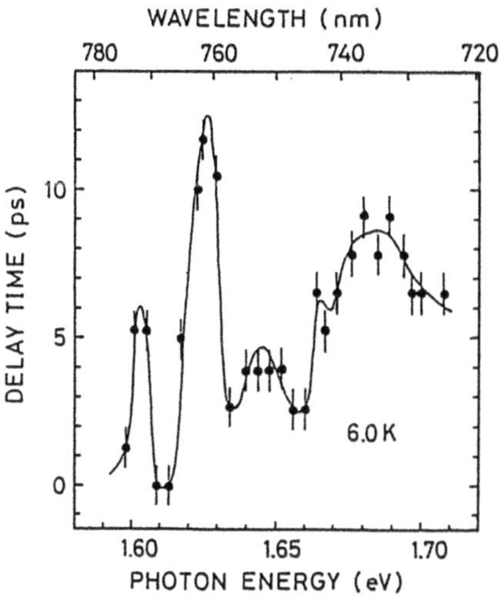

Fig. 3 Propagation delay time of the probe pulse transmitted through the waveguide as a function of photon energy or wavelength.

velocity of the laser pulse decreases around the photon energy resonant to the exciton line. The delay time is about 12 ps at the photon energy of the hh-exciton line of 1.625 eV. The filling factor of the QW in the waveguide is only 0.1 %. Simply multiplying the actual length of the waveguide, 650 μm, by the filling factor, we obtain the effective length of the waveguide, which contributes to the decrease in group velocity, as 0.9 μm. Thus the group velocity is 7 x 10^4 m/s at 6.0 K. This is less than 1/1000 of the velocity of light. Such a drastic decrease in group velocity indicates that QWEPs are formed in the QW.

Time-of-flight measurements have also been performed for light propagation perpendicular to QW layers. No noticeable change in light group velocity has been observed. The excitonic-polariton effect is quite small in this direction.

In conclusion, a drastic decrease in the group velocity of a light pulse transmitted through a GaAs quantum-well waveguide has been observed. This result yields the first experimental evidence for quantum-well excitonic polaritons.

References

1. J. J. Hopfield, D. G. Thomas: Phys. Rev. 132, 563 (1963)
2. D. Frolich, E. Mohler, P. Weisner: Phys. Rev. Lett. 26, 554 (1971)
3. R. G. Ulbrich, G. W. Fehrenbach: Phys. Rev. Lett. 38, 865 (1979)
4. J. S. Weiner, D. S. Chemla, D. A. B. Miller, H. A. Haus, A. C. Gossard, W. Wiemann, C. A. Burrus: Appl. Phys. Lett. 47, 664 (1985)
5. H. Kakibayashi and F. Nagata: Surf. Sci. 174, 84 (1986)

Second Harmonic Generation from a Monolayer on a Water Surface and Accumulated Structure on a Glass Substrate (LB Film)

K. Shirota, K. Kajikawa, Y. Sano, H. Takezoe, and A. Fukuda

Tokyo Institute of Technology, Department of Organic and Polymeric Materials, O-okayama, Meguro-ku, Tokyo 152, Japan

1. Introduction

Nonlinear optical properties in organic materials have attracted great attention because of the large nonlinearity and the fast response time. Among them, even-order nonlinear optical effects such as second harmonic generation (SHG) do not vanish only in non-centrosymmetric media. Therefore, it is very important to fabricate desired structures using molecules with a high molecular polarizability, $\alpha^{(2)}$. For this purpose, the Langmuir-Blodgett (LB) method serves as a powerful technique.

In this paper, we will present preliminary measurements of SHG from a molecular monolayer on water surface, which is sometimes called Langmuir film (L film). Importance of this study is twofold: This technique provides a reliable method for determining $\alpha^{(2)}$ /1/ and a knowledge of molecular orientation distribution /2/. Therefore, the study of L films is in turn quite useful to obtain LB films of high quality suitable for nonlinear optical devices.

2. Experiments

The sample used was a mixture of hemicyanine dye/3/,

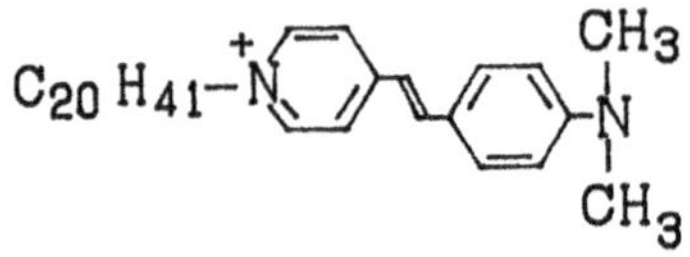

and arachidic acid, $C_{19}H_{39}COOH$, of a volume ratio of 1:2. It was dissolved in a 1 mM solution of chloroform and was spread over a water surface using a micropipet. The subphase was water containing 0.5 mM $CdCl_2$. The measurements of SHG from the L films were carried out with a Langmuir trough made of tefron after the solvent was evaporated. The surface pressure was monitored by a filter paper (Wilhelmy plate). The LB films were prepared by using another Kuhn-type Langmuir trough. After compressing to a surface pressure of 30 dyne/cm, the monolayer was deposited onto a glass slide, at a dipping speed of 7 mm/min.

The SHG measurements were performed with a 10 Hz Q-switched Nd:YAG laser operating at 1064 nm. The incident laser power was controlled by ND filters to be about 10 mJ for the L films and about 1 mJ for the LB films. The SH light from optical components was blocked by a color filter just before the sample. The p-polarized fundamental beam was cast upon water surface at an incident angle 75°, and the SH light was detected along the reflected direction.

Springer Proceedings in Physics, Vol. 36
Nonlinear Optics of Organics and Semiconductors
Editor: T. Kobayashi

For LB films, the sample was mounted 45° declined from vertical and the SH light was detected in a transmission geometry. The output fundamental beam was removed by a $CuSO_4$ solution filter. The SH light was detected by a photomultiplier tube (PMT) after passing through a 532 nm interference filter. The fundamental beam intensity was monitored by a photodiode (PD). The signals of PMT and PD were averaged by a digital storage oscilloscope, and then integrated by a microcomputer to evaluate the light intensity.

3. Results and Discussion

3.1 L films

The SH light intensity I_{SH} was measured while compressing the monolayer. A typical pressure-area (π-A) isotherm during the measurement is shown in Fig.1. The SHG measurements were made in several static conditions. During each measurement, however, the pressure tended to relax as noticed by the discontinuous drops in the π-A isotherm (Fig.1).

In Fig.2 is displayed the square root of I_{SH} against the surface number density of the dye molecules, ρ. The finite I_{SH} at zero surface number density is caused by the air-water interface. When the mixture was spread on the subphase, the observed I_{SH} increased by three times that of the pure subphase. Compression of the monolayer brought about a further increase of I_{SH}. Thus, the SH light from the L film was definitely distinguished from the other origins, although the SH light was extremely weak.

Since I_{SH} is caused by the water surface as well as from the L film, I_{SH} is expressed as

$$I_{SH} \propto (E_{SH}^{W} + E_{SH}^{L})^2, \tag{1}$$

where E_{SH}^{W} and E_{SH}^{L} are the electric fields of the SH light from the water surface and the L film, respectively. Therefore, the contribution from the L film $(I_{SH}^{L})^{1/2}$ is given by

$$E_{SH}^{L} \propto (I_{SH}^{L})^{1/2} = (I_{SH})^{1/2} - (I_{SH}^{W})^{1/2}. \tag{2}$$

Since $(I_{SH})^{1/2}$ at $\rho=0$ is equal to $(I_{SH}^{W})^{1/2}$, it is safely concluded in Fig.2 that $(I_{SH}^{L})^{1/2}$ is simply proportional to ρ as

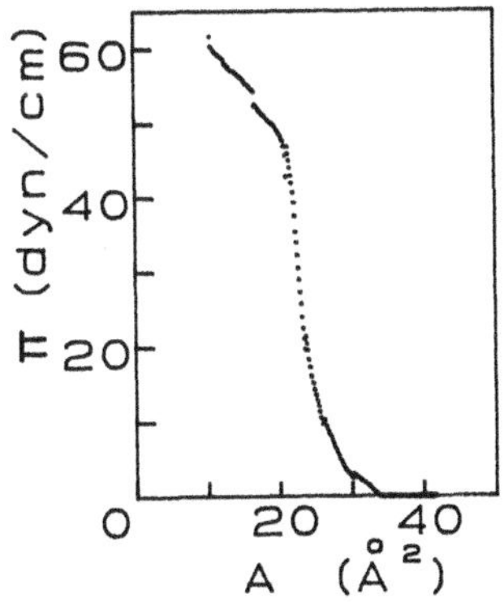

Fig.1 π-A isotherm monitored during the SHG measurement.

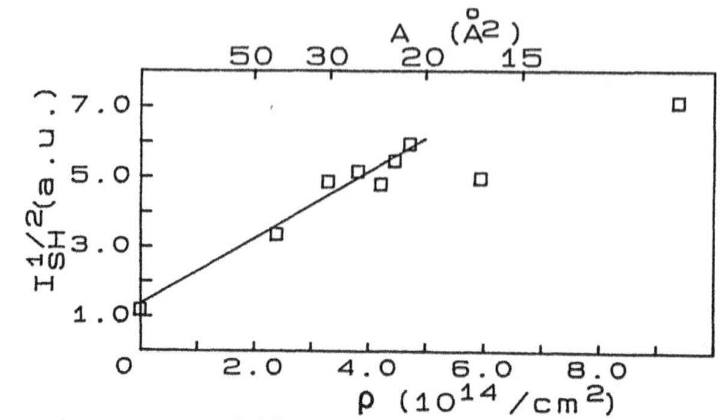

Fig.2 $(I_{SH})^{1/2}$ against the surface number density ρ of the dye molecule.

shown by a straight line except in the high ρ region. The square root of $I_{SH}{}^{L}$ is proportional to the second order susceptibility $\chi^{(2)}$. Assuming that the interaction between molecules is negligible,

$$\chi_{zzz}{}^{(2)} = \rho \langle \cos^3 \theta \rangle \alpha_{\xi\xi\xi}{}^{(2)}, \tag{3}$$

where z and ξ are along the surface normal and a certain molecular axis, respectively, and θ is the polar angle between them/2/. Therefore, the proportionality between $(I_{SH}{}^{L})^{1/2}$ and ρ suggests that the orientational distribution of the dye molecule does not change during the compression process. The same conclusion was obtained by BERKOVIC et al./2/ for 4"-n-pentyl-4-cyano-p-terphenyl.

The conventional model for the two-dimensional molecular compression process is as follows: The molecular axes randomly distribute in gaseous and liquid phases and gradually stand up packing themselves due to the compression process. According to the present findings, however, the dye molecules form islands upon deposition and these islands come together upon compression without changing the orientational distribution.

3.2 LB films

The SHG experiments were also carried out for the LB films. Using simple Y-type LB films, it was confirmed that I_{SH} for two-layer films is negligibly weak compared with I_{SH} for one- and three-layer films. It was also found that the one- and three-layer films gave rise to the same I_{SH}. These facts are reasonable from the point of view of the simple structural consideration; there exists an inversion symmetry for even-layer films and only the single layer effectively contributes to I_{SH} of odd-layer films.

It was also confirmed that no SH signal was observed for normal incidence of light on the glass substrate. This fact means that the average molecular orientation is along the glass normal.

It is well known that the structure without a center of inversion can be formed by so-called hetero Y-type LB films, i.e., an alternative deposition of two different molecules. The measure-

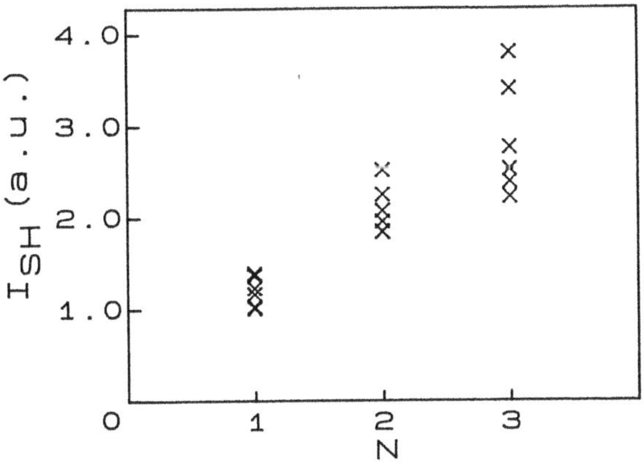

Fig.3 Dependence of the SH intensity on the number of dye layers for the hetero LB film.

ments were made for the hetero Y-type LB films composed of the same mixture and the pure arachidic acid. As shown in Fig.3, I_{SH} increases almost linearly with the number of dye layers. This result contradicts the theoretical prediction of a square law. Most of the experiments so far reported show somewhat less than the quadratic dependence/3,4/, although it was suggested that the square law seems to be held for rather large number of layers/5/. Careful measurements must be performed for a further discussion of the square law.

References

1. Th. Rasing, G. Berkovic, Y. R. Shen, S. G. Grubb and M. W. Kim: Chem. Phys. Lett. 130, 1 (1986)
2. G. Berkovic, Th. Rasing and Y. R. Shen: J. Opt. Soc. Am. B 4, 945 (1987)
3. I. R. Girling, N. A. Cade, P. V. Kolinsky, J. D. Earls, G. H. Cross and I. R. Peterson: Thin Solid Films 132, 101 (1985)
4. L. M. Hayden, S. T. Kowel and M. P. Srinivasan: Opt. Commun. 61, 351 (1987)
5. I. Ledoux, D. Josse, P. Vikakovic, J. Zyss, R. A. Hann, P. F. Gordon, B. d. Bothwell, S. K. Gupta, S. Allen, P. Robin, E. Chastaing and J. Ç. Dubois: Europhys. Lett. 3, 803 (1987)

Femtosecond Pulse Compression Using Highly Nonlinear Organic Materials

M. Yamashita, K. Torizuka, and T. Sato

Laser Research Section, Radio- and Opto-Electronics Division, Electrotechnical Laboratory, 1-1-4 Umezono, Tsukuba, Ibaraki 305, Japan

1. Intracavity pulse compression using highly nonlinear organic materials

Recently, Haus and Silberberg [1] analytically pointed out that a nonlinear refractive index (NRI; n_2) medium introduced into a CPM laser with adjustable cavity dispersion leads to shorter pulses. However, to the authors' knowledge, no one has expeirmentally investigated how the introduction of a NRI medium into a practical CPM laser influences the genaration of femtosecond pulses. In this paper, we demonstrate that the addition of high-NRI organic materials to a solution of the saturable absorber permits intracavity pulse compression in a simple CPM laser employing cavity-mirror dispersion adjustment [2].

The NRI organic materials used have the following properties: (1) a higher NRI than that of an ethylene glycol (EG) solution (n_2=3.0 × 10^{-16} cm^2/W), (2) an ultrafast time-dependent index change by electronic hyperpolarizability, (3) high solubility for EG or a mixture of EG and benzylalcohol, (4) no absorption around the lasing wavelength region, and (5) good photochemical stability in the solution. Recent studies of high-NRI organic materials enabled us to find suitable materials that satisfy these properties. The third-order molecular hyperpolarizabilities, γ, of paradimethylamino-β nitrostyrene (DMA-NS) [3] and 2'-4'-methyl nitroaniline (MNA) are much larger than that of usual solvents (by a factor of ~10^4). Therefore, from the NRI equation, $n_2=3\pi N\gamma(n^2+2)^4/81$, where n is the linear refractive index of the solvent and N is number of molecules per unit volume, we find that n_2 for both the DNA-NS solution and the MNA solution at concentrations of ~10^{-2} M/liter is larger than that of EG. This is true even through N for both DMA-NS and MNA is smaller than that for EG (the ratio of the former to the latter is 10^{-2}~10^{-3}). In this experiment, we employed a 1.1 × 10^{-2} M/liter solution of DMA-NS (n_2=5.3 ×10^{-15} cm^2/W) and a 3.8 ×10^{-2}M/liter solution of MNA (n_2=3.0 ×10^{-15} cm^2/W). The NRI's for both the DMA-NS and the MNA solutions are larger than that for EG by a factor of 20 ~10. The absorption peaks of the DMA-NS and MNA solutions are 475 and 410 nm, respectively.

The simple CPM (R6G + DODCI) laser (Fig.2-B) and background-free second-harmonic generation autocorrelation used are identical to those that we previously described [4,5]. The intracavity dispersion was adjusted by changing the incident angle of multilayer dielectric mirrors M_6 and M_7. M_6 was an angle-dependent dispersive mirror, while M_7 was an angle-independent and negligibly small dispersive mirror. The dispersion of the other mirrors except for M_1 was negligibly small. M_1 provides a constant dispersion.

Measurement of the pulse duration and of the spectrum was initially carried out for the CPM laser with the 7 × 10^{-3}M/liter solution of DODCI absorber only. The cavity-mirror dispersion, cavity alignment, and pump power were carefully adjusted while the fast-scanned autocorrelation trace and pulse spectrum were monitored on oscilloscopes. The duration of the generated pulse with the relatively asymmetrical spectrum was 73 fsec, as shown in Fig.1(a). Next, the 1.1 ×10^{-2}M/liter DMA-NS was dissolved in the DODCI solution. Consequently, the pulse duration was reduced to 59 fsec. After adjustment of the cavity-mirror dispersion, the pulse duration was further shortened to 50 fsec (Fig.1(b)).

Springer Proceedings in Physics, Vol. 36
Nonlinear Optics of Organics and Semiconductors
Editor: T. Kobayashi

The dispersion dependence of the pulse duration generated from the CPM laser with the addition of the 1.1×10^{-2}M/liter solution of DMA-NS is shown in Fig.2-A. It is found that with the increase of the second-order phase dispersion $\ddot{\phi}(\omega)$, the pulse duration decreases slowly, reaches a minimum, and then rapidly increases for $\ddot{\phi}(\omega) > 3 \times 10^{-28}$sec^2. The value of $\ddot{\phi}(\omega)$ for which the pulse duration is minimum is larger for the case considered here than when DMA-NS is not added (a case previoulsy reported by us [4, 5]), which is in agreement with theoretically derived results [1]. The rapid increase in the pulse width for $\ddot{\phi}(\omega) > 3 \times 10^{-28}$sec^2 is caused not only by the excess second-order dispersion, but also by the increase of the third-order phase dispersion of multilayer dielectric mirror M_6. This is because the third-order dispersion of M_6 rapidly increases over $\dddot{\phi}(\omega) = 1 \times 10^{-41}$sec^3 when $\ddot{\phi}(\omega) > 3 \times 10^{-28}$ sec^2. Therefore, when $\dddot{\phi}(\omega)$ becomes negligibly small, the pulse duration will increase more slowly in the region $\ddot{\phi}(\omega) > 3 \times 10^{-28}$sec^2[1].

The same intracavity pulse-compession experiment was carried out for the CPM laser with a thinner film (flared-stream jet nozzle) of a solution of a mixture of DODCI and 3.8×10^{-2}M/liter MNA (instead of 1.1×10^{-2}M/liter DMA-NS). The results showed that pulses as short as 53 fsec were shortened to 48 fsec, as shown in Fig.3. We believe that shorter pulses are generated when newly designed mirrors providing small third-order dispersion are utilized [6].

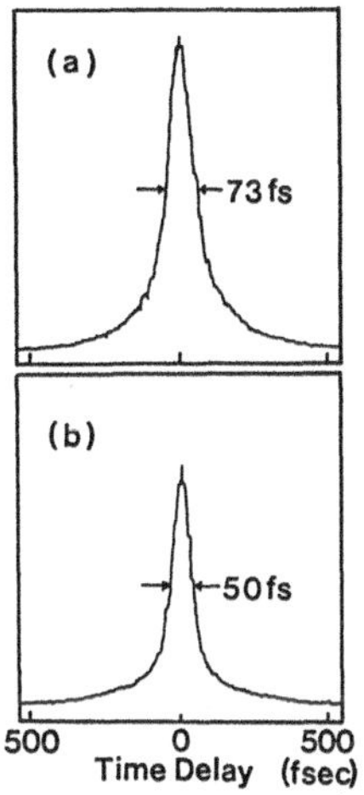

Fig.1 Autocorrelation traces from the simple CPM laser: (a) without and (b) with addition of 1.1×10^{-2}M/liter DMA-NS.

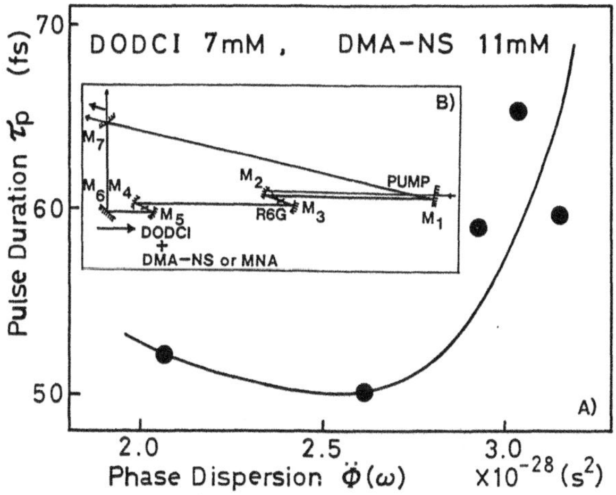

Fig.2 A) Generated pulse duration as a function of cavity-mirror dispersion $\ddot{\phi}(\omega)$ in the laser with 1.1×10^{-2}M/liter DMA-NS. B) Cavity configuration of the laser.

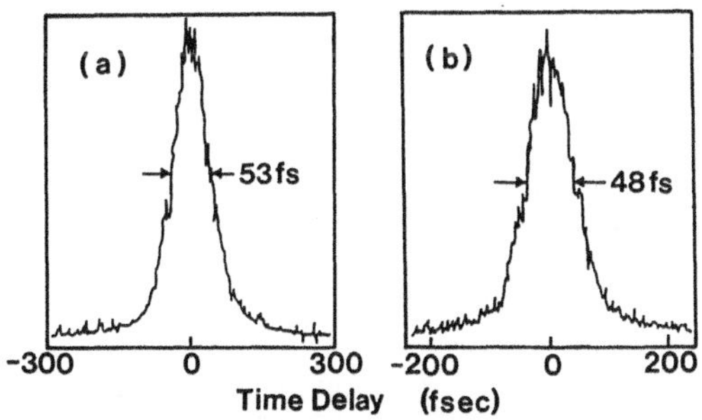

Fig. 3 Autocorrelation traces from the laser: (a) without and (b) with addition of 3.8×10^{-2}M/liter MNA.

2. Organic fibers for a femtosecond pulse compressor

The pulse-compression technique by self phase moduration (SPM) in a single-mode glass-fiber and linear-chirp compensation is one of the successful applications of optical nonlinear-effects. The compression ratio is determined by the intensity of incident pulses, the NRI and the group velocity dispersion (GVD)[7]. Highly-nonlinear oganic materials are attractive candidates for a compressor of low peak-power ultrashort pulses because of its extremely large electronic (fast-response) NRI [8] and moderate GVD. The pulse-compression making use of other core-materials than a glass has not been realized although it was pointed out in an early picosecond work by others [9]. This is because single-mode fibers of large and fast-response NRI materials did not exist. Here, we study experimentally the properties of highly-nonlinear organic fibers in the femtosecond region.

Our fibers consist of a 1 cm long quartz-glass capillary-clad and a 4 μm radius organic-liquid core. Sample (a) is a multimode fiber of pure nitrobenzene and Sample (b) is a singlemode one of a mixture of MNA, nitrobenzene and ethanol. In both of (a) and (b), electronic NRI and GVD are expected to be about 70 and 5 times of those in a glass fiber, repsectively. Fig.4 shows typical results of the background-free autocorrelation measurements. The peak-power of the incident pulse (i: a colliding pulse modelocked ring laser) in the fibers is not so strong that SPM occurs. The output-pulse of Sample (a) is broadened by not only material GVD but additional mode-dispersion. That of Sample (b) is much shorter and is really recompressed (r also) by the negative GVD of a grating-pair. The spectrum-broadening by SPM is measured by amplified and GVD-compensated 70 fsec pulses (the repetition rate: 10 pps), as shown in Fig.5. Curve (g) is the result of the singlemode glass-fiber. In Sample (a), a large asymmetric broadening to the longer-wavelength side occurs. Here, the pulse-broadening by the mode-dispersion enhances a contribution of the slow-response NRI by molecular reorientation. In Sample (b), the observed modulation is similar to that in (g) because of the fast-response NRI, where the incident peak-power is lower about one order of magnitude.

A crystal-core fiber has the advantage that NRI is completely free from the molecular orientation. In Fig.6, the SPM results in a crystal MNA-fiber (a quartz-glass clad: multimode) is shown. Even with a pulse-broadening by mode-dispersion, a symmetric spectral-broadening occurs. The peak-power for SPM is in a reasonable agreement with NRI and GVD under the condition that the crystal c-axis is in the axial-direction of the fiber, and lower than that in a glass-fiber.

In addition, our result shows that the modulated spectrum is sensitive to the time-response of nonlinear refractivity. Therefore, the measurement of the SPM in organic-fiber is also usable for the study of ultrafast molecular-dynamics.

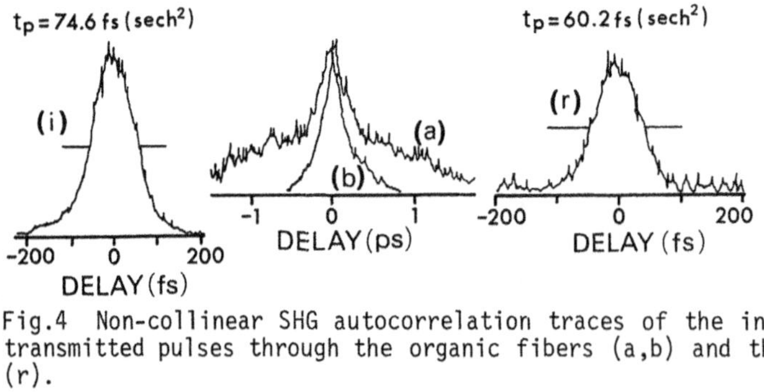

Fig.4 Non-collinear SHG autocorrelation traces of the incident pulse (i), the transmitted pulses through the organic fibers (a,b) and the recompressed pulse (r).

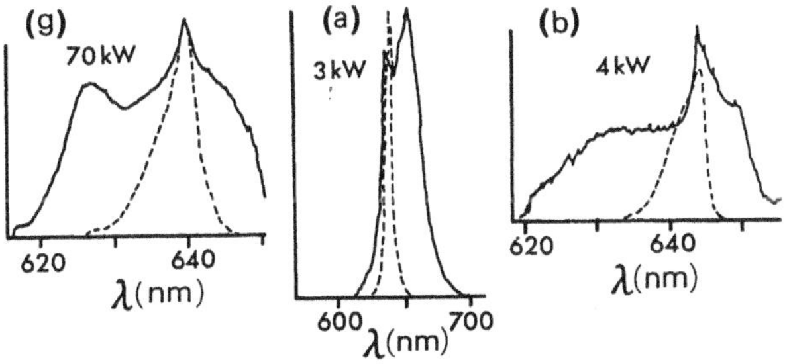

Fig.5 Spectra of the incident pulses (dashed curves) and the modulated pulses (solid curves). Peak-power of the incident pulses is also shown.

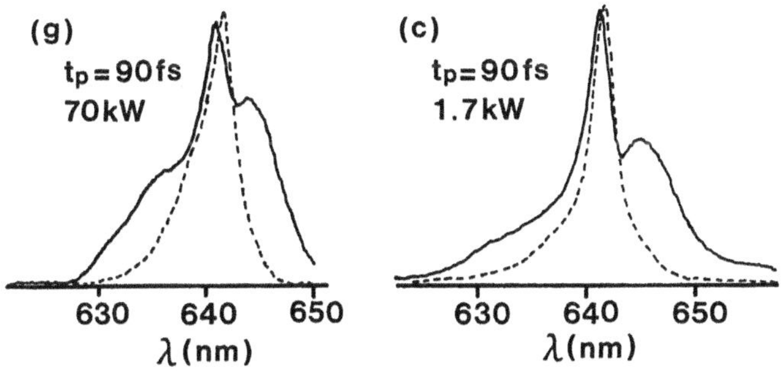

Fig.6 Spectra of the incident pulses (dashed curves) and the modulated pulses (solid curves) to/from a glass-fiber (g) and a MNA crystal fiber (c). In both fibers, core radii are 4 μm and lengths are 1 cm.

References

1. H.A.Haus and Y.Silberberg, IEEE J.Quantum Electron. QE-22, 325(1986).
2. M.Yamashita, K.Torizuka and T.Sato, Opt.Lett. 13, 24(1988).
3. J.L.Oudar, J.Chem.Phys. 67, 446 (1977).
4. M.Yamashita, M.Ishikawa, K.Torizuka and T.Sato, Opt.Lett. 11, 504(1986).
5. M.Yamashita, K.Torizuka and T.Sato, IEEE J.Quantum Electron. QE-23, 2005 (1987).
6. M.Yamashita, S.Kaga, K.Torizuka and T.Sato in Ultrafast Phenomena VI, (Springer-Verlag, Berlin, 1988) to be published.
7. W.J.Tomlinson et al., J.Opt.Soc.Am. 81, 139(1984).
8. B.F.Levine and C.G.Bethea, J.Chem.Phys. 63, 2666(1975).
9. E.P.Ippen, C.V.Shank and T.K.Gustafson, Appl.Phys.Lett, 24, 190(1974).

Picosecond Third Harmonic Generation in β-BaB_2O_4 and Calcite

A. Penzkofer, P. Qiu, and F. Ossig*

Naturwissenschaftliche Fakultät II-Physik, Universität Regensburg, D-8400 Regensburg, Fed. Rep. of Germany

*On leave from Shanghai Institute of Optics and Fine Mechanics, Academia Sinica, Shanghai, Peop. Rep. of China

1. Introduction

Phase-matched third harmonic generation has been achieved in metal vapors [1,2] inert gases [3], organic dyes [4-6], liquid crystals [7] and birefringent crystals [8-14]. Third harmonic light may be generated by the direct third-order nonlinear interaction, $\nu_L+\nu_L+\nu_L\to\nu_3$, due to the third-order nonlinear susceptibility $\chi^{(3)}_{THG}$, or it may be generated by cascading the second harmonic generation, $\nu_L+\nu_L\to\nu_2$, and the frequency mixing, $\nu_2+\nu_L\to\nu_3$. The second harmonic generation and the frequency mixing are due to the second-order nonlinear susceptibilities $\chi^{(2)}_{SHG}$ and $\chi^{(2)}_{FM}$. Phase-matching, $\Delta k=0$, is necessary for efficient third harmonic light generation. In birefringent crystals it is achieved by crystal orientation and angle tuning. In crystals with inversion center only direct third harmonic generation is allowed while in crystals without inversion center cascading and mixed (direct and cascading) third harmonic generation are possible. A double phase-matching of the second harmonic generation and the frequency mixing requires two separately oriented crystals in series. The subsequent phase-matched second harmonic generation and phase-matched frequency mixing is most efficient and is widely used [15]. The various phase-matching schemes for third harmonic generation in negative uniaxial birefringent crystals are summarized in Table 1.

Table 1 Single and double phase-matched angle-tuned generation of third-harmonic light in negative uniaxial birefringent crystals ($n_e<n_o$). D = direct. C = cascading. D+C = mixed. IC = inversion center.

Crystal	Process	Phase-matching	Interaction		Contribution
single crystal, single phase-matched					
with IC	D	$\Delta k_{THG}=0$	type I	$o_L o_L o_L\to e_3$	$\chi^{(3)}_{eff,THG}$
			type II	$o_L o_L e_L\to e_3$	
			type III	$o_L e_L e_L\to e_3$	
without IC	C	$\Delta k_{SHG}=0, \Delta k_{FM}\neq 0$ a)	type I	$o_L o_L\to e_2$	$\chi^{(2)}_{eff,SHG}$
			type II	$o_L e_2\to e_2$	
	C	$\Delta k_{SHG}\neq 0, \Delta k_{FM}=0$	type I	$o_2 o_L\to e_3$	$\chi^{(2)}_{eff,FM}$
			type II	$o_2 e_L\to e_3$	
			type III	$e_2 o_L\to e_3$	
	D+C	$\Delta k_{THG}=\Delta k_{SHG}+\Delta k_{FM}=0$	type I	$o_L o_L o_L\to e_3$	$\chi^{(3)}_{eff,THG}+\chi^{(2)}_{eff,cas}$
			type II	$o_L o_L e_L\to e_3$	
			type III	$o_L e_L e_L\to e_3$	
two crystals, double phase-matched					
without IC	C	$\Delta k_{SHG,1}=0$ and	type I	$oo\to e$	1: $\chi^{(2)}_{eff,SHG}$ and
		$\Delta k_{FM,2}=0$	type II	$oe\to e$	2: $\chi^{(2)}_{eff,FM}$

a) Cascading third harmonic generation with $\Delta k_{SHG}=0$ is less efficient compared to $\Delta k_{FM}=0$ [13].

Springer Proceedings in Physics, Vol. 36
Nonlinear Optics of Organics and Semiconductors
Editor: T. Kobayashi

In this paper the single phase-matched third harmonic generation in single crystals of calcite and β-BaB_2O_4 is studied. Calcite is a negative uniaxial crystal with inversion center (trigonal system, space group $R\bar{3}c$, point group $\bar{3}m$). β-BaB_2O_4 is a newly developed negative uniaxial crystal without inversion center [16,17] (trigonal system, space group R3, point group 3; a higher symmetry of R3c and 3m is stated in [18]). The large effective second harmonic coefficients, the wide transparent waveband (190 - 3500 nm), and the high damage threshold make BBO very important for second harmonic generation and frequency mixing in the ultraviolet spectral region [19-24]. β-BaB_2O_4 was applied successfully in the second harmonic generation of femtosecond pulses [25,26].

Table 2 gives a list of crystals that have been applied to the phase-matched third harmonic generation in a single crystal.

Table 2 Realized phase-matched third harmonic generation in single crystals. All crystals are negative uniaxial. D = direct. D+C = mixed.

Crystal	Class	Interaction	Laser	λ_L [nm]	References
Calcite	trigonal, $R\bar{3}c$	D	Q-switched	694.3	8,9
			Q-switched	1060	11,14
			Mode-locked	1054	this work
KDP	tetragonal, $\bar{4}2m$	D+C	Q-switched	1064	11
ADP	tetragonal, $\bar{4}2m$	D+C	Q-switched	1060	10
			Mode-locked	1060	10
β-BaB_2O_4	trigonal, R3(R3c)	D+C	Mode-locked	1054	this work

2. Fundamentals

The geometrical arrangement of phase-matched third harmonic generation in a single crystal is sketched in Fig.1. The angle Θ between the crystal fixed z-axis (optic axis) and wave-vector $\vec{k}_L(\|\vec{k}_3)$ is adjusted to phase-matching. α_L and α_3 are the walk-off angles between the ray directions of extraordinary and ordinary polarized light.

Collinear phase-matching of mixed or direct THG requires

$$\Delta k = k_{e3} - k_{aL} - k_{bL} - k_{cL} = 0 \; ; \qquad (1)$$

a,b,c indicate the polarizations o or e of the fundamental waves. In the case of type-II phase-matching it is $k_{3e}-2k_{oL}-k_{eL}=0$. The wave-vectors are related to the refractive indices n by $k=2\pi n\nu/c_0$ where ν is the frequency and c_0 is the vacuum light velocity. The ordinary refractive index n_o is independent of crystal orientation. The extraordinary refractive index depends on the polar angle Θ by

$$n_e(\Theta) = \frac{n_o n_e}{(n_e^2\cos^2\Theta + n_o^2\sin^2\Theta)^{1/2}} \; ; \qquad (2)$$

n_o and n_e are the principal refractive indices.

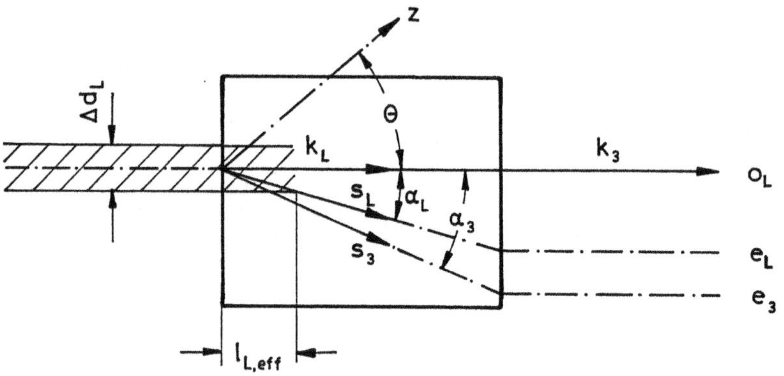

Fig.1 Schematic geometrical arrangement

The walk-off angle α_L limits the overlap length of the o und e pump laser components in the case of type-II and type-III phase-matching to

$$\ell_{L,eff} \simeq \frac{\Delta d_L}{2\alpha_L} \tag{3}$$

where Δd_L is the pump pulse beam diameter. The walk-off angle α_3 allows third harmonic generation over the whole crystal length but amplifying interaction between pump and third harmonic light occurs only within an effective length

$$\ell_{3,eff} \simeq \frac{\Delta d_L}{2\alpha_3} \quad . \tag{4}$$

For femtosecond pulses the temporal overlap of the o and e pump pulse components of type-II and type-III phase-matched crystals may be limited by the group velocity dispersion. The group refractive index is

$$n_g = \frac{n}{1 - \frac{\nu}{n}\frac{\partial n}{\partial \nu}} \quad . \tag{5}$$

The time delay between ordinary and extraordinary rays is given by $(\delta t/\delta \ell)_{oLeL} = [n_{goL} - n_{geL}(\Theta)]/c_0$ and the overlap length is limited to

$$\ell_o \simeq \frac{\Delta t_L}{(\delta t/\delta \ell)_{oLeL}} \quad . \tag{6}$$

The duration of the generated third harmonic pulse is given approximately by

$$\Delta t_3 \simeq [\tfrac{1}{3}\Delta t_L^2 + (\delta t/\delta \ell)^2_{e3oL}\ell_I^2]^{1/2} \tag{7}$$

where ℓ_I is the shorter length of ℓ, ℓ_o, and $\ell_{L,eff}$.

The energy conversion efficiency of third harmonic generation is given by [12,13]

$$\eta_E = \kappa \ell^2 I_{OL}^2 |\chi_{eff}|^2 \frac{\sin^2(\Delta k\ell/2)}{(\Delta k\ell/2)^2} \quad f(\Delta d_L, \Delta\Theta_L, \Delta\nu_L, \Delta t_L) \tag{8}$$

κ comprises constant factors. The function f takes care of the reduction of conversion efficiency due to the finite beam diameter Δd_L, the divergence $\Delta\Theta_L$, the spectral width $\Delta\nu_L$, and the pulse duration Δt_L of the pump pulse.

3. Crystal data

The dispersion of the principle refractive indices n_o and n_e of calcite [27] and BBO [19] are depicted in Fig.2. The transmissions T are shown in Fig.3. The type-I, II, and III phase-matching angles of direct THG in calcite and of mixed THG in BBO are plotted in Fig.4. The corresponding walk-off angles α_L and α_3 are diagrammed in Fig.5.

The angular dependence $n_E(\Theta)/n_E(\Theta_{PM})$ of BBO is shown in Fig.6a for λ_L = 1.054 µm ($\Delta\Theta_L$=0, $\Delta\nu_L$=0, Δd_L=∞). $\Delta\Theta_{1/2}$ is the FWHM of the angular detuning curve.

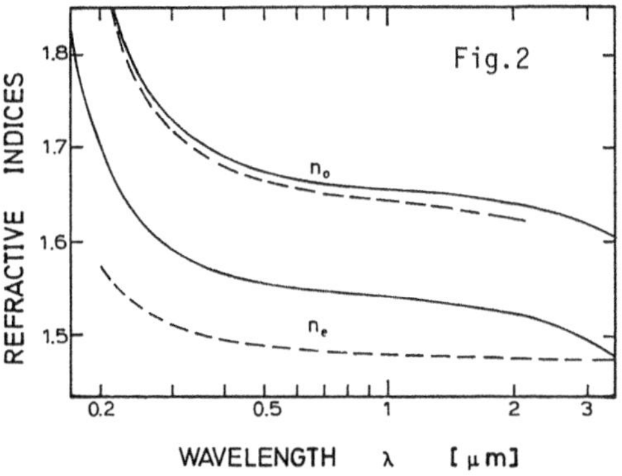

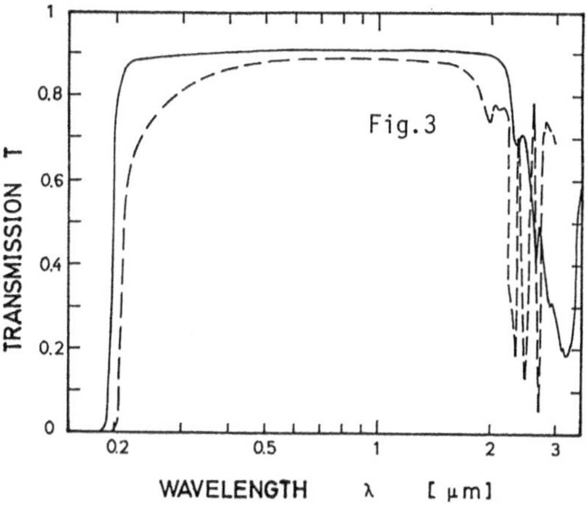

Fig. 2 Refractive indices of β-BaB_2O_4 (solid)[19] and calcite (dashed)[27].

Fig.3 Spectral transmission of β-BaB_2O_4 (ℓ = 6 mm, solid) and calcite (ℓ = 33 mm, dashed).

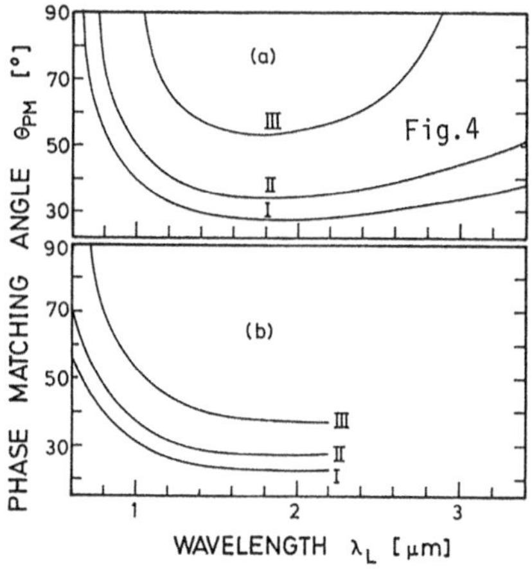

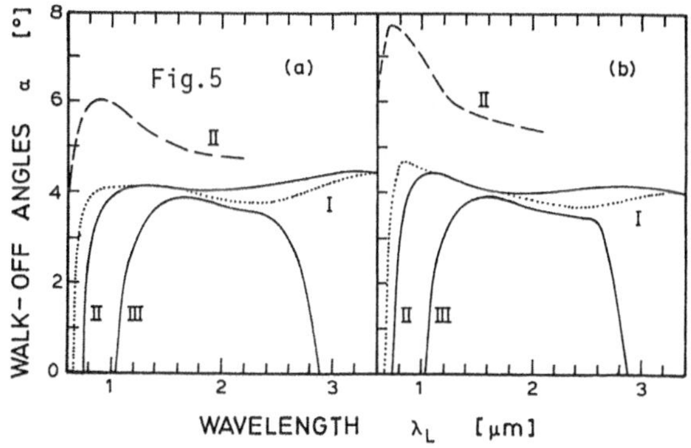

Fig.4 Type-I, II, and III phase-matching in β-BaB_2O_4 (a) and calcite (b).

Fig.5 Walk-off angles α_L(a) and α_3(b) of BBO (solid and dotted) and calcite (dashed, only type-II is shown).

$\Delta\Theta_{1/2}$ is inverse proportional to the crystal length ℓ. $\Delta\Theta_{1/2}\ell$ versus wavelength is plotted in Fig.7 for calcite and BBO. The internal divergence of the pump laser radiation, $\Delta\Theta_{L,int}$, should be less than $\Delta\Theta_{1/2}$ in order to avoid remarkable loss of efficiency (external divergence angle $\Delta\Theta_L \simeq n_{oL}\Delta\Theta_{L,int}$).

The frequency dependence $\eta_E(\nu)/\eta_E(\nu_L)$ at a fixed angle is similar to the angular dependence at a fixed wavelength. $\eta_E(\nu)/\eta_E(\nu_L)$ of BBO is depicted in Fig.6b ($\Delta\Theta_L=0$, $\Delta\nu_L=0$, $\Delta d_L=\infty$). Phase-matching is adjusted to $\tilde{\nu}^{-1} = \lambda_L = 1.054$ µm. $\Delta\nu_{1/2}$ is the FWHM of the spectral detuning curve. $\Delta\nu_{1/2}$ is inverse proportional to the crystal length ℓ. $\Delta\tilde{\nu}_{1/2}\ell$ versus wavelength is displayed in Fig.8. The spectral width of the pump laser $\Delta\nu_L$ should be less than $\Delta\nu_{1/2}$ in order to avoid remarkable loss of efficiency.

The group velocity dispersion limits the temporal overlap (Eqs.6 and 7). The curves of $(\delta t/\delta\ell)_{oLeL}$ and $(\delta t/\delta\ell)_{e3oL}$ are depicted in Fig.9a and 9b, respectively.

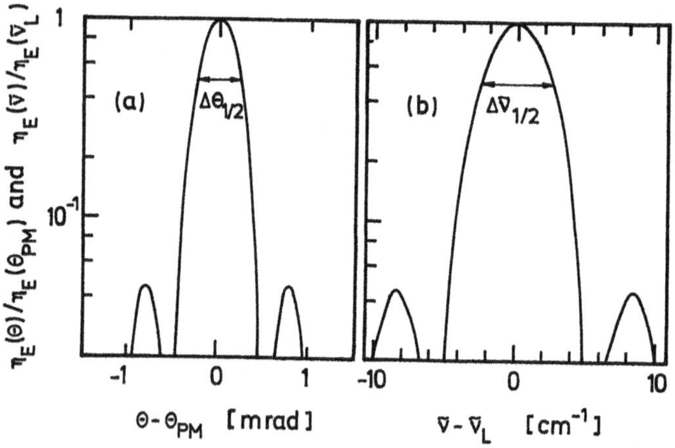

Fig.6 Angular (a) and spectral (b) detuning curves of β-BaB_2O_4 at λ_L = 1.054 µm for type-II phase-matched mixed THG (ℓ = 0.72 mm).

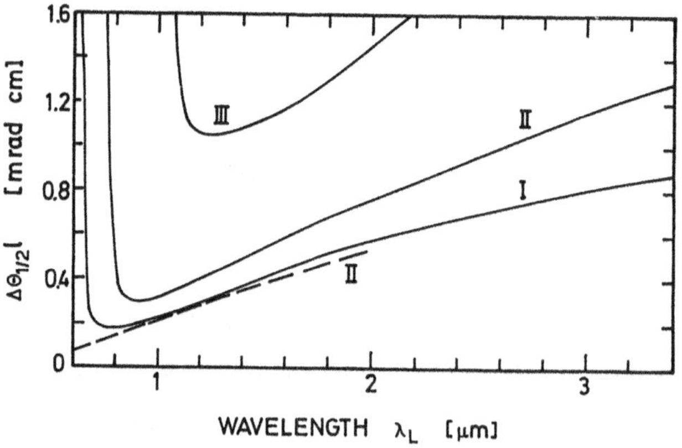

Fig.7 Halfwidth of angular tuning curves for BBO (solid) and calcite (dashed, only type-II). Mixed THG.

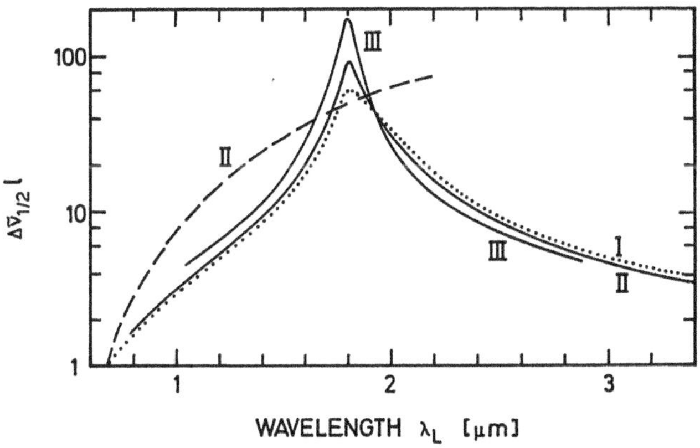

Fig.8 Halfwidth of spectral tuning curves for BBO (solid and dotted) and calcite (dashed, only type-II is shown). Mixed THG.

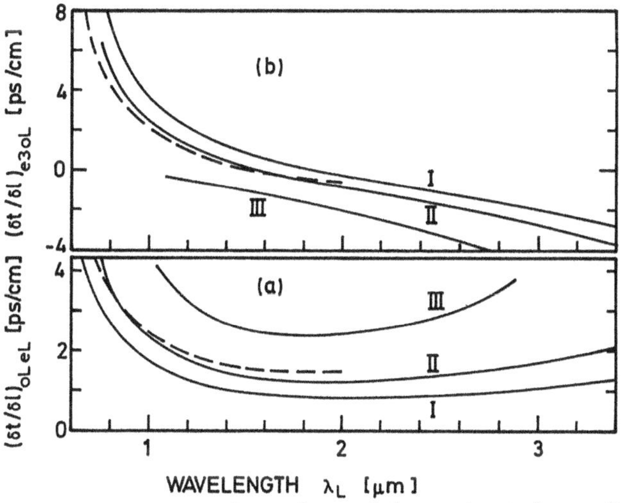

Fig.9 Time delays $(\delta t/\delta \ell)_{oLeL}$ (a) and $(\delta t/\delta \ell)_{e3oL}$ (b) for mixed THG in BBO (solid) and direct THG in calcite (only type-II is shown).

4. Experimental

The experimental setup is shown in Fig.10. Single picosecond light pulses of a passively mode-locked Nd-phosphate glass laser ($\Delta t_L \simeq 5$ ps, $\lambda_L = 1.054$ μm) are used as pump pulses. The energy conversion efficiency of third harmonic light versus input pump pulse peak intensity is measured and the angular detuning curves are determined. The calcite crystal is 2 cm long and the length of the β-BaB_2O_4 crystal is $\ell = 7.2$ mm. In some calcite measurements a cylindrical lens is inserted to generate a line-focus which increases the light intensity at the crystal without increasing the relevant laser divergence $\Delta\Theta_L$ in the plane spanned by the optic axis and the light propagation direction.

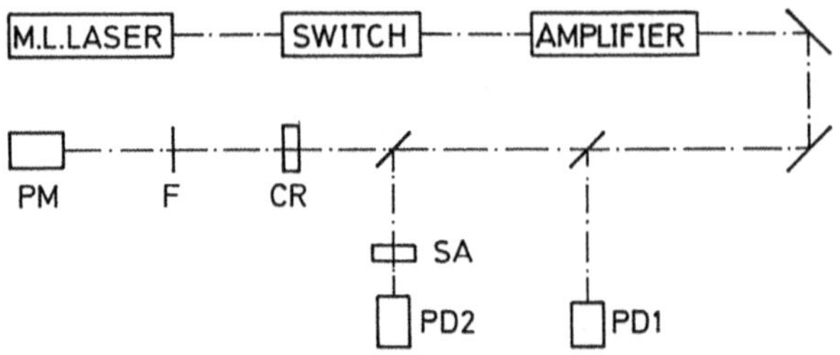

Fig.10 Experimental setup. PD1, PD2, photodetectors. SA, saturable absorber (Kodak No.9860) for peak intensity detection [28]. CR, crystal. F, filter. PM, photomultiplier.

5. Results

Type-II phase-matched mixed (BBO) and direct (calcite) THG are investigated. The angular detuning curves $\eta_E(\Theta)/\eta_E(\Theta_{PM})$ of BBO and calcite are shown in Fig.11. For BBO the spectral width of the pump laser is $\Delta\tilde{\nu}_L = 10\ cm^{-1}$. In the case of calcite two curves are depicted for $\Delta\tilde{\nu}_L = 10\ cm^{-1}$ and $\Delta\tilde{\nu}_L = 40\ cm^{-1}$ (self-phase modulated pulses).

The phase-matched THG energy conversion efficiency versus pump pulse peak intensity is depicted in Fig.12. At the highest intensities applied conversion efficiencies of $\eta_E \simeq 0.01$ ($I_{0L} = 5\times10^{10}\ W/cm^2$) and $\eta_E \simeq 8\times10^{-5}$ ($I_{0L} = 10^{11}\ W/cm^2$) have been obtained for BBO and calcite, respectively. The damage threshold of calcite is $I_{th,C} > 10^{13}\ W/cm^2$ and the damage threshold of BBO is $I_{th,B} \approx 10^{12}\ W/cm^2$ [18,22] for single pulses of 5 ps duration. At pump pulse intensities slightly below the damage threshold very high conversion efficiencies are expected in both crystals.

The effective nonlinear susceptibilities $\chi^{(3)}_{eff}$ are determined by comparison of the measured energy conversion efficiencies η_E with calculations (Eq.5). The obtained values are listed in Table 3 together with other crystal parameters. A

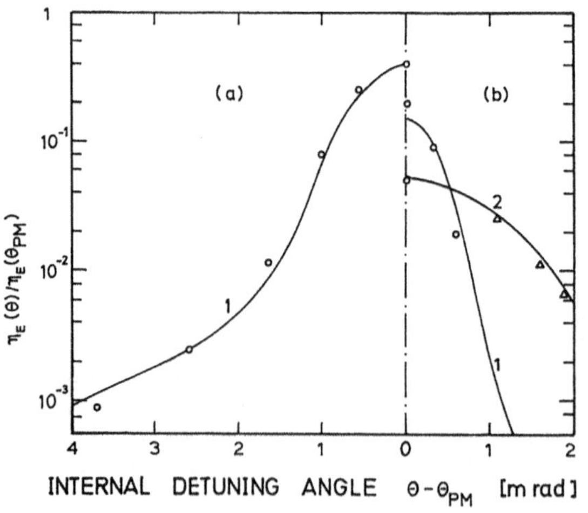

Fig.11 THG conversion efficiency versus detuning angle for BBO of 0.72 cm length (a) and calcite of 2 cm length (b). $\Delta\Theta_L = 5\times10^{-4}$ rad. (1,o), $\Delta\tilde{\nu}_L = 10\ cm^{-1}$. (2,Δ), $\Delta\tilde{\nu}_L = 40\ cm^{-1}$. Type-II phase-matching.

detailed analysis of the effective susceptibilities indicates that $\chi^{(3)}_{eff,THG}$ and $\chi^{(2)}_{eff,cas}$ are of the same order of magnitude for BBO [13].

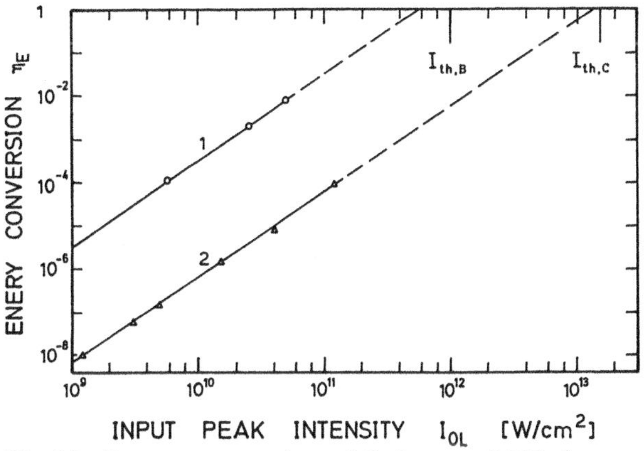

Fig.12 Energy conversion efficiency of BBO (curve 1,o; ℓ = 7.2 mm) and calcite (curve 2,Δ; ℓ = 2 cm). Type-II phase-matching. $\Delta\Theta_L = 10^{-4}$ rad. $\Delta\tilde{\nu}_L$ = 20 cm^{-1}. The damage thresholds $I_{th,B}$ (BBO) and $I_{th,C}$ (calcite) are indicated.

Table 3 Phase-matched third harmonic generation of picosecond pulses of a Nd-phosphate glass laser in calcite (ℓ = 2 cm) and β-BaB_2O_4 (ℓ = 7.2 mm). Δt_L = 5 ps, λ_L = 1.054 µm.

Parameter		Calcite	β-BaB_2O_4
System		trigonal	trigonal
Point group		$\bar{3}$m	3 (3m)
Space group		R$\bar{3}$c	R3 (R3c)
Process		D	D + C
Phase-matching		type-II (ooe→e)	type-II (ooe→e)
Θ_{PM}	[°]	35.96	47.4
$\Delta\Theta_{1/2}\,\ell$	[rad cm]	2.3×10^{-4}	3.4×10^{-4}
$\Delta\tilde{\nu}_{1/2}\,\ell$		8.8	3.7
α_3	[°]	6.75	4.45
α_L	[°]	5.85	4.05
$(\delta t/\delta\ell)_{e3oL}$	[ps cm^{-1}]	1.7	2.0
$(\delta t/\delta\ell)_{oLeL}$	[ps cm^{-1}]	2.2	2.1
$\chi^{(3)}_{eff,THG}$	[m^{2}V^{-2}]	3×10^{-24}	6.4×10^{-23}
$\chi^{(2)}_{eff,cas}$	[m^{2}V^{-2}]	-	6.6×10^{-23}
$\chi^{(3)}_{eff}$	[m^{2}V^{-2}]	3×10^{-24} (2.1×10^{-16} esu)	1.3×10^{-22} (9.2×10^{-15} esu)
η_E		8×10^{-5} a)	0.01 b)
I_{th}	[W cm^{-2}]	$\gtrsim 10^{13}$	$\sim 10^{12}$
$\eta(I_{th})$		→ 1	→ 1

a: $I_{OL} = 10^{11}$ W/cm^2. b: $I_{OL} = 5\times10^{10}$ W/cm^2.

References

1. J.F. Reintjes: In Laser Handbook, ed. by M. Bass and M.L. Stitch, Vol.5 (North-Holland, Amsterdam, 1985) Ch.1.
2. C.R. Vidal: In Tunable Lasers, ed. by F.L. Mollenauer and J.C. White (Springer, Berlin, Heidelberg, 1987) p. 57.
3. A.H. Kung, J.F. Young, S.E. Harris: Appl. Phys. Lett. 22, 301 (1973).
4. P.P. Bey, J.F. Guiliani, H. Rabin: IEEE J. Quant. Electron. QE-7, 86 (1971).
5. J.C. Diels, F.P. Schäfer: Appl. Phys. 5, 197 (1974).
6. A. Penzkofer, W. Leupacher: Opt. Quant. Electron. 20, 222 (1988).
7. J.W. Shelton, Y.R. Shen, Phys. Rev. Lett. 26, 538 (1971).
8. R.W. Terhune, P.D. Maker, C.M. Savage: Appl. Phys. Lett. 2, 54 (1963).
9. P.D. Maker, R.W. Terhune: Phys. Rev. 137, A801 (1965).
10. C.C. Wang, E.L. Baardsen: Appl. Phys. Lett. 15, 396 (1969).
11. S.A. Akhmanov, L.B. Meisner, S.T. Parinov, S.M. Saltiel, V.G. Tunkin: Sov. Phys. JETP 46, 898 (1977).
12. A. Penzkofer, F. Ossig, P. Qiu: Appl. Phys. B, to be published.
13. P. Qiu, A. Penzkofer: Appl. Phys. B45, 225 (1988).
14. A.P. Sukhorukov, I.V. Tomov, Sov. Phys. JETP 31, 872 (1970).
15. D. Eimerl, IEEE J. Quant. Electron. QE-23, 575 (1987).
16. C. Chen, B. Wu, A. Jiang, G. You: Sci. Sinica (Ser.B) 28, 235 (1985).
17. A. Jiang, F. Cheng, Q. Lin, C. Chen, Y. Zheng: J. Cryst. Growth 79, 963 (1986).
18. D. Eimerl, L. Davis, S. Velsko, E.K. Graham, A. Zalkin: J. Appl. Phys. 62, 1968 (1987).
19. K. Kato, IEEE J. Quant. Electron. QE-22, 1013 (1986).
20. K. Miyarzaki, H. Sakai, T. Sato: Opt. Lett. 11, 797 (1986).
21. P. Lokai, B. Burghardt, D. Basting, W. Mückenheim, Laser und Optoelektronik 19, 296 (1987).
22. H. Schmidt, R. Wallenstein: Laser und Optoelektronik 19, 302 (1987).
23. R.S. Adhav, S.R. Adhav, J.M. Pelaprat: Laser Focus 23/9, 88 (1987).
24. J.T. Lin, C. Chen, Laser Focus 23/11, 59 (1987).
25. Y. Ishida, T. Yajima: Opt. Commun. 62, 197 (1987).
26. K.L. Cheng, W. Bosenberg, F.W. Wise, I.A. Walmsley, C.L. Tang: Appl. Phys. Lett. 52, 519 (1988).
27. American Institute of Physics Handbook, 3rd ed., ed. by D.E. Gray (McGraw-Hill, New York, 1972) p. 6-20.
28. A. Penzkofer, D. von der Linde, and A. Laubereau, Opt. Commun. 4, 377 (1972).

Index of Contributors

GPSR Compliance

The European Union's (EU) General Product Safety Regulation (GPSR) is a set of rules that requires consumer products to be safe and our obligations to ensure this.

If you have any concerns about our products, you can contact us on ProductSafety@springernature.com

In case Publisher is established outside the EU, the EU authorized representative is:

Springer Nature Customer Service Center GmbH
Europaplatz 3
69115 Heidelberg, Germany

Batch number: 10761697

Printed by Printforce, the Netherlands